Transport Without Wheels

Transport Without Wheels

Edited by
E. R. LAITHWAITE
Professor of Heavy Electrical Engineering, Imperial College, London

Elek Science London

First Published in 1977 by
Paul Elek (Scientific Books) Ltd.,
54–58 Caledonian Road, London N1 9RN

ISBN 0 236 40066 5

Text set in 10/11 pt. IBM Press Roman, printed by photolithography, and bound in Great Britain at The Pitman Press, Bath

Contents

Contributors

M. BARTHALON Compagne d'Énergetique Linéaire, France

F. T. BARWELL Department of Mechanical Engineering, University College of Swansea, UK

K. M. CHIRGWIN Airesearch Manufacturing Company, California, USA

A. J. ELLISON Department of Electrical Engineering, The City University, London, UK

E. R. LAITHWAITE Department of Electrical Engineering, Imperial College, London, UK

K. MATSUI Railway Technical Institute, Japanese National Railways, Japan

F. C. MOON Department of Theoretical and Applied Mechanics, Cornell University, USA

S. NONAKA Department of Electrical Engineering, Kyushu University, Japan

K. OBERRETL Department of Electrical Engineering, Dortmund University, Germany

J. C. SABONNADIÈRE Laboratoire d'Électrotechnique, Institut National Polytechnique de Grenoble, France

T. SAIJO Railway Technical Institute, Japanese National Railways, Japan

G. P. SCELZO PRT Systems Corporation, Illinois, USA

G. R. SLEMON Department of Electrical Engineering, University of Toronto, Canada

T. UMEMORI Railway Technical Institute, Japanese National Railways, Japan

K. YOSHIDA Department of Electrical Engineering, Kyushu University, Japan

Editor's Preface

In October 1974, the Institution of Electrical Engineers held an international conference at their headquarters in Savoy Place, London, on the subject of linear motors. During my year as Chairman of the IEE Power Division I had been responsible for the initial suggestion to hold such a conference, and had therefore been appointed its organising chairman. I was able to contribute a truly international flavour by inviting delegates from France, Germany, Japan, Canada and the U.S.A. The conference lasted four days, of which two were devoted to high-speed transport applications.

At the first session on high-speed transport I was approached by Elek Books Ltd to suggest the names of authorities from among the delegates who might be prepared to contribute a chapter to a book on linear motor propulsion for high-speed vehicles. Having provided a list of possible contributors I realised that difficulties might arise in that the authors might have some constraint put upon them not to disclose this or that new idea, for the world of high-speed transport became highly competitive in the late 1960s. In the event, my fears were unfounded and everyone wrote as up-to-date an account as possible so that the book would not appear out-of-date by the time it was published, for the technology described is a rapidly-changing one where a clearly preferred system has still to emerge. Some authors indeed sent updated material at the final proof stage.

In such a climate, opinion is probably more important than fact, and certainly makes more interesting reading when you find (as you will) conflicting opinions between one author and another. Lewis Carroll's 'Alice' thought 'What is the use of a book without pictures or conversations?' In my turn I can add 'What is the use of a book on high-speed transport without arguments?' Of pictures there are plenty, and we are indebted to many organisations for allowing their authors to use them.

The two years since the conference have seen a financial recess in the world generally, and in such a climate exotic projects like conveying people over the ground at 400 k.p.h. are the first to be slashed. That not all the projects described have so suffered is an indication that at least one of the forms of transport described in these pages will emerge as a commercial system. In the years to come the text of this book will be seen to have been prophetic and inspired in some places and quite definitely wrong in others.

In the formative years of rotating electrical machines there can be seen some very definite themes running through the years between 1831 and 1900. The most important of these was a systematic study of topology. The number of different ways of linking a magnetic circuit with one or more electric circuits is larger than most engineers imagine. The Victorians tried them all, and by trial and error rejected some and promoted others. Since those days, electrical technology has been advanced by the discovery and availability of electronic valves, and more recently of solid state devices capable of handling megawatts of power. 'Power electronics' was born in the 1950s, and continues to influence rotary machine design, especially for adjustable speed drives. The discovery of superconductors presents a challenge to the designer to exploit their properties in electromagnetic devices and it is interesting to note that almost the first exploitation of niobium-tin involved a return to 'year one' (1813) and to Faraday's disc dynamo structure.

When a linear motor first showed commercial promise the topological gates were re-opened, for *every* kind of rotary motor can be 'linearised'. We can therefore take out and dust the museum pieces that the Victorian pioneers rejected, for they might just fit a linear motor's requirements in ways their original inventors might never have conceived. The main difference is that in rotary machines the d.c. motor came first, thanks to a most unfortunate meeting in 1832 between one Hippolyte Pixii (who invented the basic modern alternator, no less) and the great Ampère, the latter of whom showed Pixii how to fit a commutator and get 'battery-like current'. The world waited a further 56 years for Tesla and the induction motor, which kind was to dominate the world of electric drives to the extent of over 95% of its total driving power. At least we began linear motors in the 1940s with that knowledge and developed induction devices first. But, as these pages will show, linearised versions of the synchronous motor, the d.c. machine and the reluctance motor are beginning to infiltrate the high speed world, ably aided and abetted by power electronics. One might, of course, reflect that if Ampère had never met Pixii and the alternator had moved forward the invention of the induction motor to the 1930s we just might have had a brilliant young engineer of the 1970s inventing the commutator and being told by industry that it was 'not commercial and impractical'!

The linear motor pioneers have been no exception to those who have to bear the slings and arrows of industrial vested interests, national and international politics and their accountants. I am indebted to a former postgraduate of Professor F. T. Barwell, Dr. Parkinson, for permission to publish the cartoon opposite. Anyone who heard the evidence given in public before the Select Committee on Science and Technology of the House of Commons in 1973 in relation to the closure of the UK Tracked Hovercraft project will appreciate this cartoon all the more.

Another difference between linear motor pioneering today and rotary machine developments of a century ago, which for me is by far the most important, is that the Victorians did it all for fun – and to engage in friendly rivalry with each other. No one (except perhaps Tesla) had high-minded thoughts of patriotism, or of humanitarianism. There were no prizes for reducing pollution and noise – rather the reverse! Today we live in a world of rapidly changing values, developing countries, monetary crises and political complexities such as never troubled Victorian England. To say that the engineering world is in a state of flux would be a bad pun, but the accountants moved in, in earnest, on the engineers after World War II and we have not recovered our poise ever since.

I am delighted to have been instrumental in putting together a book such as this, for I know that is it a book full of optimism written by men who recaptured some of that adventurous optimism of the nineteenth century – tempered a little by the commercial necessities, but still with that twinkle of delight that comes in childhood at the discovery of a new toy and happily stays with a few of us for the rest of our lives.

Here are no pessimists. This is the work of men of honest endeavour, not caring to look too far to the right or left in case they get tangled in the barbed wire of politics

'The whole business is economically unsound, gentlemen. With a train of this length and forty miles of track, we find that only 0.0568 per cent of the track will be in use at any given time, representing a constant idle investment of 99.9432 per cent.'

and policies. Here is high technology, no less in quality than that which put men on the moon. By their essays we can see the authors think and by appreciating their efforts can recapture the spirit from which engineers inherited their title, from the French 'Les ingènieurs' – the ingenious ones.

E.R.L.

1

Advances in Transport Without Wheels

F. T. BARWELL

1.1 Overcoming transport congestion

The development of transport during the central fifty years of this century has been dominated by the internal combustion engine which, because of its high power/weight ratio and because the energy in the form of petroleum fuel was so readily transported, conferred a freedom of movement unknown to previous modes. Previously, the steam engine and its fuel were so bulky that it was applicable only to massive situations such as ships or, where land transport was required, only on specially constructed tracks such as railroads. Electric power, in many ways the ideal, was only available on fixed routes (trams, trolley buses, railways), where wires could be stretched or in areas restricted by the capacity of the comparatively crude storage batteries available.

The half century is noteworthy for the liberation of transport from fixed facilities, notably the development of air transport, and the ubiquitous use of pneumatic tyres on roads varying from unmade lanes to motorways. The construction of motorways with fixed lanes and entry and exit points can be seen as the beginning of a reversion to a more disciplined form of transport movement involving large-scale investment in the route.

With the growth of population and moreover the growth in standards of living, whereby larger proportions of the public make greater demands on transport both for personal transport and the transport of goods for consumption, the freedom conferred by the internal combustion engine begins to result in congestion which is particularly marked in city thoroughfares and airports. Figure 1.1 shows an overcrowded bus in Calcutta and Figure 1.2 an electric suburban train in Bombay, which exemplify the world-wide pressure on existing means of transport. In addition to the absolute growth in population, there is a world-wide tendency for people to live in or about larger and larger centres. In developing countries the tendency is for people to crowd into the centres of the larger cities, but under more prosperous conditions such as those encountered in Western Europe and the United States the predominant shift of population has been to the low and medium density suburban fringes around urban centres[1]. Nearly half the population of the USA is now located in such areas and the trend is expected to continue. This has resulted, in many instances, in a reduction in the travel to and from the city centre, and future urban public transportation must augment peripheral and cross-town facilities.

A source of controversy exists between those individuals who consider that greater emphasis should be placed on public transport so as to lessen the environmental and

Figure 1.1 Overcrowded bus in Calcutta.

Figure 1.2 Overcrowded train in Bombay.

the community disruption caused by complete reliance on private automobiles and those other citizens who, by contrast, consider that existing road facilities are insufficient to enable the community to grow economically and for the individual to choose from a sufficiently wide range of social, recreational and employment opportunities. The resolution of this controversy has led to emphasis on what is known as 'balanced transportation', which implied some investment in various forms of public transport whilst not excluding appropriate highway development.

In contrast to the situation in Asia, inter-city travel in North America is dominated by the automobile. The airlines are beginning to capture travel beyond 200 miles, but even for trips of over 1000 miles the automobile share is about 66% of the total, with bus and rail carrying only about 3%. Although the automobile will continue to play a major role in inter-city transport, public policy is to transfer trips in the 200–500 mile range to upgraded rail and bus services, and to air for trips over 500 miles.

In contrast to the freedom of movement enjoyed by the motor car and the aeroplane, railways represent guided transport in its traditional form. It can be shown that disciplined movement under the control of signals and the coupling of vehicles into trains enable much greater capacities and higher speeds to be obtained than when vehicles are controlled as individual entities by drivers, each of whom adopts his own following and guiding strategy. However, during the immediate post-war period, the role of the railways was considered by most authorities to be the conveyance of heavy

Figure 1.3 German multiple unit train.

freight, particularly coal and steel. It was thought that this would be sufficient to provide revenue for the survival of the industry and that passenger and high quality freight would continue to be diverted to road and air transport. As a result of World War II and its aftermath, equipment was in a poor condition and much replacement had become overdue. The announcement by the late Lord Robertson in 1956 therefore, of a modernisation plan which envisaged investment of some £1500 million over a period of 10 years, provided an opportunity for a re-assessment of the role and re-equipment of the railways in Great Britain.

Even those most stalwart in their confidence in the future of railways were somewhat amazed by the decision of the Japanese to construct an entirely new railway designed to offer unprecedented speed and capacity. However, such a railway was constructed and has been an outstanding success both socially, financially and technically, although the normal speed of operation is 200 km h^{-1} rather than 250 km h^{-1} which was proposed during the early stages of the development. Other technically advanced countries followed Japan in achieving the speed of 200 km h^{-1}

but using existing routes. Figure 1.3 shows a modern German multiple unit train.

Wales and the West of England are served by high-speed trains, in this case diesel propelled. These operate, at least, at 125 m.p.h. between Paddington and the Severn tunnel. France, notable in achieving a world speed record of 206 m.p.h. with an electric train as long ago as 1955, operates numerous high-speed trains, notably the 'Mistral', has exported gas-turbine driven trains to the United States and has constructed prototype trains capable of speeds up to 180 m.p.h. British Rail's Advanced Passenger Train which has similar objectives, will first be used as an electric train operating between London and Glasgow.
Glasgow.

Exciting as these developments on the railway are, their impact upon the total transport pattern is likely to be limited and it is probable that conventional railway will remain an important minority carrier. The main growth of traffic will continue by road and air, although there is at present a set-back in the growth rate due to financial difficulties primarily associated with the supply of petroleum. This set-back is likely to be temporary, however, and problems of congestion of airports (a temporary respite has been earned here by increasing size of aircraft rather than the number of flights) and the growing congestion of motorways will pose great problems. The important factor then is the *capacity* of a route provided by a given investment. The capacity of a route can be determined from simple Newtonian dynamics provided certain assumptions are accepted as being necessary for safety [2]. If a vehicle must be separated from its predecessor so that it can stop at any time without colliding therewith, then a single line motorway can handle 2000 vehicles per hour at a moderate speed. This, of course, excludes passing, because passing can only occur if the faster lanes are not themselves loaded to capacity.

The penalising effect of braking distance can, however, be reduced by coupling vehicles together into trains. The optimum speed is that at which the spacing of trains becomes equal to their length and, assuming normal practice, a single line of railway can carry between 40 000 and 60 000 passengers per hour, whereas a six line motorway would have a capacity of 27 000 passengers in each direction, on the assumption that the average car load is $1\frac{1}{2}$ passengers. Thus a double track railway can carry at least as many people in trains as a six-line motorway. It is reasonable to assume that the capital expenditure to provide a railway will be much lower than that of a motorway for a given capacity. This lesson has been learned in a number of important cities where motorway construction has been halted. A notable example is provided in Toronto where a large area was cleared to provide a radial link joining the peripheral motorway with the city centre. Some millions of dollars were expended before public protest caused the work to be halted. Apparently, the normal protests of the local people whose homes were to be destroyed in the interests of urban motorways were supplemented by protests throughout the city on environmental grounds. The motorway project has been replaced by an extension of the underground railway now under construction.

As far as densely populated industrial areas are concerned, for example the north-east corridor of the United States between Boston and Washington, congestion of road and air facilities has already become serious, although the full capacity of the railway is not exploited and a substantially greater number of inter-city rail passengers could be carried. It is therefore proposed [3] during the next 20 years to invest about 1 500 000 000 dollars in providing exclusive track, in eliminating level crossings and in introducing improved trains. The state of the art would not be greatly extended and speeds of about 120 m.p.h. would be considered adequate.

1.2 Innovative technology

The last decade has been characterised by extensive studies in many countries in which attempts have been made to evolve new technology of guided transport, ranging from attempts to provide personal means of transport for very short distances up to 100 metres to the full inter-city range. A classification in ascending order of operational capacity is as follows:

1. Dual mode (buses with electronic controls)
2. PRT (personal rapid transport)
3. Light rail (tram-cars)
4. Medium rail (underground)
5. Heavy rail (suburban and inter-city)

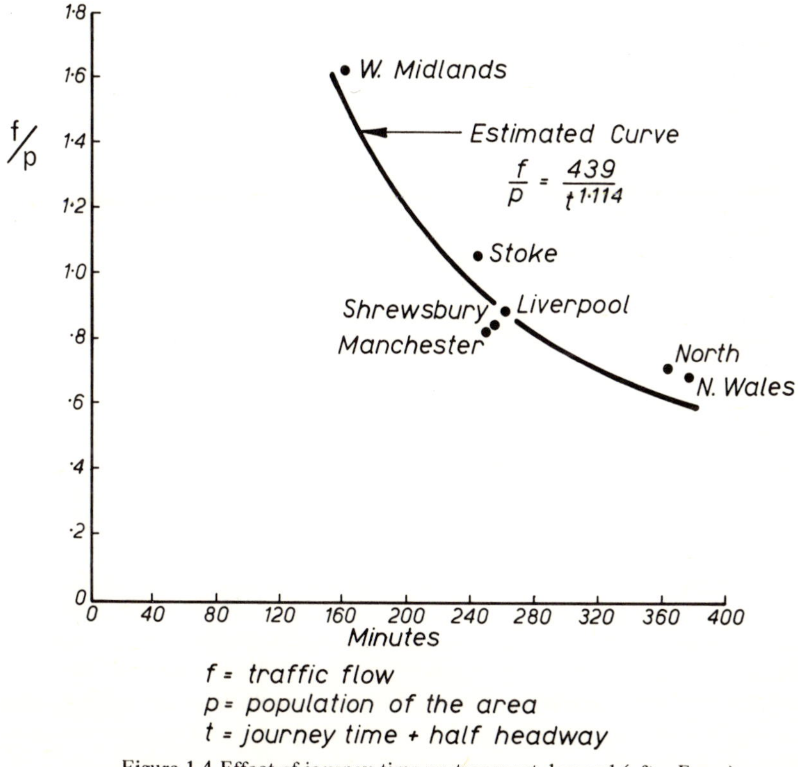

Figure 1.4 Effect of journey time on transport demand (after Evans).

Whilst it can be accepted that guided transport can provide capacity more cheaply than motorways on a route basis, justification for its use must depend on its ability to attract sufficient traffic to warrant the investment in the track. All other things being equal, increased demand follows reduced journey time, as was indicated in the studies of Evans[4] (Figure 1.4).

There appears to be no physical reason why the speed of the conventional railway system based on steel wheels and steel rail should not be extended. It is difficult, however, to decide whether to devote effort in extending this much beyond the present practice of 125 m.p.h. to perhaps double this value, or whether new alternatives such as the 'Hovertrain' might be more flexible, less expensive and more acceptable from the amenity point of view. A major limitation of the conventional railway, referred to in modern jargon as the 'dual rail' system, is its reliance on adhesion for traction and braking. This is particularly important, because, in determining the spacing of the signals, it is necessary to assume a braking distance corresponding to the worst conditions of adhesion likely to be encountered. This represents a major limiting factor.

Figure 1.5 Light electric vehicle in Mannheim.

Whilst electric railways provide a high quality, high-capacity service, modern environmental standards require expensive construction, if not in tunnels, then in screened alignments with expensive attention to amenity preservation. The only cheaper alternative available in most cities is the bus, which is noisy, causes pollution and, at least when operated on the public highway, has a very limited capacity. The need has been expressed [5], for some mode of transport to complete the system; a mode which would be able to provide economically a closer net than can be provided by the electric railway with the costly construction associated with high capacity. Light electric vehicles are candidates for this application. These have evolved from the traditional tram-car, which has been eliminated in some countries but which has been maintained and developed in others, particularly in Germany. Figure 1.5 shows a tram-car in a street in Mannheim passing through a shopping district from which other wheeled traffic has been excluded.

Another objection to steel wheels running on steel rails is noise [6], particularly in urban situations. This led to consideration of the pneumatic tyre which has been used with great success on the Paris underground railway and in Montreal. The pneumatic tyre suffers from the disadvantage, however, that friction losses are increased which leads to difficulties regarding the dissipation of heat in tunnels. A more compact arrangement is that adopted by Westinghouse for their transit expressway. Whilst this is intended for local use, nevertheless it illustrates the principle of using pneumatic tyres both for support and, separately, for guidance. A demonstration was provided in the South Park of Pittsburgh featuring a prototype which was intended for use in that city, Figure 1.6 (a and b). Unfortunately the construction is held up by political

Figure 1.6 (a) Track for transit expressway.

(b) Station and vehicle.

argument between two levels of local government. The system has, however, been installed and is giving satisfactory results, at Seattle airport.

1.3 Air cushion support and guidance

Any vehicle/track system must fulfil six important functions: namely support, guidance, propulsion, braking, current collection and communication for control. In the conventional railway system, massive wheels are required to make contact with a fixed railhead. This arrangement calls for a high quality of construction and maintenance of both track and vehicle suspension. The area of contact between an individual rail and a corresponding wheel approximates to the size of a sixpence, as is shown in Figure 1.7. This means that there is intensive pressure at the point of

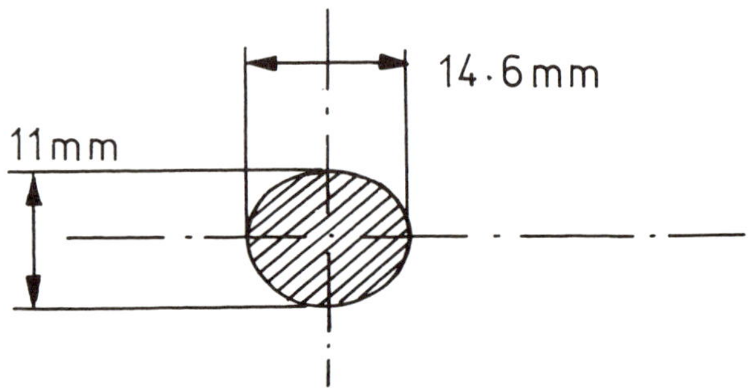

Figure 1.7 Contact between wheel and rail.

contact and therefore steel of the highest quality must be used to resist deformation and fatigue. Ultimately the load has to be distributed on the rails through sleepers and ballast to reduce the intensity of loading to that which can be withstood by the natural sub-soil as shown in Figure 1.8. Whilst much progress has been made in recent years in improving the high-speed potentialities of the track, notably by continuous welding of rail and experimenting with long concrete slabs to replace sleepers, nevertheless the rolling of a wheel on a rail must inevitably produce noise and ground-borne vibration, giving rise to environmental disturbances to which greater attention is now being given than was the case when railways were first developed. The controversy surrounding the proposal for the high-speed link to the Channel Tunnel is evidence of this. An attractive alternative to this elaborate technical exercise would be to use air cushions to distribute the load over a wide area of track surface, so avoiding the expensive construction necessary to withstand the concentrated load of wheels on rails/roads. The susceptibility to small-scale track irregularities would also be eliminated by the air cushion because there would no longer be any necessity for a massive component to follow surfaces intimately. Figure 1.9 shows four possible methods of air support. The system in Figure 1.9(a), originally proposed by the Ford Motor Company, is unlikely to commend itself for high speed working because it would require a very perfect track to allow the necessary fine clearances to be employed. However, it has been suggested by the National Engineering Laboratory, that if one of the components is made of compliant material this objection may be overcome. Even if unsuitable for

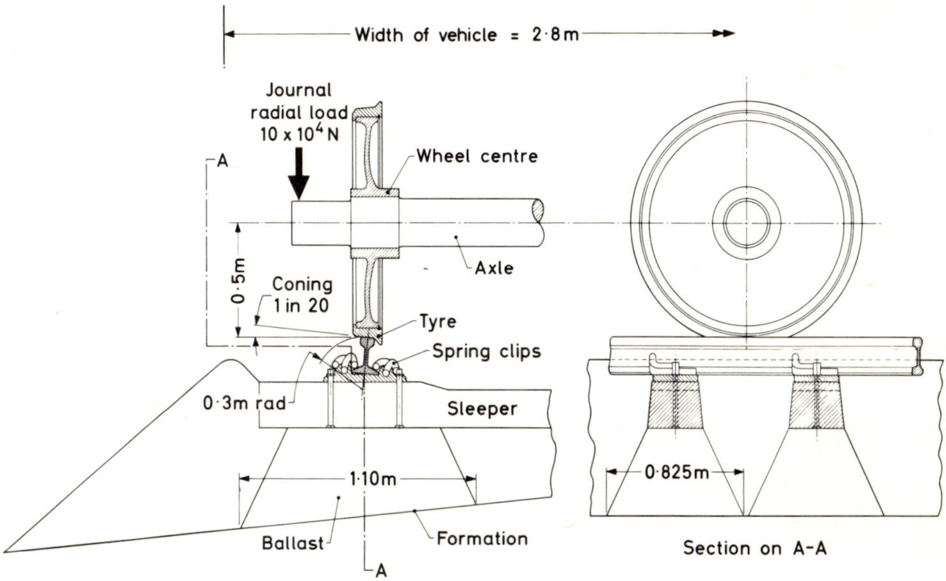

Figure 1.8 Construction of conventional railway track.

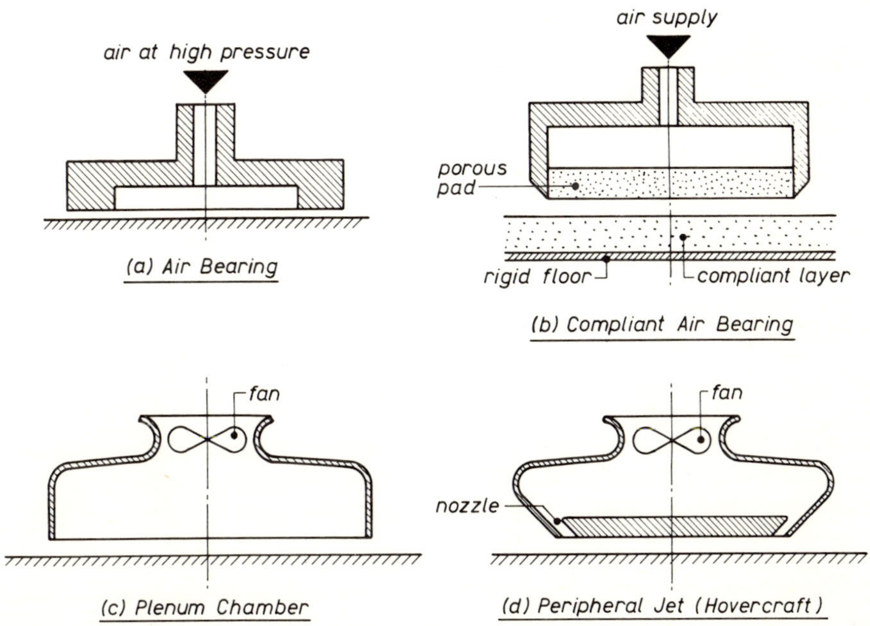

Figure 1.9 Forms of air suspension.

support of the main vehicle, this system offers promise for support of the linear motor to which reference will be made later. The method shown in Figure 1.9(c) has been developed by M. Bertin in France (see Figure 1.10). Figure 1.9(d) represents the peripheral jet system employed in the United Kingdom in the development of the Hovercraft.

Figure 1.10 Aerotrain.

1.4 Propulsion by linear motor

Vehicles are usually driven through the agency of the friction between the wheel and the fixed surface, be it rail or road, and variation of this friction with weather, speed, load, etc., is an embarrassment to all transport operators. If, however, a vehicle floats on an air cushion, it is no longer possible to adopt this method and accordingly alternatives have to be provided. One of the simplest alternatives which was selected some 15 years ago by British Railways was the linear motor. Figure 1.11

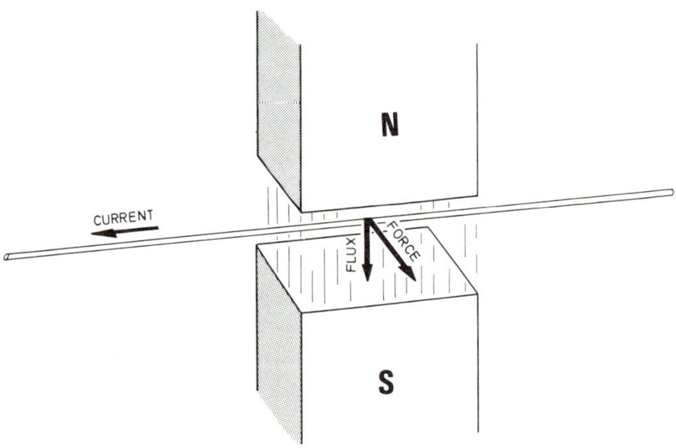

Figure 1.11 Faraday's Law.

shows the basic relationship between electrical and mechanical energy. The output is a force which is mutually at right angles with the flux and the current flowing. Figure 1.12 shows a feasible mechanical system in which the situation is reversed with the magnet being carried on the vehicle and the current directed by brushes through a plate which is fixed relative to the earth. Such a system would be uneconomic unless very great currents were used or a number of units were put in series.

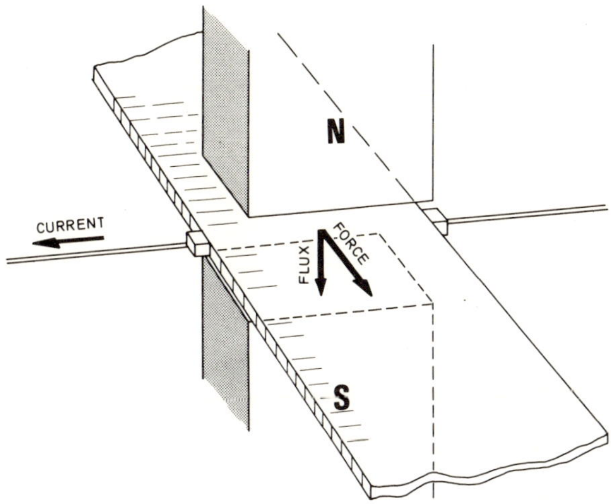

Figure 1.12 Homopolar linear motor

The use of alternating current enables more realistic designs to be achieved. For example, the synchronous linear motor [7] consists of a powerful magnet carried on the vehicle. It interacts with a set of coils laid into the track. As shown in Figure 1.13 these coils are fed with alternating current, each being displaced laterally from its neighbour so that at any instant in time the current in one phase is declining, in another near to its maximum and in the third increasing. Thus, in effect, the current is being progressed along the track at a speed which is proportional to the supply frequency multiplied by the overall pitch of the coils. This is equivalent to a single d.c. coil being progressed along the track at synchronous speed. Thus the magnetic field is required to produce an adequate flux to interact with the track coils. This is achieved by taking advantage of the principle of super-conductivity in which a coil is mounted within a liquid helium bath so that current once generated will continue to flow almost indefinitely.

A test facility for the investigation of synchronous linear motors has been set up at the Canadian Institute of Guided Ground Transport at Queen's University, Kingston, Ontario, Canada. It consists of a test wheel, Figure 1.14 (a) to (d), having a diameter of 7.6 metres and an axial depth of 1.5 metres. Apart from the bosses and the spokes, which are made of aluminium, it was constructed entirely of wood and was claimed to be round within a tolerance of 1/16th of an inch. The coils laid on the outside formed the primary of the linear motor, the cryogenic magnet being held by a stationary dynamometer. The peripheral speed of the drum was 100 km h^{-1}, which

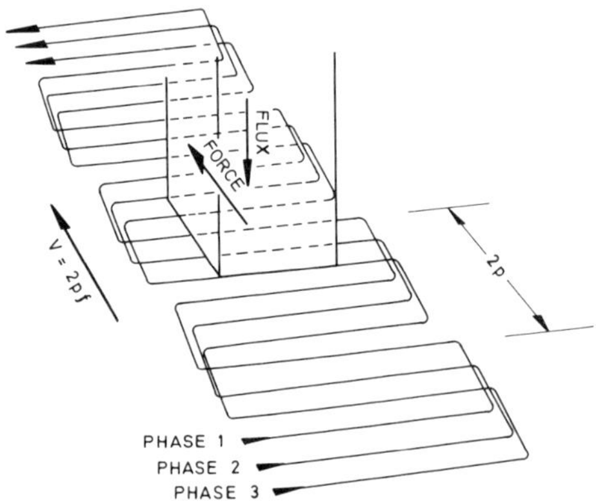

Figure 1.13 Synchronous linear motor.

is about one-fifth of the speed envisaged in a full-scale application of this type of motor.

Linear motors can, of course, exist in as many manifestations as rotary machines, but by far the greatest progress to date has been made with the linear induction machine which has the advantage of requiring a very simple reaction member on the track, the wound coils being carried on the vehicle.

If a 3-phase winding is placed on the vehicle so as to produce a flux-wave which moves at the velocity of $2pf$ in one direction this will induce an e.m.f. which will result in a circulating current as shown in Figure 1.15.This current will react with the flux to produce a force tending to propel the vehicle in the direction indicated in the arrow. This force will accelerate the vehicle until a speed of $-2pf$ is reached. At this speed there will no longer be any relative motion between the moving flux field and the reaction rail so that in the absence of interaction of current and flux the force will fall to zero. This is known as the synchronous speed.

If as a result of some external force the vehicle is propelled leftwards at a greater speed than the synchronous speed, the relative motion between the field and the reaction rail would be re-introduced but in the opposite section of motion, thus providing a braking action. This form of braking can be described as hypo-synchronous in contrast to the situation where field and motion are in opposite directions. This is known as 'plugging'. Hypo-synchronous braking can be achieved by reducing the supply frequency in consonance with the speed of the train so that the requisite amount of negative slip exists. Such braking is of course regenerative, the system acting as an induction generator.

(a)

(b)

Figure 1.14 Test facility for linear synchronous motor
(a) General view.
(b) Test drum.

(c)

(d)

Figure 1.14 (contd.) (c) Cryogenic magnet.
(d) Detail of field winding.

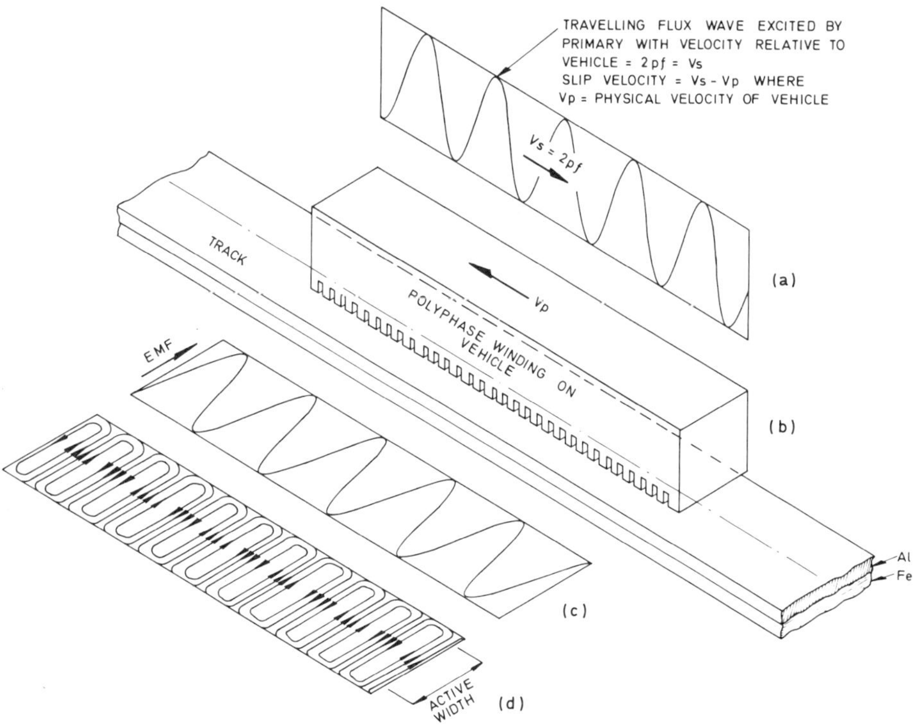

Figure 1.15 Principle of linear induction motor.

1.5 Tracked Hovercraft

It seemed natural that the two British inventions, the linear induction motor and the hovercraft should come together as a solution to the problem of providing high-speed guided transport. The British Hovercraft Corporation exhibited a model at Browndown in 1966 which embodied all the features of subsequent construction, excepting that the reaction rail was mounted vertically.

The next stage was the construction of a full-scale vehicle and track by a company specially set up for the purpose called Tracked Hovercraft Limited, under the aegis of the National Research Development Corporation. Chiefly because of the difficulty of devising points and crossings, but also because the elevated reaction rail was regarded as being hazardous should it become broken or detached, single sided linear motors were introduced in which a continuous iron member was laid in the track to complete the magnetic circuit. This was covered by a thin aluminium plate which afforded a low resistance path for the circulating currents as sketched in Figure 1.15. Because, at high speeds, the value of p would be rather large for a motor which was fed with power at commercial frequencies of 50 or 60 Hz, a modified design was evolved known as the 'transverse flux machine' in which the flux path was extended laterally beyond the aluminium reaction rail so that flux flowed in a plane at right angles to the direction of motion. This, of course, has no effect on the configuration in the zone of interaction between flux and induced current, the flux at this point being orientated vertically, whether the remainder of the circuit is of the longitudinal or transverse flux configuration. The project was closed however, before the first full-scale transverse flux machine was built.

The track is shown in Figure 1.16 and one of the vehicles in Figure 1.17. Just over

Figure 1.16 Hovercraft track at Earith.

one mile of track had been constructed and some interesting experiments carried out at speeds up to 107 m.p.h.[8]. The extension of the track to 2.5 miles was in hand when the Government decided that there would be no more financial support for Tracked Hovercraft Limited. A number of universities approached the Science Research Council for support to extend and operate the track in order to settle some vital questions about the choice of systems. The Government unfortunately refused the requests in spite of the firm recommendation of the Select Committee on Science and Technology [9]. The rest of this article will therefore be concerned with work carried out in other countires, notably the USA, Canada and Germany. Important developments have also taken place in Japan[10].

Figure 1.17 Research test vehicle.

There can be no doubt that the main stimulus for innovation in transport technology arose from the initiative of President Kennedy (see Section 2.1). This was continued by President Johnson who set up the Department of Transportation (henceforth referred to by the initials D.O.T.). An important report prepared at the Massachusetts Institute of Technology [11] established that some form of guided transport was the only conceivable solution to the problems of congestion anticipated in the North Eastern corridor of the United States.

1.6 Tests at Pueblo.

The D.O.T. has spent large sums in supporting research and development in transport including the establishing of the Transportation Systems Center at

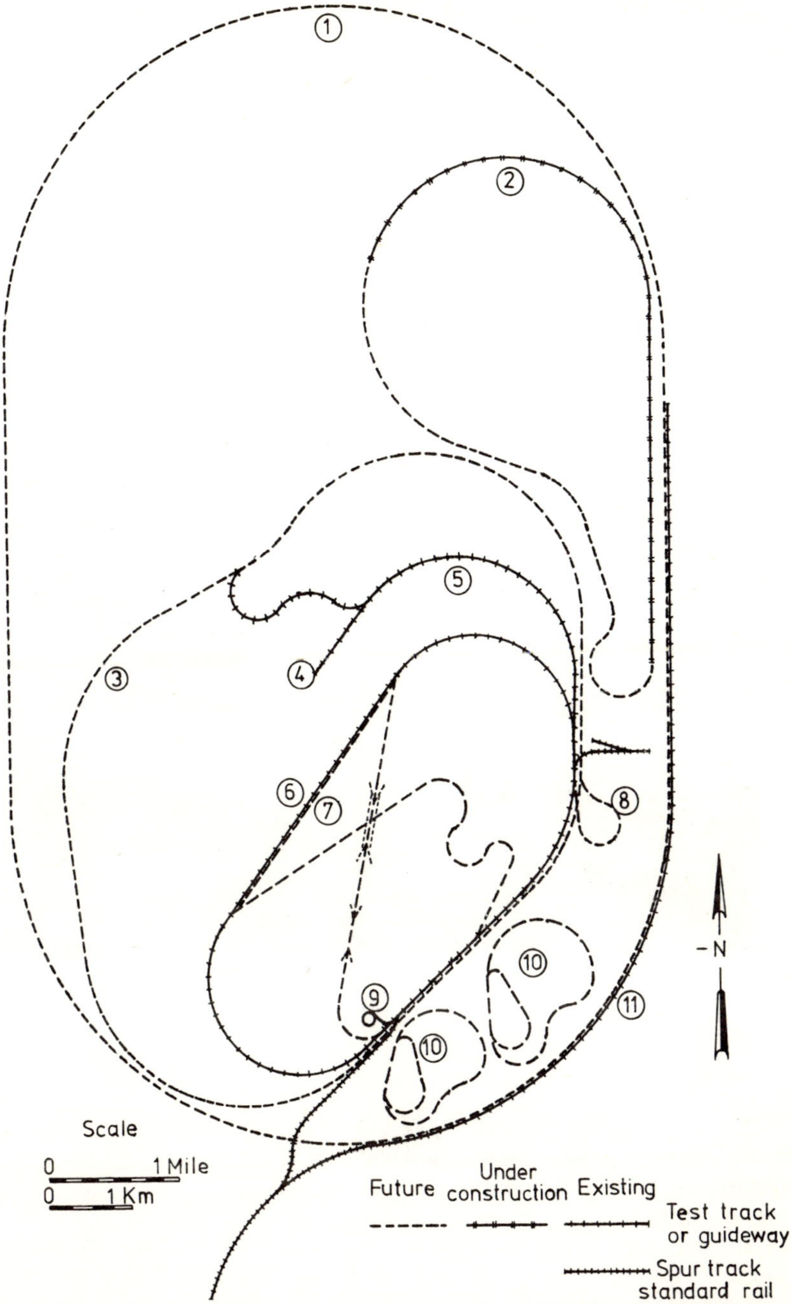

Figure 1.18 Transportation Test Center, Pueblo, Colorado. (1) TLRV guideway (300 m.p.h.); (2) PTACV (150 m.p.h.); (3) Railroad (160 m.p.h.); (4) Impact facility; (5) Train dynamics track; (6) Transit (80 m.p.h.); (7) Perturbed track; (8) Baloon track; (9) Screech loop; (10) Dual mode; (11) LIMRV track (250 m.p.h.).

Cambridge, Massachusetts, and the Transportation Test Center at Pueblo, Colorado, the lay-out of which is shown in Figure 1.18.

At the time of writing, political support for high-speed ground transport is lacking and financial support is no longer available for effective progress. Nevertheless, a good deal of successful testing has taken place, as indicated in Table 1.1. The LIMRV (linear induction motor research vehicle) was devised for testing linear motors on conventional track. To simplify the test arrangement this vehicle was fitted with a 2240 kW 2-shaft gas turbine which was connected to an alternator which gave 3000 kVA at a maximum frequency of 173 Hz. So that the tests could be carried out on the limited lengths available, thrust was provided by two jet engines (Figure 1.19) and additional braking was provided by drag parachutes. Successful test running has been recorded up to speeds of 255 m.p.h.[12].

Figure 1.19 Jet engines used to augment thrust of LIMRV.

The running of the vehicle, which was supported on Budd 'Pioneer' trucks embodying an air suspension system, was particularly good, cinematographic records indicating the virtual absence of 'hunting' even at the highest speed. Figure 1.20 shows an arrangement of track including the central reaction rail.

Experiments with air cushion support on the Hovercraft principle were carried out at Pueblo using the vehicle known by the initials TLRV (tracked levitated research vehicle). Figure 1.21 shows this vehicle under construction at the Grumman works at Hixville, Long Island. The ducts for supplying air to the various support and guidance hoverpads can be seen in the photograph. The guideways for the TLRV are shown in Figures 1.22 and 1.23.

Figure 1.20 LIMRV track.

Figure 1.21 TLRV under construction at Hixville.

Figure 1.22 View of TLRV track taken from inside.

Figure 1.23 Exterior view of TLRV track.

Figure 1.24 Inverter.

It was intended that power should be drawn from the supply system at the conventional frequency and converted on the vehicle to correspond with the speed of the train. Figure 1.24 shows the inverter. Figure 1.25 shows a synchronous condenser. A synchronous condenser is necessary for to correct the low power factor which has so

Figure 1.25 Synchronous condenser.

Figure 1.26 Guidance air cushion for TLRV.

far been associated with experimental axial-flux linear motors. Figure 1.26 shows the guidance air cushion including the emergency pads which are lined with anti-wear material. At the time of writing, a linear motor and a short section of reaction rail are in course of installation. So far the vehicle has operated only as fast as the reaction from a gas turbine exhaust will permit. Nevertheless, at over 90 m.p.h. it is reported to ride particularly well, even when artificial irregularities were incorporated in the track.

A vehicle known by the initials PTACV and which is basically Bertin's aerotrain, was originally intended to serve an airport, operating at a maximumm speed of 67 metres per second (150 m.p.h.), but because the plans for the airport were discon-

Figure 1.27 PTACV general view.

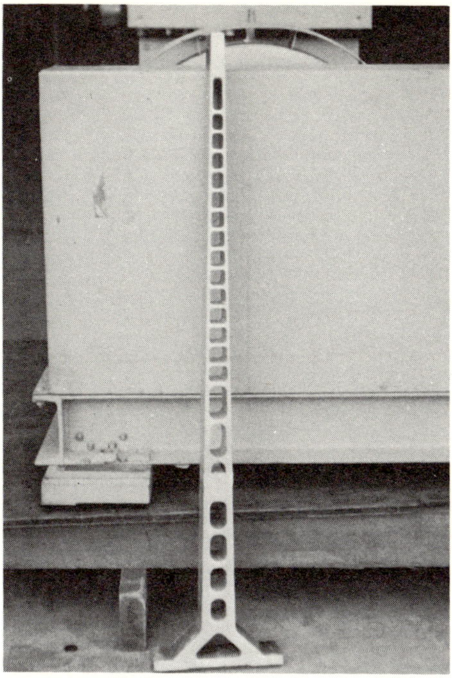

Figure 1.28 PTACV reaction rail section.

tinued, the project has been evaluated by the Federal Railroad Administration at Pueblo. As will be seen from Figure 1.27, the vehicle is guided and propelled from a central aluminium rail which is shown in greater detail in Figure 1.28. Care is taken

Figure 1.29 PTACV view of track.

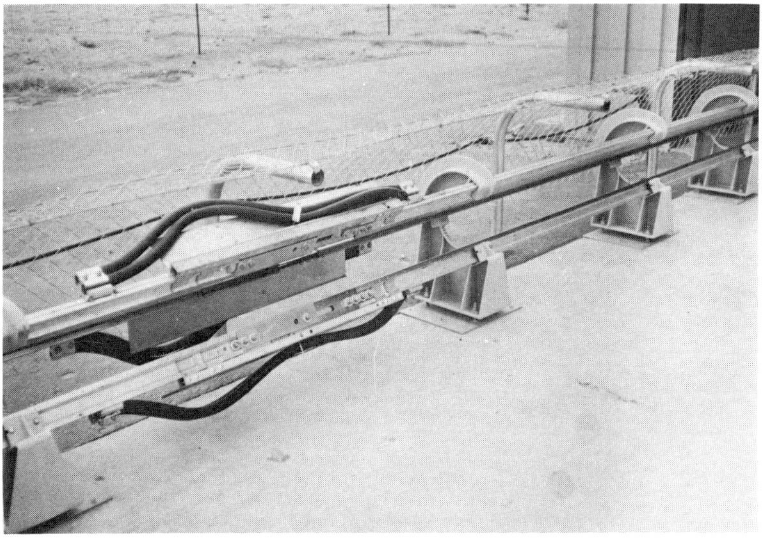

Figure 1.30 PTACV current supply system.

Figure 1.31 PTACV interior view

to install the reaction rail at a controlled temperature (60°F) so that thermal stresses are minimised. Provided this operation was carried out correctly, no buckling has occurred even in the high daytime temperatures experienced in the Colorado desert. Similarly, no tensile failure has occurred in the extremely cold winter conditions. (The site of the test track is one mile above sea level). Figure 1.29 shows a general view of the track and Figure 1.30 illustrates the current supply system. Three

Table 1.1
High-speed Vehicle Ground Tests at Pueblo and Earith

System / Feature	LIMRV	TLRV	PTACV	RTV 31
Projected speed (m.p.h.)	250	300	150	300
Means of suspension and guidance	Wheel on rail	Air cushion with peripheral jets. Trough section	Air cushion plenium chamber with central reaction rail	Air cushion with peripheral jets, box girder
Power system	Double sided LIM augmented by jet	Turbine exhaust reaction only	Double sided LIM (start by rollers)	Single sided LIM
Length of track (miles)				
Projected	6	22	6	7
Actual	6	5	6	1.2
Maximum speed attained (m.p.h.)	255	94	102	107
Principal Contractor	Garrett	Grumman	Rohr	Tracked Hovercraft

rails are provided, one for each phase. They surround and support a current collecting fitting which is carried on an arm projecting from the vehicle.

The vehicle rides well at 100 m.p.h. and is expected to meet all specifications at 150 m.p.h. once the installation of power conditioning equipment has been completed. Of the three high-speed vehicles at Pueblo this is the only one which has been fitted out as a demonstration of a passenger carrying vehicle, the interior arrangements being as shown in Figure 1.31. Table 1.1 summarises the results of tests on high-speed vehicles at Pueblo and Earith.

1.7 Magnetic suspension

The main effort in Britain and the USA as far as full-scale experimentation has been concerned has been the use of the Hovercraft principle. However, more recently, magnetic suspensions have come into the picture. Some work is in progress at the University of Sussex [13]; their vehicle is illustrated in Figure 1.32. The University of Warwick is also working in this field but on another system [14]. The primary distinction between the two systems is that one is based on the magnetic attraction of a d.c. solenoid for a ferrous armature and the other is based on repulsion between a heavy current in one member and an induced current in the other.

Figure 1.32 Experimental vehicle installed at University of Sussex.

The attraction system was the one chosen by the firm of Krauss Maffei for demonstration in Toronto. There was a ferrous rail and an extension of the vehicle projected beneath this carrying a direct current magnet which, by pulling upward on the rail, lifted the vehicle. Similar arrangements were available for guidance. This system was essentially unstable. Consider that there was a certain force between the vehicle and the track. If this force were less than that due to the weight of the vehicle, the gap would widen, the magnetic circuit would be extended, the magnetic flux would fall off and the vehicle would fall to the ground. Conversely, if the magnetic force were greater than the gravitational force, the vehicle would lift, thereby closing the air gap enabling the force to become still greater. Thus there would be a tendency either for the vehicle to fall away or, for it to rise to the greatest height thereby eliminating the gap between the magnet and the track. If a solenoid were used to provide the whole of the magnetic flux, or to moderate the flux provided by permanent magnets, a control system would be required which measured the value of the air gap at any instant of time and moderated the applied current in order to keep the separation within acceptable limits. When this was considered by British Rail some 15 years ago, it was decided that the power required for such a control system would be so great that the necessary equipment presented unacceptable weight and volume penalties on any vehicle. It is a tribute to the development of control engineering during the intervening period, (this is a subject in which the University of Sussex is eminent) particularly in the application of solid state devices, that these demonstrations indicate a measure of feasibility.

The city of Toronto is one of the most successful in developing a co-ordinated transport system so as to avoid the impact on the centre of the city by the motor car.

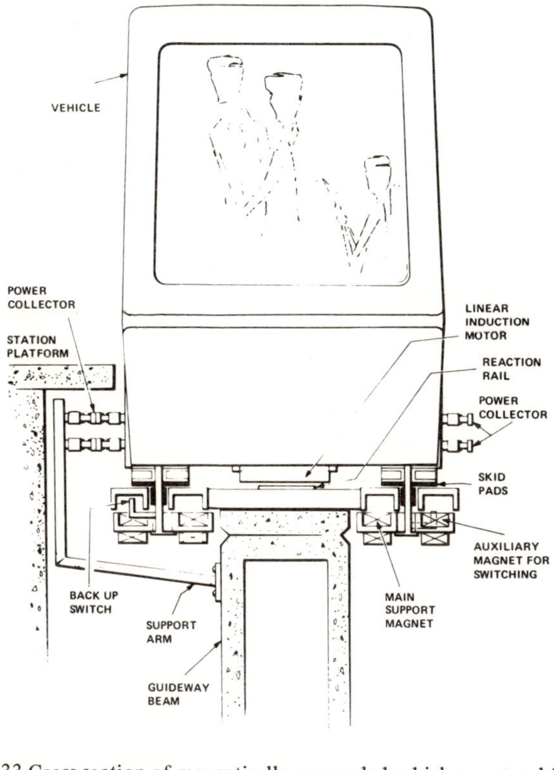

Figure 1.33 Cross section of magnetically suspended vehicle proposed for Toronto.

They successfully developed underground systems as well as bus, trolley-bus and tram systems but they felt the need for something intermediate between the two, something which was environmentally compatible with city conditions but cheaper and more flexible than conventional undergrounds. As the result of a world-wide study and a comparison of many tenders, they finally chose that of the German firm of Krauss Maffei, embodying the magnetic suspension system. Figure 1.33 shows a cross-section of the proposed vehicle. It will be noted that there were two horizontal coils which could be switched in order to provide a directional control at junctions entirely from the vehicle. In other words, static switching would be employed, which has important connotations from the point of route capacity. The German Government was supporting the firm of Krauss Maffei in this development and it was a great shock at the beginning of December, 1974, when it decided not to continue its support into 1975 on the grounds that the application to urban work was not necessary and that development should be confined to high speeds. This, among other things, led to the firm withdrawing from the contract in Toronto.

After the experience with the 'Tracked Hovercraft' system this was indeed depressing news. However, the contract was a very businesslike one and Krauss Maffei has already passed to the Canadian Government a letter of credit for 1 million dollars towards the compensation. A statement made by the Minister of Transportation and

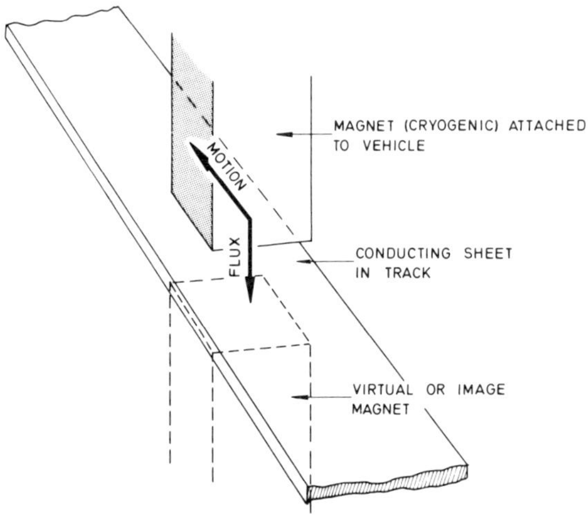

Figure 1.34 Principle of magnetic levitation.

Communications in the Ontario legislature contained the following:

'First the Ontario Transportation Development Corporation will now develop and will now own the intermediate capacity technology using existing Krauss Maffei research and development data for which the West German government and Krauss Maffei have invested 30 million dollars. This valuable asset is available to Toronto without cost. We have also negotiated the right to an exclusive royalty–free licence in Canada and a royalty–free non-exclusive licence in most of the rest of the world. Second our contract with Krauss Maffei has now been terminated and Ontario will be refunded all the money it has spent so far on a demonstration product at C.N.E.'.

The Toronto authority has completed an eighteen months programme. The first three months were devoted to determining operational requirements and the remaining period to the evaluation of a technical specification. The sum authorised for this work was 6.1 million dollars. It was concluded that their particular requirements would be best served by vehicles running on steel wheels but propelled and braked by linear induction motors.

The other system, the repulsion system, is inherently stable but has the disadvantage that it only works when the vehicle is going at speed. Therefore, rather like an aircraft, the vehicle would have to start up from a stationary position using wheels which may be retracted as operational speed is approached. The repulsion system shares with the air cushion the advantage of stability because, as the load is raised the air gap is decreased and the magnetic force increases until a balance is achieved. Moreover, it is feasible to employ large air gaps with the system thereby ensuring a better ride with less severe track alignment requirements. Repulsion must be achieved by magnetic induction. Figure 1.34 shows the interaction of flux with a plane consisting of a conducting sheet. The currents, induced in the sheet by relative motion of the magnet, are equivalent to the existence of a magnet of similar polarity to the real magnet and situated as though it were a mirror image thereof. The repulsion effect can therefore be regarded as the reaction between the real and virtual magnets.

The number of ampere-turns necessary to provide a reasonable gap is very large indeed. This is made possible by using the cryogenic system as shown in Figure 1.35

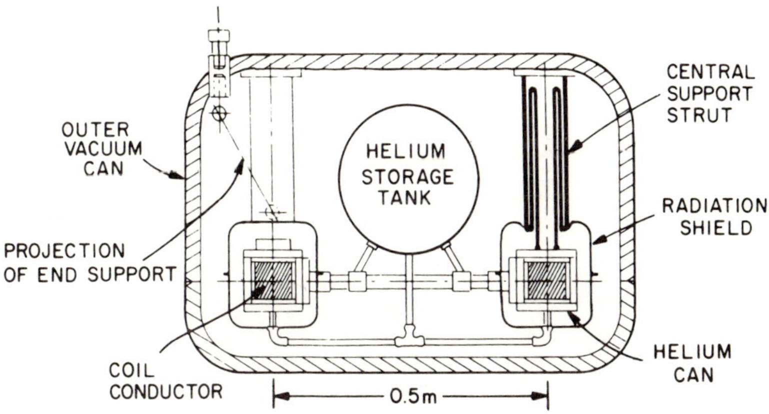

Figure 1.35 Basic configuration of supercooled magnet.

wherein the electro-magnet which is supplied with power from an external source, is kept at extremely low temperatures so that many thousands of amperes can flow with negligible I^2R loss. One of the demerits of the system is that the action of inducing levitation results in a force tending to retard the vehicle. Research is aimed at reducing this to a minimum, but it appears from the results so far obtained that the electro-magnetic drag increases rapidly to a peak at about 100 m.p.h. and then falls off as speed is raised. Aerodynamic drag becomes significant at these higher speeds whilst magnetic drag becomes less important at the higher speeds, so that the overall power requirement can be kept within bounds.

The D.O.T. has expressed interest in the repulsion system because it appears to make less demands for precision of the track and a consortium of German firms is working seriously on the topic. Figures 1.36 and 1.37 show a test track which has

Figure 1.36 Experimental vehicle, supported by a magnetic field, on the test track at Erlangen.

Figure 1.37 Test track at Erlangen showing current supply rails.

been constructed at Erlangen. It will be noted that the track has been canted at a steep angle so as to enable relatively high speeds to be obtained on a track of moderate radius.

1.8 The future of high-speed ground transport

From the technological point of view there are, therefore, at least four main systems of support and guidance to choose from: the conventional wheel on rail, air cushion support, attraction magnetic support and repulsion (cryogenic) magnetic support. As regards propulsion, the conventional wheel can be driven by rotary motors through friction at the wheel rims or by linear motor. All the other types require a linear motor in some form or another. These can of course be combinations such as conventional wheeled support, magnetic guidance and linear motor propulsion. The linear motor may also take numerous forms.

Most important of all, a decision must be made as to the social and economic implications which must govern the installation of any new system. The scale of the investigation, to say nothing of the investment required for actual installation, is such as to necessitate government involvement. Moreover, the type of government involvement required is something which must embrace a time-span of some 15 to 20 years. This is so much longer than the life-span of a parliament that the requisite political basis for a decision would appear to be lacking. The British government decision relating not only to Tracked Hovercraft but to Maplin Air Terminal and the Channel Tunnel, and the German government's reaction to the Krauss Maffei development for Toronto cause doubt as to whether, in fact, a sufficiently long view is being taken.

The universities have a responsibility here, insofar as fundamental studies are concerned and there is a growing, healthy interest in all aspects of transport in a number of universities. Such effort must, however, be abortive in the absence of full-scale facilities whereby the interaction of the various facets of the system can be evaluated on a realistic scale. As regards public demand, it is likely that with the continued development of speed in the air, people will become irritated by the speeds available on the surface, particularly if these are reduced spasmodically by the effects of congestion. Figure 1.38 is based on a diagram produced in 1965[15] but has been up-dated to take into account the Concorde. It will be noted that it had a prophetic quality and one which indicates the supremacy of air from the point of view of speed. Quite clearly the problems of airport location imply that for the shorter journeys, inherently slower transport modes may result in a shorter journey time as is indicated in Figure 1.39, which is the result of recent German studies. These indicate high-speed ground transport (HSGT) is likely to be competitive within a range of distances from 200 to 500 km. This is based on the assumption that the average time required to reach an airport and to board a plane is 1.8 h, whereas the HSGT stations would all be within an hour of the originating point of any journey.

The number of questions to be answered in formulating a forward policy for high speed transport is so great that only a very massive study can produce any quantification. One such study which has been carried out for the German Transport Ministry examines to what extent it will be possible to relieve road congestion by channelling goods and

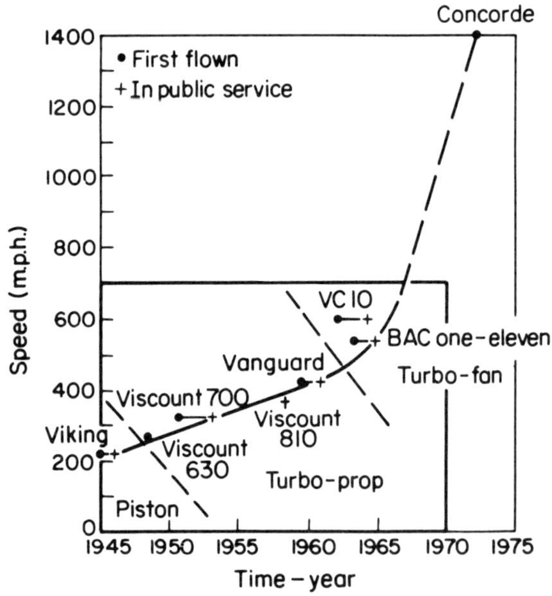

Figure 1.38 Growth of speed in commercial flight.

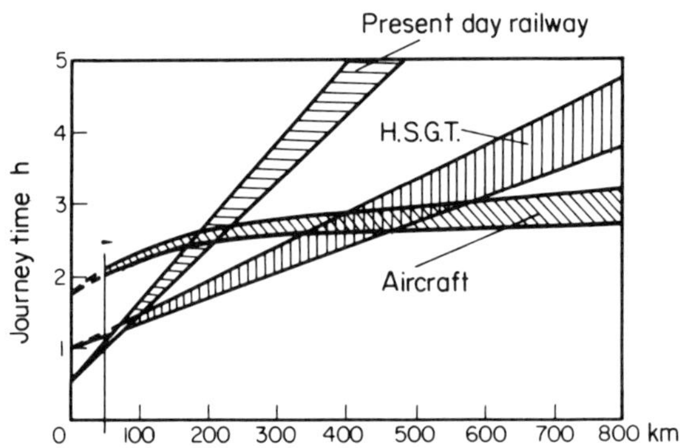

Figure 1.39 Range of application of HSGT.

passenger traffic. The report describes a high capacity electric railway (Hochleistungsschnellbahn) which will henceforth be referred to by the initials HSB and includes a cost benefit analysis, the results of which should allow the responsible authorities to assess the total benefit of an HSB system to the community and especially to investigate the economic effect on other forms of transport. The studies commenced in 1966 and the report is indicative of the thoroughness with which study took place. In fact it has cost 7 million German marks and has involved 150 000 man hours.

A great many assumptions were made as regards route, traffic carried and so forth, and these were eventually reduced to the four systems embodied in Figure 1.40. Here in System I, all traffic, lorries, cars and individuals were carried on a rail system driven by linear motors. System II involved the same technology but the traffic was separated into different classes of trains; cars and lorries going forward at 257 km h^{-1} whilst passengers went forward at 375 km h^{-1}. System III, which would appear to be the most sensible, envisaged cars and lorries going forward by conventional techniques at 225 km h^{-1}, whereas passengers went forward on an air-cushioned linear motor propelled system at 500 km h^{-1}. System IV, is for a combined system but with quite conventional technology. Some consideration was given to the size of the vehicle and it was decided that a vehicle which was 4.3 metres wide and 5.6 metres high would be the ideal solution insofar as it would carry two cars, one above the other, or a single lorry. The cross-section of the proposed vehicle is compared with the British and American load gauges in Figure 1.41.

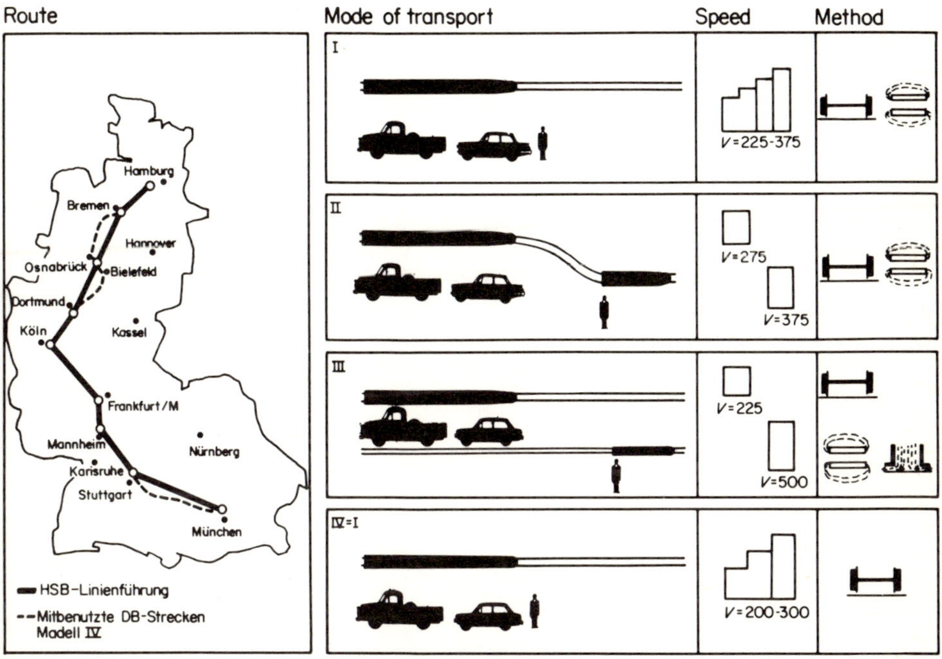

Figure 1.40 Four systems of HSGT.

Because of the time taken to load wheeled vehicles, intermediate stops would be avoided and trains would operate between two stations only. There would be nine stations so that 72 trains would be needed to complete a single traffic cycle. The system finally selected consisted of trains 400 metres long operating at hourly intervals at a speed of 300 km h^{-1}.

Cost estimates were prepared for various systems. For the conventional trains operating at 225 km h^{-1} the capital cost was £2800 million, whilst a system using

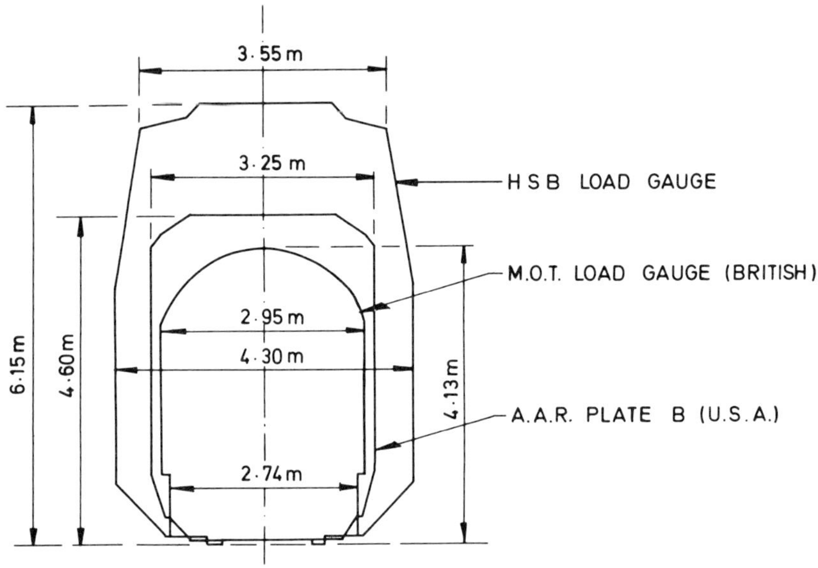

Figure 1.41 Comparison of HSB with British and American load gauges.

air-cushion or magnetic suspension capable of 500 km h^{-1} would cost £4400 million.

The result of a cost benefit analysis is illustrated in Figure 1.42, in which the cost of providing a service is plotted against speed. Receipts, shown here as 'balance' are

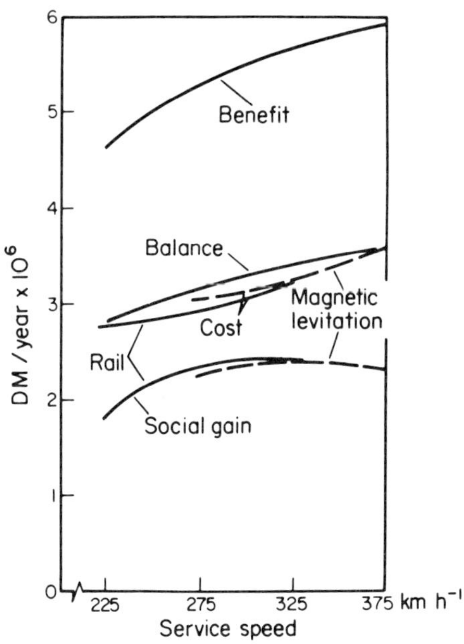

Figure 1.42 Results of cost benefit analysis of HSB.

arranged to just marginally cover the cost, but the benefit to the community in lost time, saving in investment on alternative modes such as motorways, energy savings and so forth are placed on the highest line labelled 'benefits'. Subtraction of cost from benefit gives the social gain thereof. Though the benefits are greatest at the highest speeds studied involving magnetic levitation, the costs are also higher leaving a slight reduction in the social gain. The conclusion from the diagram was that conventional rail at 300 km h^{-1} would be the ideal from the point of view of the community.

The report emphasises that forward predictions are uncertain. Economic prediction relating to the year 1980 might be only 80% correct and technical estimates only 30% correct. Whilst, therefore, the cost benefit studies appear to be a useful way of avoiding colossal blunders, they are no substitute for imagination and technical courage. Surely Brunel[16] would have gone for the highest speed possible and he would have been right.

References

[1] Storey, W. E. (1975) To-day's R & D. Building a Base for Better Transportation, U.S. Department of Transportation, April 15th, 1975.

[2] Barwell, F. T. (1973) *Automation and Control in Transport*, Pergamon Press, Oxford.

[3] Improved High Speed Rail for the North East Corridor, U.S. Department of Transportation, January, 1972.

[4] Evans, A. W. (1968) 'A study of inter-city travel between London, the West Midland and the North West of England', University of Birmingham Department of Transportation and Environmental Planning Publication, No. 38, January, 1968.

[5] Sobermann, R. M. (1975) 'Choice for the future, summary report', Metropolitan Toronto Transportation Plan Review, Toronto, Canada, January, 1975.

[6] Nayak, P. R. (1972) 'Contact vibrations of rolling discs', *J. Sound and Vibration,* 22, pp. 297.

[7] Slemon, G. R., Turton, R. A., Burke, P. R. and Dewan, J. B. (1974) ' "Analysis" and control of a linear synchronous motor for high speed ground transport', IEE Conference, Publication No. 120, pp. 163–148.

[8] Fellows, T. G. (1974) 'High speed ground transport', *Railway Engineering Journal,* 3, No. 2, pp. 4–20.

[9] Select Committee of Science and Technology, Third Report, 'Tracked Hovercraft Ltd., Session 1972–3'. London. H.M. Stationery Office, July 1973.

[10] Fujiwara, S., Fujie, I. and Usami, Y. 'Studies on linear motors in the Institute of J.N.R.' IEE Conference Publication No. 120, pp. 131–136.

[11] Siefert, W. W. (1965) 'Survey of technology for high speed ground transportation, Massachusetts Institute of Technology'. June 15, 1965.

[12] Chirgwin, K. M. (1974) 'Test result from the U.S. linear inducation motor research programme', IEE Conference Publication No. 120, pp. 236–261 + addendum

[13] Jayawant, B. V., Hodkinson, R. L., Wheeler, A. R. and Warton, R. J. (1974) 'Transducers and their influence in the design of magnetically suspended vehicles', IEE Conference Publication No. 117, pp. 200–206

[14] Abel, E., Corbett, B. E., Mulhall, R. E. and Rhodes, R. G. (1974) 'Levitation and propulsion of guided vehicles using superconducting magnets', IEE Conference Publication No. 120, pp. 223–229

[15] Barwell, F. T. (1962) 'Some speculation on the future of railway mechanical engineering', *Proc. I. Mech. E.,* **176**, pp. 61–106
[16] Rolt, L. T. C. (1957) *Isambard Kingdom Brunel*, Longmans Green, London. (Also in Penguin edition).

2

Linear Electric Motor Propulsion Systems

K. CHIRGWIN

2.1 The background

Both of the programmes which will be described in this chapter trace their history back to an activity called the North East Corridor Project, which was set up during the Kennedy administration in the US Department of Commerce (there was no Department of Transportation at that time). The North East Corridor Project was charged with studying the future transportation needs of the highly urbanised area along the Atlantic Coast encompassing the cities of Washington, New York, and Boston.

The North East Corridor Project contracted with Massachusetts Institute of Technology (MIT) to perform a study and to make recommendations. The MIT report recommended that a technology development programme be set up to examine the use of linear induction motors as a means of propulsion for high-speed, guided ground vehicles. Both the linear induction motor research vehicle (LIMRV) programme and the tracked levitated research vehicle (TLRV) programme form part of the implementation of this recommendation.

2.2 History of the linear induction motor research vehicle (LIMRV) programme

The LIMRV programme started in March 1966 with the contract award to The Garrett Corporation to study the linear induction motor and its feasibility for high-speed ground transportation [1]. In October 1967, design and fabrication began on a full-scale linear induction motor (LIM) capable of a continuous output of 1860 kW at 400 km h^{-1}. Early in 1968, work commenced on the design and fabrication of a vehicle in which to test this LIM. An instrumentation and remote control system was added later that year. The official roll-out of the completed vehicle took place in Los Angeles in December 1969.

After about a year of low-speed checkout and testing on a 400 m test track in California, the vehicle together with its instrumentation and control van were transferred to the new Transportation Test Center near Pueblo, Colorado, where the US Department of Transportation had built a special 10 km high-speed test track.

In May 1971, dedication ceremonies took place at the Transportation Test Center near Pueblo. As part of these ceremonies, the LIM vehicle, with the then Secretary of Transportation the Honorable John A. Volpe on board, was operated at speeds up to 150 km h^{-1} (Figure 2.1).

In August 1974, the vehicle first exceeded its designed speed of 400 km h^{-1} on the same 10 km test track at Pueblo. Testing on this track at speeds up to 400 km h^{-1} is expected to continue to the end of 1976. In 1977, it is planned that the vehicle will be used for testing another (single-sided) LIM configuration.

Figure 2.1 Linear induction motor research vehicle (LIMRV).

2.2.1 *The LIM*

The many different configurations that are possible with a LIM, and some of the advantages and disadvantages of these configurations, have been discussed elsewhere [2]. Suffice it to say here that the configuration chosen for the test vehicle is categorised as a double-sided, short-primary LIM. The arrangement is shown diagrammatically in Figure 2.2.

The LIM's main characteristics are listed in Table 2.1. Figure 2.3 shows the two motor primaries in the inverted U-frame mounted on an assembly dolly. Figure 2.4, which is an underside view of the vehicle, shows the motor complete with its guidance and air cooling subsystems.

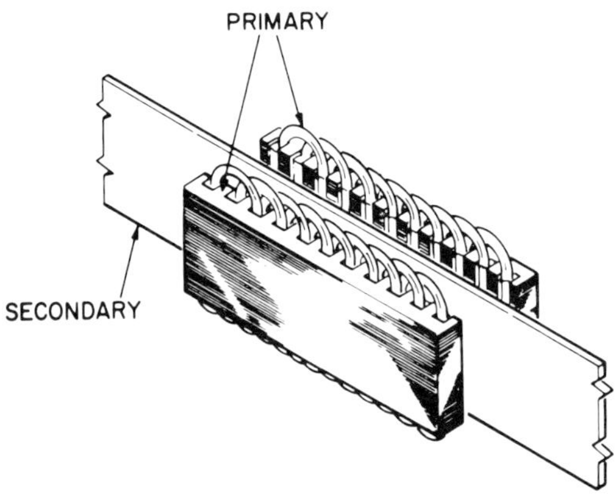

Figure 2.2 Double-sided, short-primary LIM.

Table 2.1
Salient features of the 1860 kW linear induction motor for a 400 km h^{-1} rail vehicle

Item	Salient features
Design point rating	The design point is 1860 kW or 16 650 N thrust continuous at 400 km h^{-1}, which requires 1000 V, 3-phase, 173 Hz, 2000 A, 3500 kVA input. NOTE: At lower speeds, the power, voltage, and frequency are reduced, while the thrust is maintained at 16 650 N continuous.
Motor dimensions	The stack height is 254 mm, the core length is actually 3.81 m (3.556 m effective); the pole pitch is 356 mm, and the primary-to-primary air gap can be adjusted from 28.6 mm to 66.7 mm. Air gap design point is 38.1 mm.
Weight	The electromagnetic weight is 1542 kg; the total weight including support, guidance, connecting cable, and housing is 2950 kg.
Winding	Double-layer diamond-coil, lower-coil ends bend at 90 degrees to minimize height. Windings are 10 pole, 3-phase, 15 slots/pole, 2/3 pitch. Each primary is connected in series Y, and the two primaries are connected in parallel. Ten of the slots at each end are half filled.
Guidance	The LIM is attached directly to the vehicle in heave and pitch. For the direction of travel, there is limited LIM-to-vehicle movement (for thrust measurement). In lateral, roll, and yaw motions, the LIM is free to move with respect to the vehicle but is guided to the reaction rail by the guidance wheels.
Cooling	The LIM is air cooled by separate blowers for each primary. Cooling air passes over the top-end turns through cooling holes in the core, continues to flow over the lower-end turns, and then exhausts to the atmosphere.

Figure 2.3 1860 kW, 400 km h^{-1} LIM.

Figure 2.4 LIM with guidance wheels installed on vehicle.

2.2.2 *The vehicle*

The vehicle (Figure 2.1) was built primarily to provide a means for testing the LIM under real-world conditions. Because of this primary purpose, certain short cuts were permitted in order to concentrate on a programme to test the motor in as timely a manner as possible, rather than attempt to solve all the problems at the same time.

For example, the vehicle was supported and guided on more-or-less standard railway bogies running on standard gauge railway track, whereas revenue vehicles will probably have exotic suspension and guidance systems. An onboard prime mover power source (2230 kW gas turbine) was used, whereas revenue vehicles will almost certainly use wayside electric power.

In the interest of personnel safety, the vehicle was arranged for remote monitoring and control from an instrumentation van positioned near the test track; however, facilities are provided for operating the vehicle from a cab in the front of the vehicle This same van is used to record and partially process a majority of the test data generated by the vehicle.

Table 2.2 lists pertinent data on major components of the vehicle. Figure 2.5 is a ghost view of the vehicle and shows the location of most of these components. Figure

Table 2.2
Salient features of the 400 km h^{-1} rail vehicle that uses a 1860 kW LIM

Item	Salient features
Vehicle dimensions and weight	The vehicle, which includes a cab for a driver and engineer, is 16.6 m long, 2.13 m high, 3.0 m wide. Dry weight is 22t.
Main turbine alternator	The prime mover is a 2230 kW free-turbine gas turbine (G.E. T64) that can operate up to 13 600 r.p.m. and provides a torque of 1827 N m maximum. A 2.613-to-1 stepdown gearbox is incorporated. The 4 pole synchronous alternator has a maximum 1000 V, 3-phase, 173 Hz, 2000 kW power output at 5200 r.p.m.
Auxiliary power supply	A 142 kW gas turbine supplies (1) compressed air for starting the main turbine and for pneumatic brakes, (2) 120/208 V, 3-phase, 400 Hz power for blowers and excitation of the main alternator, (3) 115/230 V, single-phase, 60 Hz power for air conditioning, lights, and instrumentation, and (4) 28 V d.c. power for actuators and instrumentation.
Bogies	Two railway bogies (Budd – Pioneer III) constructed for 1435 mm gauge rail; each has a 2.29 m wheelbase and 965 mm wheels (so the axles clear the reaction rail). All spring rates and damping coefficients are adjustable by substitution or change of air pressure.
Brakes	The three types of vehicle brakes are (1) dynamic, using LIM and 2000 kW onboard resistor that is ram air cooled, (2) aerodynamic, using two doors deployed in the airstream, and (3) friction, using eight ventilated disk brakes that are electro-pneumatically operated.
Instrumentation	Instrumentation is used for (1) gathering, processing, and telemetering of test data, (2) onboard control and monitoring of vehicle operation from the cab, and (3) remote control and monitoring of the vehicle.

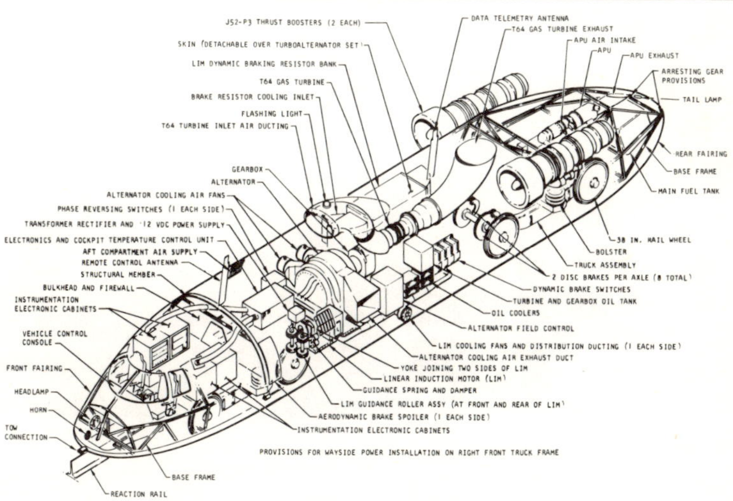

Figure 2.5 General arrangement of equipment on LIMRV.

2.6 is a view looking down on the vehicle with its centre cover removed and shows the main turbine and alternator installed in the vehicle.

Figure 2.6 Turbine alternator installed in vehicle.

2.2.3 *The reaction rail*

The reaction rail, or the long, stationary, secondary member of the LIM, is a comparatively simple piece of hardware, but because it is as long as the railway, it becomes a major cost item. The programme required that a considerable amount of effort be devoted to minimising the cost of both producing the reaction rail and installing it on the track.

The material finally chosen was 6061-T6 aluminium alloy. The form is a hollow extrusion in an inverted T-shape 559 mm high, 15.9 mm thick, and 127 mm wide at the base. Except at the base, the thickness of the metal wall is 3.3 mm.

The extrusions come in lengths of about 30 m, which are welded in the field to form a continuous reaction rail with no expansion joints. The continuous rail is installed in the centre of the track on aluminium tie plates. The rail is accurately aligned by means of shims placed in the tie plates. The rail is pretensioned by heating it to about 140°F and then spiking it down to the tie plates while still hot. This arrangement keeps the rail straight and avoids buckling, even on the hottest day.

2.2.4 *Testing of the LIMRV*

The first tests performed on the LIMRV were checkout and functional tests designed to assure that all the many vehicle subsystems were operating in a safe and satisfactory manner. When these tests had been completed, the testing proceeded to speed-upgrading tests, which were concerned mainly with wheel, rail and vehicle dynamics, and were designed to assure that the vehicle could proceed safely from a low to a higher speed, until finally the designed speed of 400 km h^{-1} was achieved.

While some incidental testing of the LIM was performed during both the functional and the speed-upgrading tests, serious systematic testing of the LIM did not start until the speed-upgrading tests had been completed. At the time of writing, this LIM testing phase has almost been completed.

During the remainder of 1976, testing is proposed which will explore LIM end effects with less than the whole LIM energised. When these tests are complete, it is planned to convert the vehicle, the LIM, and the test track so that a single-sided LIM configuration can be tested.

Presentation of the tremendous amount of test data that has been generated by this programme is beyond the scope of this article. Small sections of the test data have been published [3, 4]. The complete data will be presented in Department of Transportation publications which are presently in preparation and which it is hoped will be available in 1977 [11, 12, 13].

2.3 History of the tracked levitated research vehicle (TLRV) programme

In 1968, the US Department of Transportation embarked on a second high-speed technology programme. Because of what had been accomplished by the LIMRV programme by this time and because of other advances in the technology, this programme did not need to take the same short cuts that had been necessary with the LIMRV programme. It was recognised, however, that this programme would leave many problems still unsolved.

It was decided that the vehicle should be levitated and guided without wheels, and that its suspension system should be designed to explore effective ways of providing a high level of passenger comfort in high-speed guided ground vehicles.

Also, it was decided that the propulsion system should be all electric and should pick up power from a wayside distribution system. To overcome drag at high speed, a considerable power rating is required. It was anticipated that, at least initially, testing would have to be performed on a track of limited length, so high acceleration and deceleration were considered important. This requirement and limitations of the vehicle levitation system dictated a low weight for the propulsion system.

To obtain high deceleration during braking, the propulsion system is used to provide a major part of the braking effort. The power represented by this electrical braking is large and the most convenient sink for this power is the wayside electrical distribution system. Accordingly, the propulsion system is designed to have regenerative braking, with a maximum regenerative braking capability equal to the maximum power capability during motoring.

Figure 2.7 Tracked levitated research vehicle (TLRV).

In 1969, a contract was awarded to Grumman Aerospace Corporation to design and build a TLRV, and in 1970 a contract was awarded to The Garrett Corporation to design and build a second-generation LIM propulsion system for this vehicle that included on-the-vehicle power conditioning. In 1971, another contract was awarded to The Garrett Corporation to design and build a wayside electric power pick up for use with this vehicle and its power conditioning system. The completed vehicle was unveiled at the Grumman Plant in April, 1972. It is now at the Transportation Test Center near Pueblo, Colorado, and has completed low-speed levitation tests (see Figure 2.7).

The LIM and the onboard power conditioning equipment have completed static laboratory tests both in California and at Pueblo. The equipment is now installed on the vehicle in Pueblo and has been operated on the vehicle at speeds up to 65 km h^{-1}. The wayside power pickup system has been tested at speeds up to 500 km h^{-1} on a test sled at the rocket test facility at China Lake, California. It successfully transferred the equivalent of 17 MW of power at this speed. The wayside and vehicle parts of this system are now installed in Pueblo and have been operated at speeds up to 65 km h^{-1} [5].

Because of the present critical shortage of funds for long-range development, this TLRV programme has been redirected; it is planned to convert it to a propulsion system development programme for more conventional (lower speed) vehicles.

2.3.1 *The LIM*

The second generation LIM built for the propulsion of the TLRV uses the same double-sided, short-primary configuration shown in Figure 2.2 that was chosen for the LIMRV programme. This LIM's main features are listed in Table 2.3, and Figure 2.8 shows three views of the motor.

Table 2.3

Characteristics	Value
Weight	2030 kg (including air cushions)
Thrust	22 200 N continuous, 33 300 N overload
Power	2980 kW continuous
Voltage	4120 V r.m.s. line-to-neutral, 3-phase ungrounded neutral
Frequency	Variable to 165 Hz at 480 km h^{-1}
Primary current	530 A at rated thrust and speed, 700 A overload
Power factor	57.5% at rated thrust and speed
Efficiency	80.5% at rated thrust and speed
Reaction rail	6061-T6 Al alloy, 6.35 mm effective thickness, 0.60 m high
Air gap (electrical)	38 mm primary-to-primary

As explained in Section 2.3, a very low weight for both the LIM and the power conditioning system is required. Weight savings were made by many different means, and the largest contributor was the direct water cooling of the windings of the electrical machines using water circulated through hollow copper conductors. The cooling water manifolds are visible in Figure 2.8(c). When compared with the LIM for the LIMRV, this motor has a power-to-weight ratio that is 2.3 times larger.

To be able to use a small air gap in the LIM and still avoid contact between the moving LIM primaries and the fixed reaction rail, the support and guidance of the LIM is separated from that of the vehicle, and there is essentially complete freedom of movement between the vehicle and the LIM in all degrees of freedom except the direction of travel. The LIM has four circular air cushions for levitation that react against the guideway floor and four rectangular air cushions for the guidance that react against the vertical sides of the reaction rail.

(a)

(b)

(c)

Figure 2.8 Linear induction motor: (a) end view with reaction rail; (b) stack assembly; (c) housing assembly.

2.3.2 *The test vehicle*

The vehicle for the TLRV programme is shown in Figure 2.7 with the near wall of the guideway removed, exposing the levitation and guidance cushions and the friction brake blocks. A ghost view of the vehicle shown in Figure 2.9 indicates the LIM position and the power conditioning equipment position.

The propulsion system comprising the LIM and the power conditioning equipment for this vehicle were designed as two identical systems of half rating. This was done for a number of reasons, which included the ability to use a back-to-back laboratory test set-up allowing the power conditioning system to be tested at full power.

Because of a shortage of funds, only a single half system was built and installed on the vehicle. The arrangement shown in Figure 2.9 is for this single half system. It was planned to build the second half system at a later phase of the programme. When this happened, it was planned to replace the present boiling water heat sink for the cooling system with an air-to-water heat exchanger at the front of the vehicle. This would make room in the centre compartment of the vehicle for the second set of the power conditioning equipment. The second LIM would be mounted under the vehicle just ahead of the existing one shown in Figure 2.9.

Table 2.4 lists data relating to the propulsion system of the vehicle.

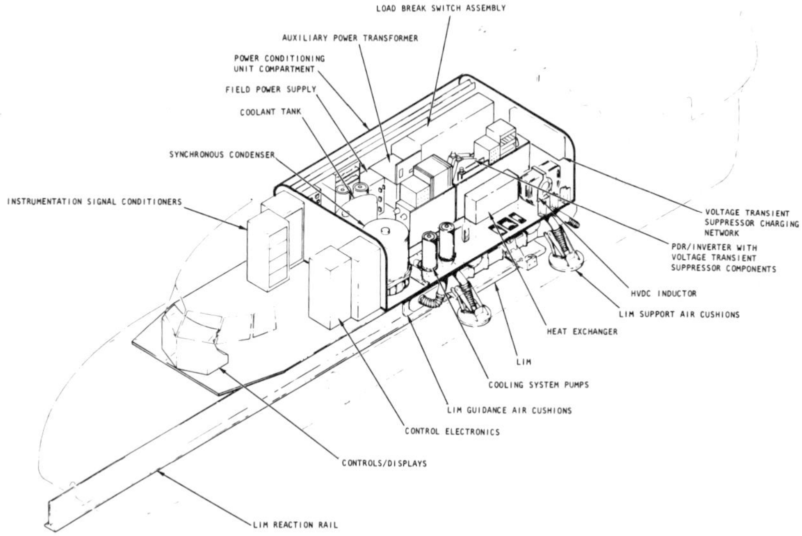

Figure 2.9 Linear induction motor propulsion system installed in the TLRV.

Table 2.4

Propulsion system specification (complete system)

Weight inside vehicle	8500 kg
Total weight	12 400 kg
Thrust for acceleration and deceleration	66 600 N
Continuous thrust	44 400 N
Maximum speed	480 km h^{-1}
Continuous power output	6000 kW
Power supply	8.2 kV, 3-phase, 60 Hz
Stopping distance	3200 m from 480 km h^{-1}

2.3.3 *The power conditioning system*

A simplified schematic of the power conditioning system adopted for the TLRV is shown in Figure 2.10. Because of size and weight constraints, it is not possible to use an input transformer in the power conditioning system. Therefore, it is necessary to operate the LIM and all the power components of the propulsion system at a maximum voltage almost equal to the 8.2 kV input voltage.

During drive operation, the input converter is a phase delay rectifier (PDR), which uses thyristors as the switching element. It converts the relatively fixed three-phase

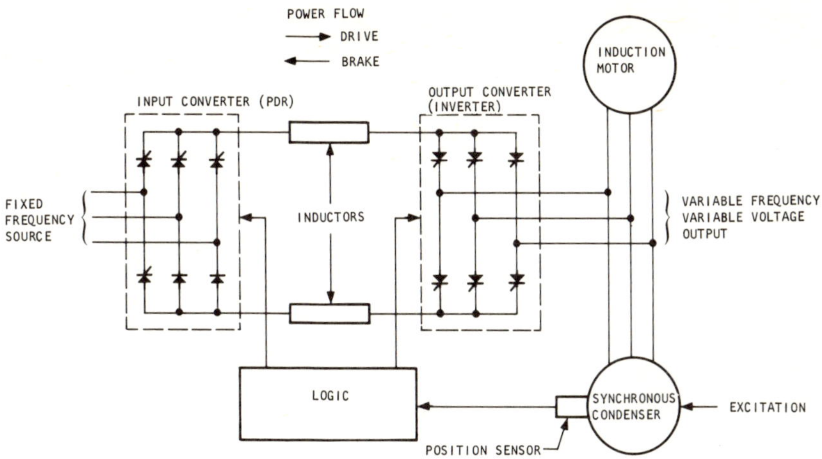

Figure 2.10 Propulsion system simplified schematic.

a.c. voltage input to a controlled variable voltage d.c. This d.c. power is fed by way of the coupled smoothing inductors to the output converter, which, during drive operation, is an inverter that also uses thyristors as the switching element. It converts the variable voltage d.c. input to variable voltage, variable frequency (approximately constant volts per Hertz), 3-phase a.c. output power. This variable frequency a.c. power is fed to the LIM and to a synchronous condenser in parallel. The functions of this (variable speed) synchronous condenser are to: supply the lagging reactive kVA required by the LIM, supply the lagging reactive kVA required by the inverter, provide three-phase sinusoidal voltages that commutate the (line commutated) inverter, and absorb most of the current harmonics generated by the inverter.

The inverter is shown in Figure 2.11. The phase delay rectifier is similar to this inverter. Figure 2.12 shows the synchronous condenser.

2.3.4 *The wayside power pickup*

The wayside power pickup for the TLRV is designed to pick up 17 MW of power at 8.2 kV, 3-phase, 60 Hz, while travelling at speeds up to 480 km h^{-1}.

A view looking into the open end of the wayside installation is shown in Figure 2.13. This view shows the three hat-shaped power rails mounted with 120 degree spacing on a 3/4-circle support insulator. Each power rail carries a copper strip pointing towards the centre. The white object visible in the centre is the collector. This is captured within the rails, and it is a passive device supported and guided from the copper strips of the three power rails.

Figure 2.14 shows the double-ended, missile-like collector, in the right foreground, and the movable telescoping tow arm and the vehicle attachment. Three pairs of current collecting brushes and one pair of guidance brushes are visible in one phase on each end of the collector. Each of these brush pairs pinch on the copper strip of the power rail.

Figure 2.11 TLRV inverter.

The conventional pantograph-type collector relies on upward pressure between the collector elements and the overhead wire to maintain contact. At high speeds, the dynamics of this arrangement may result in separation and consequent arcing between the wire and the collector. The arrangement used with the TLRV collector, where the collector brushes pinch on either side of the copper strip, is capable of maintaining contact at much higher speeds than the conventional arrangement. This results from the fact that misalignment in the plane of the copper strip has a negligible effect on contact pressure. Misalignment perpendicular to this plane, together with the inertia of the collector brushes, cause a decrease of contact pressure on one side and an increase in contact pressure on the other side, consequently, one or another of these collector brushes will tend to be in contact.

Contact with at least one brush will be maintained up to some very high speed, where one brush has not regained contact after a disturbance before the other brush is separated by another disturbance in the opposite direction. The 500 km h^{-1} test demonstrated that contact was in fact maintained at all times.

(a)

(b)

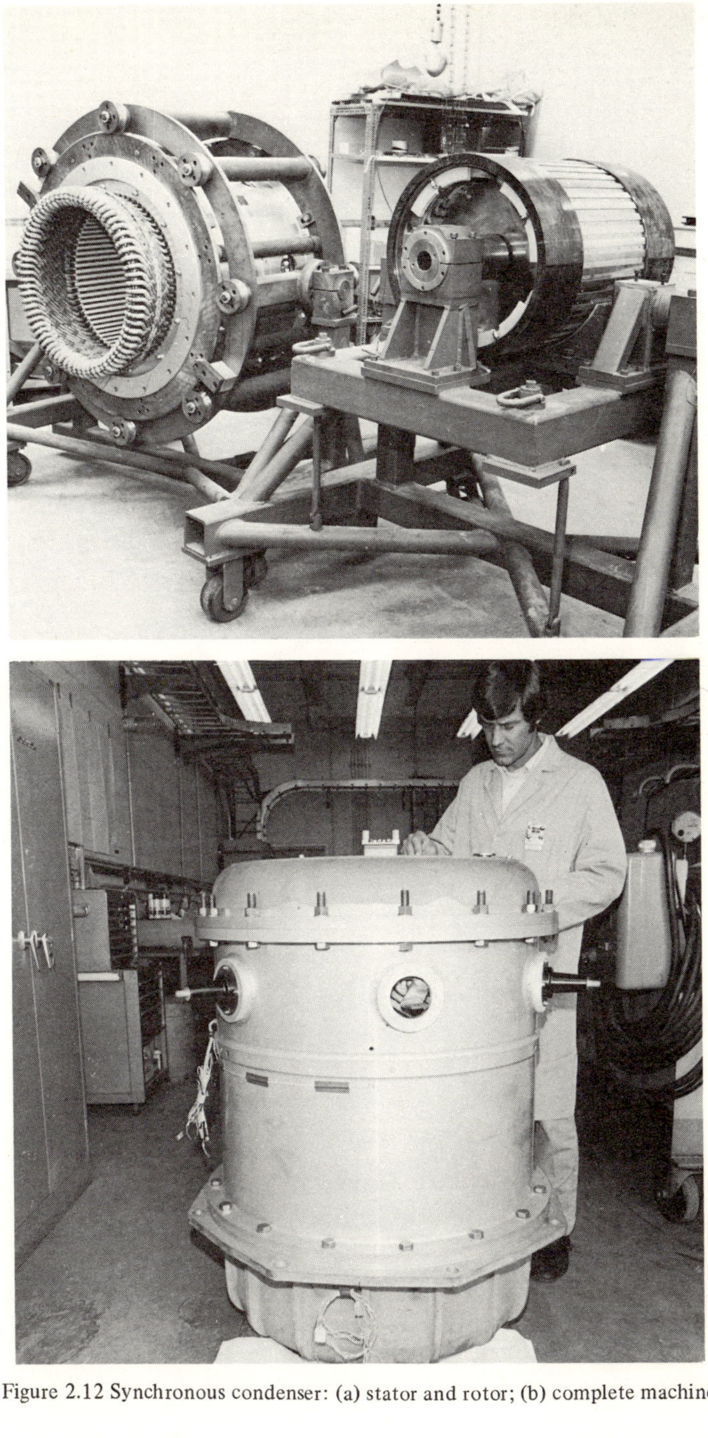

Figure 2.12 Synchronous condenser: (a) stator and rotor; (b) complete machine.

Figure 2.13 Wayside power pickup system.

Figure 2.14 Wayside power collector.

2.4 Reasons for the choices made

The field of propulsion by linear electric motors is new and uncharted. There are many choices that must be made with no precedent as a guide, so all the important aspects of these choices must be carefully considered. This tends to make the decision process complicated. In an article of this length, it is not possible or desirable to go into the details of the tradeoffs and interactions that entered into each of the decisions made. Instead, the more important factors in the more important choices are briefly discussed below.

2.4.1 *Active vehicle versus active guideway*

With an active guideway system, the linear electric motor mounted in the guideway may be either a linear induction motor or a linear synchronous motor. (The linear synchronous motor together with its associated variable frequency power conversion equipment could be categorised as a linear brushless d.c. motor if the input power to the conversion equipment is d.c.) The passive vehicle part of this arrangement carries some form of d.c. magnet for a linear synchronous motor or some form of short-circuited conductor for a linear induction motor. In either case, the cost of the vehicle part of the propulsion system is a small part of the total, and it is the guideway part that largely determines the cost and economic viability of the system.

With an active vehicle system, the vehicle carries the motor and the power pickup and may also carry the power conversion equipment. The guideway has the reaction rail, the wayside power distribution system, and it may have the power conversion equipment. So there are items having substantial cost effect both on the vehicle and in the guideway.

To have the most meaningful tradeoff between an active vehicle and an active guideway system, it is necessary to know in detail the requirements of the application, then to determine the best active vehicle system to fit these requirements and the best active guideway system to do the same, and then compare these two best systems. The information needed to make this kind of tradeoff is obviously not available at present.

Because of the nonavailability of this type of information, the MIT North East Corridor study made a judgement that with the train frequencies that seemed likely, the active vehicle system appeared to be a more attractive economic solution. This choice, then, had been made before either of the LIM programmes had started.

2.4.2 *Fixed versus variable frequency*

Obviously, it is preferable to use a fixed frequency equal to that of the electric utility system (50 or 60 Hz), if at all possible. A propulsion system must operate at variable speed, so fixed frequency implies operation at large slips and therefore large losses for at least some part of the time. With conventional fixed frequency, rotating synchronous or induction motors, operation at large slips occurs during starting and can be tolerated for a few seconds provided it is not repeated too often. In a vehicle application, this starting period may last for time periods as long as 5 minutes, and for even longer times if there are civil or other speed limits lower than maximum speed. Such times would be quite impossible for conventional rotating machines of several thousand kilowatt rating.

With linear machines of this size, the secondary heating is not a problem above some speed because sufficient cold reaction rail enters the motor. But problems of primary winding heating, heating of the reaction rail at low speeds and of large energy consumption remain.

With a fixed frequency linear induction motor, these remaining problems can be reduced in severity by the use of a pole-changing motor[6]. With a variable frequency supply to the linear motor, none of these problems occur. The only disadvantages are the size, weight, and cost of providing the variable frequency power. For both of the programmes discussed, it was decided to incur these costs and use variable frequency.

2.4.3 *Maximum operating frequency*

Having accepted the need for a variable frequency supply to the LIM, the choice of the maximum operating frequency is open. In both programmes, this frequency was chosen to give a LIM pole pitch about one and one-half times the LIM stack height. This, with a coil of 2/3 pitch, gives a coil throw about equal to the active length of a coil side, which has been found to be about optimum for conventional rotating electrical machines.

2.4.4 *Double-sided Versus Single-sided LIM Configuration*

At the start of the LIMRV programme, so little was known about the single-sided LIM that there was no choice but to use the double-sided configuration. However, by the time the TLRV programme came along there was a choice.

Because the double-sided LIM uses both sides of the reaction rail, it needs a smaller but thicker reaction rail, and results in a more compact motor arrangement. The single-sided LIM needs a larger area of reaction rail and needs some amount of back iron behind the reaction rail to complete the magnetic circuit. Because this iron is probably as long as the railway, it is a considerable cost penalty. The double-sided LIM always maintains an essentially constant air gap, and the position of the reaction rail within this air gap has a negligible effect on motor performance. With a single-sided LIM, the air gap depends upon the LIM levitation system employed and is probably quite variable. These air gap variations will cause variations in motor performance.

For the TLRV propulsion system, it was decided to accept the less convenient guideway arrangement rather than deal with the new problems raised by the single-sided LIM. However, it is planned to convert the LIMRV to a single-sided configuration when its double-sided tests have been completed.

2.4.5 *Linear synchronous versus linear induction motor*

For high-speed vehicle propulsion applications, both linear synchronous and linear induction motors will have main air gap lengths that are very large when compared with a conventional rotating machine of the same type and rating. This means that an unusually large number of ampere-turns will be needed to magnetise their main flux paths. Producing these ampere-turns in a d.c. field winding, as it is done in a linear synchronous motor, is preferable to providing them in the primary winding, as it is done in a linear induction motor. This is because the synchronous machine's d.c.

field has only copper losses, whereas in the induction machine the magnetisation of the main flux path represents substantial lagging reactive kVA in the primary winding.

However if, as in TLRV, one has a conventional rotating synchronous machine in parallel with the linear induction motor, then the combination behaves in many respects as a linear synchronous motor, and the reactive kVA are produced much more effectively in the rotating synchronous machine than they can ever be in a linear synchronous machine. If the choice is between a linear synchronous motor and a combination linear induction motor with parallel rotating synchronous condenser, then factors other than reactive kVA are dominant. In the case of TLRV, one of the important factors was that with a double-sided LIM there is essentially zero interaction between its guidance system and the production of thrust.

2.4.6 *Line commutated versus forced commutated inverter*

Inverters used to power induction motors in variable speed industrial applications are usually of the voltage fed, forced commutated type. Because of the high power (6 MW), high voltage (almost 10 kV d.c.) and the size and weight constraints of the TLRV application, a current fed, line commutated inverter was used. This is the simplest of all inverters and it has the ability to ride through a commutation failure and resume normal operation after a cycle or two, simply by a change in thyristor firing times. It is the type of inverter used on the receiving end of high voltage d.c. transmission systems, so there is considerable operating experience at very high power levels and very high voltages. The choice of this type of inverter does require that the load contain a synchronous machine that would not be needed with a forced commutated inverter.

2.4.7 *LIM Guidance*

The simplest arrangement of the LIM in the vehicle is to rigidly attach the LIM to the vehicle. This requires that the clearance between the LIM primaries and the reaction rail is large enough to accommodate both the vehicle dynamic motions with respect to its guideway elements and the misalignments between the reaction rail and these vehicle guideway elements, without producing physical contact between the moving motor primary and the stationary reaction rail.

For the LIMRV, it was practical to do this for vertical (heave and pitch) motion, but to do it for lateral motion would have required an unacceptably large LIM air gap. Accordingly, on this vehicle the lateral, roll and yaw motions of the LIM were separated from those of the vehicle, and the LIM was guided from the reaction rail itself in these degrees of freedom by means of guidance wheel sets at each end of the LIM which pressed against both sides of the reaction rail and centred the reaction rail in the LIM air gap (see Figure 2.4).

For the TLRV, the vertical motions of the vehicle body were so large (about 300 mm) that the extra height of the reaction rail needed was unacceptably large. Accordingly, on this vehicle all motions of the LIM, except the direction of travel, were separated from those of the vehicle, and guidance of the LIM from the floor of the guideway and from the reaction rail was provided by the use of eight air cushions (see Figure 2.9).

2.5 Speculation for the future

Conventional steel wheel on steel rail technology presently is capable of providing passenger transportation at speeds as high as 250 km h^{-1} and this will probably be extended to 300 km h^{-1} in the foreseeable future. Because of the high power requirements occasioned by air drag at sea level atmospheric pressure, it seems unlikely that speeds above 500 km h^{-1} will be economical for new high-speed ground transportation systems. Accordingly, these speculations assume the propulsion system will have a maximum operating speed in the range of 300 to 500 km h^{-1}.

A number of countries are doing development work on a high-speed transportation concept that uses superconducting d.c. magnets on the vehicle and aluminium plates in the guideway to provide levitation and guidance. Propulsion is provided by a variable frequency linear synchronous motor with its primary in the guideway and reacting with d.c. superconducting magnets on the vehicle [7] (see Chapters 4 and 5).

Expending effort to develop such an active guideway system implies either that an active vehicle system is not feasible (because of inability to pick up power at the high speed, for example), or that an active guideway system is likely to be preferable from an economic point of view. The TLRV programme will demonstrate that an active vehicle system is feasible and particularly that adequate amounts of power can be picked up from wayside at speeds up to 500 km h^{-1}. The question of the relative economic merits of active vehicle and active guideway schemes is not answered by the programmes discussed here, because they are all based on the premise that an active vehicle is preferable from an economic point of view.

A number of propulsion schemes suitable for an active vehicle are discussed briefly below. They are based on the work done to date on the LIMRV and TLRV programmes. Some of these speculations are extensions of these programmes using conventional technology.

Since equipment using superconducting technology is being suggested for active guideway schemes, it seems appropriate to speculate on the effects this technology might have on active vehicle propulsion schemes. This is a relatively new technology in engineering terms. Briefly, the state of the superconducting art is: superconducting d.c. magnets are well-developed and available, large superconducting unipolar d.c. machines have been built and operated, superconducting a.c. power transmission cables are under test, and a.c. generators with normal stators and rotating d.c. superconducting fields are under test. Machines with superconducting a.c. windings are not yet feasible. In large ratings, the a.c. generators having d.c. superconducting fields have a power density as high as ten times that of a conventional generator [8].

2.5.1 *TLRV system*

The TLRV programme will demonstrate an active vehicle system, suitable for 480 km h^{-1} speed, with wayside power pickup, an onboard power conditioning unit and a LIM propulsion motor. The propulsion system weight is about 0.2 kg per N of thrust. This system is probably viable as it stands. A number of improvements are possible.

2.5.1.1 *Higher voltage* The spacing of the substations supplying power to the wayside power distribution system could be increased, with a consequent reduction of electrification costs, if the wayside voltage could be raised above 8.2 kV.

At 8.2 kV input voltage, the LIM and the synchronous condenser have all coils electrically in series and hydraulically in parallel, which gives as many water connections to the winding as there are coils. For the same power rating and the same water pressure, the number of water connections to the winding would increase as the square of the voltage increases. This limits the voltage increase that is possible with the present direct cooling arrangement.

2.5.1.2 *Single-phase pickup* The cost of the wayside distribution system could be reduced by using single-phase pickup in place of the present 3-phase pickup. The single-phase arrangement would have a single high-voltage power rail and a grounded current return through the reaction rail. The collector would no longer be captive, so some other arrangement would be needed, and there would need to be some form of contact with the reaction rail.

The power conditioning unit would have the input converter changed from 3-phase to single-phase input. The d.c. smoothing inductors would need to be increased in size. There would be no change to the remainder of the propulsion system.

2.5.2 *Variable frequency wayside power*

In a system employing vehicles that have a propulsion system similar to that on the TLRV, it is possible to place the power conditioning equipment in the wayside substation and transmit variable voltage, variable frequency power to each train separately. This has the advantage of reduced weight, space, and complexity of the equipment on the vehicle, but has the disadvantage that there must now be more than one substation set of equipment for each train on the system – except, perhaps, where trains always cruise at maximum speed. At shorter train headways, such a system may be preferable to a straight TLRV system.

Because of the higher frequency and because of the very poor power factor of the LIM, it is even more desirable to increase the wayside voltage with this system, but the same voltage limit exists with the direct-cooled LIM. Because of wayside voltage drop problems, it might be advisable to have the synchronous condenser onboard the vehicle to correct the power factor to unity or leading. Fixed capacitors placed at intervals along the wayside distribution system may also be advantageous.

2.5.3 *Superconducting Kando system*

The original Kando system, built for locomotives of the Hungarian state railways, used a conventional synchronous machine that was both a transformer and a phase converter in that it provided a low voltage, 3-phase, 50 Hz output from a high voltage, single-phase, 50 Hz input[9].

Using present technology with rotating d.c. superconducting field winding and oil cooled stator winding, a superconducting Kando machine having a 25 or even 50 kV single-phase, 60 Hz input and a 3-phase, 60 Hz output at a few thousand volts probably can be built for a weight that is compatible with mounting onboard the vehicle.

No torque, other than that corresponding to rotational losses, is developed by any Kando machine. This gives the superconducting Kando machine an advantage over a superconducting a.c. generator in that the shaft carries very small torque and hence will produce less heat leak into the cryogenic space.

The propulsion system would be completed by a LIM using direct-cooled primary conductors (see also Chapter 11). Since this LIM would have a long pole pitch, a transverse flux machine[10] would probably be used. To reduce slip losses and reaction rail heating, one or more stages of pole changing would be needed[6].

2.5.4 *Superconducting Kando frequency multiplier system*

An extension of the superconducting Kando system is to have the single-phase Kando synchronous motor mechanically drive a synchronous generator with d.c. superconducting field winding, which has more poles than the motor. This new machine will have the same 25 or 50 kV single-phase, 60 Hz input and the same low voltage, 3-phase, 60 Hz output, but will have in addition a low voltage, 3-phase output at a higher frequency, which is 2, 2-1/2, or 3 times the input frequency.

The arrangement of the Kando machine and the synchronous generator would be such that the main fluxes of the two machines are superimposed upon each other in the same air gap. This can be done without penalty since there is no iron in the magnetic circuit to saturate. Such an arrangement eliminates the need for a shaft to transmit torque between the two machines. Instead there are tangential forces between both stator and rotor windings of the two machines which are simply provided for in the construction of the windings.

This superconducting machine combined with a normal pole pitch, direct-cooled LIM similar to the TLRV LIM, but with perhaps one stage of pole changing, will complete the propulsion system.

2.5.5 *Superconducting LIM*

If superconducting wire suitable for operation at low a.c. frequencies and very high current densities becomes available, then it will be possible to make the LIM itself into a superconducting machine.

Because there is no longer any need for iron teeth, such a LIM can be wound for high voltage. It will operate at 60 Hz, so it will have a long pole pitch, and a configuration similar to a transverse flux machine may be advantageous, even though there may be no iron present. If iron is present, it will probably be only a flux shield or a flux return path.

To limit slip losses and reaction rail heating, it may be necessary to use one or more stages of pole changing.

Because of the long pole pitch and because of the very high current densities possible in superconductors, it may be practical to build the LIM with an air gap large enough to accommodate all the vehicle motions. This would eliminate the separate LIM guidance system. It would, however, increase the kVAR requirement of the LIM.

Because electrification and power pickup are difficult as 3-phase power at voltages as high as 25 kV, it is desirable to operate the propulsion system from a single-phase input. This requirement could be met by the use of a superconducting Kando machine used as a phase converter: this can now be an even smaller machine than the ones discussed above because superconductors can now be used in the stator windings as well. Alternatively, the requirement could be taken care of by using a single-phase LIM with some form of phase splitting on at least the largest number of poles. The phase splitting device could be a conventional capacitor. Perhaps a synchronous condenser using superconducting stator and rotor windings would be competitive with a conventional capacitor when there must be liquid helium on the vehicle.

The power factor of the superconducting LIM is no better than that of a conventional LIM and will be worse if the air gap is increased, so power factor correction is probably needed either on wayside or on the vehicle or both. The on-the-vehicle part might use a doubly superconducting synchronous condenser.

References

[1] Lee, C. H. (1967) 'Study of linear induction motor, its feasibility for high speed ground transportation', U.S. Clearinghouse, Springfield, Va., Report PB-174866, 1967.

[2] Laithwaite, E. R. and Nasar, S. A. (1970) 'Linear motion electrical machines', *Proc. IEE*, 58, No. 4.

[3] Chirgwin, K. M. (1971) 'Low speed operation of the LIM test vehicle', ASCE-ASME Transportation Engineering Conference, July 1971.

[4] Chirgwin, K. M. (1974) 'Test results from the U.S. linear induction motor research vehicle program', IEE Conference on Linear Electric Machines, October 1974.

[5] Dannan, J. H., Day, R. N., and Kalman, G. P. (1973) 'Linear linduction motor propulsion system for high speed ground vehicles', *Proc. IEEE*, 61, No. 5.

[6] Eastham, J. F. and Balchin, M. J. (1974) 'Pole-change linear induction motor IEE Conference on Linear Electric Machines, October 1974.

[7] Atherton, D. L. and Eastham, A. R. (1975) 'Superconducting magnetic levitation and linear synchronous motor development – the Canadian program', Cryogenic Engineering Conference, July 1975.

[8] Mole, C. J., Brenner, W. C., Haller, H. E. (1973) 'Superconducting electrical machinery', *Proc. IEEE*, 61, No. 1.

[9] Halacsy, A. A. (1970) 'An early breakthrough in railway electrification', *Electronics and Power*, 16, No. 4, April 1970.

[10] Andree, H. (1904) German Patent No. 181014.

[11] Powell, R. B. (1977) 'Linear Induction Motor Electrical Performance Test', National Technical Information Service (N.T.I.S.), in press.

[12] Powell, R. B. (1977) 'Linear Induction Motor Electrical Braking Test', N.T.I.S., in press.

[13] Powell, R. B. and McConville, J. H. (1977) 'Linear Induction Motor Research Vehicle Reaction Rail Current and Airgap Flux Distribution Test', N.T.I.S., in press.

3

The Astroglide System

G. P. SCELZO

Introduction

Technology has provided the solutions to many of society's problems. It is also often blamed for them. Some people would have us believe that there is too much technology today and that this very abundance is the basis for so many of our ills.

> *'Our main business is not to see what lies dimly ahead in the distance. But to do what lies clearly at hand.'* *Thomas Carlyle*

We believe the solutions to problems lie in the intelligent application of technology and that a misunderstanding of technology's role in man's advancement can only lead to a diminution of his ability to provide solutions.

PRT Systems Corporation has committed its entire resources and pioneering transportation experience to expand transit technology with innovative monorail transit systems. Astroglide represents the first major urban transportation alternative to be presented to cities within the last fifty years. We have developed a technologically advanced computerised and pollution-free system, incorporating scientific advancements especially formulated for Astroglide – linear induction propulsion and magnetic levitation.

Our culture has become very dependent on transportation. Almost everything we are accustomed to in daily living is produced at some distance from where we live and is made accessible through some means of transportation. Approximately one quarter of our fuel energy is currently expended in providing this daily circulation on which we are relying for our food, clothing and other requirements and to bring us to our places of employment, health, educational and recreative facilities. Transportation, in fulfilling the requirements of these functions, has become the largest single user of human, land and energy resources.

The reduction of energy used by transportation is necessary for the maintenance of our society. How long can we continue to pay the price of urban blight, excessive fuel consumption and a depressed construction industry? In this day of man on the moon, we must find better ways to transport people on earth.

People move through our city streets at a pace slower than they did in horse and buggy days. It costs more to collect fares than it does to run trains and buses. The lack of transportation locks people into ghettos with noise and the filth which robs them of human dignity.

In the effort to improve urban environment, urban mobility is a vital consideration. Urban transportation is and must be, integral to the whole national effort to improve our cities and is, in fact, inseparable from it. Transit not only serves the city, but shapes it. People live in cities where they can get to and from home and work. Urban transportation should serve people and not constrain them. However, contemporary transportation is characterised by deficiencies in the areas of urban pollution, congestion, quality of service and the efficient use of land. Air, noise and aesthetic pollution, typical to all current modes of urban transportation, contribute to degradation of the quality of the urban environment.

The Office of Emergency Preparedness has recently issued a report citing the dangers of destroying our nation's energy. The report states: 'The nations' limited fuel resources are being dangerously depleted, and the United States is becoming increasingly dependent upon foreign oil and gas supplies'. Pointing to the energy diet of the United States, the Office said that by 1980 energy consumption should increase by 39% and double by 1990 what it was in 1971. The report states that the nation's energy gobbling can be cut 25% in 20 years if Americans are willing to adopt motor fuel rationing and utilise advanced technology in mass transportation.

With the increased use of the automobile as the primary mode of travel, mass transit facilities have declined. Dependency on the automobile has greatly contributed to the blighting effects of congestion, dislocation of people and business, and air pollution in the cities. In addition, lack of adequate, efficient and economical public transit limits the mobility of non-drivers. It is estimated that nearly one-third of the urban population suffers serious disadvantages from being served inadequately or not at all by mass transit systems.

The Astroglide system

These facts present tremendous opportunities for companies engaged in urban transportation. The Astroglide system is a demand-responsive transportation system. It fulfils a demand because the cars are small compared to trains or buses, and because the service offered is on direct demand from the passenger. The car waits for the man; not he for the car. It is rapid because the cars operate on elevated guideways, free from other traffic so that even at speeds of 30 m.p.h. or less, trip time is considerably less than today's frustrating journeys in congested areas via car or bus. Astroglide is fully automatic and computerised, making use of small cars operating at frequent intervals all day long, economically feasible. The net result is vastly improved service at operating costs which are lower than any conventional form of mass transit. It is designed to move people, baggage, cargo, mail, freight, materials and supplies on a horizontal line, efficiently and automatically.

The new Astroglide system (Figure 3.1) is far superior to the monorail installation Braniff International has in use at Dallas, Love Field. Astroglide is completely automated. It requires no driver. It is precisely pre-programmed to move from one location to another, whether it is 1000 feet or 20 miles, by pushing a button. It is as simple as that.

Astroglide is designed and fabricated to be a functional working piece of equipment. The Astroglide fiberglass shell construction is light-weight and durable. It can be operated with single cars or in multiples. For example, an airport installation could have passenger, baggage and cargo units incorporated into the same system.

Astroglide is safe and reliable and provides the long-needed missing link in transportation by combining an economical balance of service and cost. The basic

Figure 3.1 Astroglide passenger car.

Astroglide design has built-in back-up systems with the ultimate in reliability and redundancy to give safe, sure operation.

Astroglide flexibility provides the freedom to meet existing conditions rather than change the conditions to fit the system. The architect, designer or planner can design the vehicle shape and configuration to suit each specific need.

Astroglide can go up grades, down grades, above ground or under ground.

It can travel suspended from rails or on smooth floors. It can turn around sharp corners; as small as a 15 ft radius or in the smooth arc of a 1500 ft radius. It can be built over, around and through buildings. And as natural growth requires, Astroglide systems can be extended or modified without serious interruption to services.

Braniff International transit service

Early in 1970, Braniff International inaugurated a new transit service at Love Field, Dallas, Texas. The system was (Figure 3.2) tailored to the needs of Braniff International. The system at Love Field was the only one of its type. It connected their portion of the air terminal and a parking lot located some 4200 ft away from the terminal. The objective was to exploit parking space far beyond tolerable walking distance and also to make access to Braniff more attractive than other airlines. It was

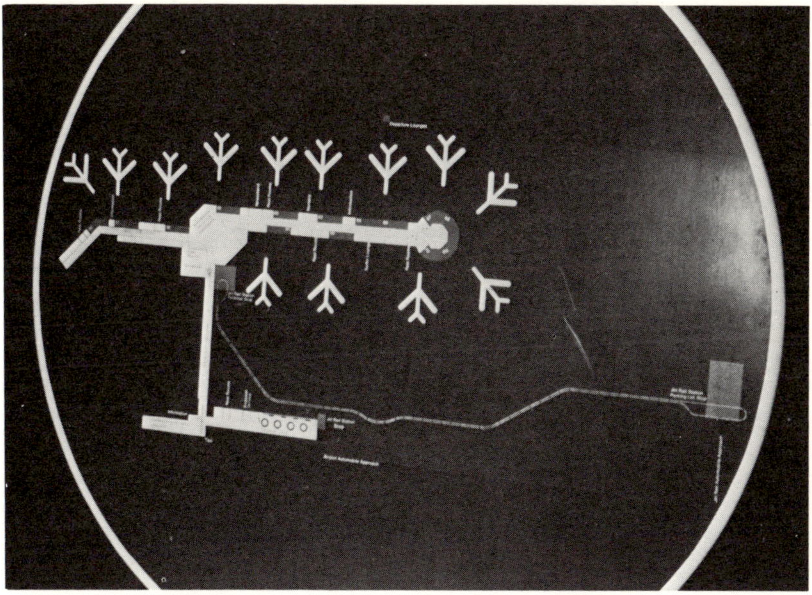

Figure 3.2 Astroglide system guideway diagram.

reported that over 47% of their passengers used the system.

The system employs a single closed loop. Switches and sidings are incorporated at both ends of the loop for empty vehicle storage and at one end for maintenance and cleaning. One terminal of the route is at a building in the parking lot and the other is in the terminal near the aircraft loading gates. A single intermediate station is located on the line to the parking lot at a point near the former baggage retrieval area. The guideway is an overhead monorail about 8400 ft in length located some 20 ft above grade in double guideway configuration with loops at each end.

Vehicles normally carry up to 10 passengers with six seated and four standing and up to 14 with crowding. Minimum headways are 10/20 seconds and maximum capacity is 2000 passengers per hour in each direction. The system has ten vehicles, each car has a total gross vehicle weight (GVW) of 6000 pounds.

Maximum speed is about 15–17 m.p.h. with an average near 13 m.p.h. Waiting is usually brief and total travel time is about 4–5 minutes. The overall speed surpasses that of the auto and bus in traffic.

Annual operating costs were $240 000 per year including about $10 000 for power. Operating costs were 45¢ per vehicle mile, 4.5¢ per passenger mile. (Astroglide LIM operating costs are 30¢ per vehicle mile and 2.5¢ per passenger mile.) Patronage was at least 2.5 million riders in the last year of service and at least 10 000 000 in the entire period of service.

Braniff International was very pleased with the reliability of the system during the five year service period. Five employees were required to maintain the system. Two employees were always available for emergencies but they performed other duties unless called.

The successful operation of our country's first commercial monorail system has been highly publicised during the past five years. The safety record of the system is excellent. One incident occurred under manual control and caused only minor damage to an empty car. There were no accidents involving passengers.

The entire fleet accumulated about 500 000 vehicle miles per year and individual vehicles travelled about 50 000 miles per year.

Development of Astroglide for urban transportation

However, even though this system for Braniff International Airways in Dallas, charted a new course for transportation, its potential for cities was not recognised by municipal governments until recently. Great technological changes have been made in the system. Astroglide now provides dramatically reduced maintenance requirements, ease of installation, flexibility to accommodate the conditions, ease of expansion and smooth and quiet pollution-free operation.

When the system was converted over to linear propulsion we removed and discarded two rotary induction traction motors, two eddy-current variable speed drive units, two rear axle differentials and four gear box assemblies for a total weight (Figure 3.3) reduction of 2300 pounds. This makes provision for five more passengers in the present 12 foot car and provides a stretch version of the car to 17 feet in length carrying 25 passengers, and all this was due to a dramatic weight reduction. The LIM

Figure 3.3 Close-up showing old mechanical drive system.

(Figure 3.4) has also dramatically increased the speed of the car from 17 m.p.h. to over 40 m.p.h. Braking of the car is accomplished by reversing the LIM's thrust. Lubrication is no longer required.

Figure 3.4 Close-up showing new LIM propulsion system.

The linear induction motor

Nearly seventy years ago, the concept of a linear induction motor for vehicle propulsion was proposed by Zeyden[1]. It is illustrated in Figure 3.5. The most common form of electric motor is the induction motor. It has a set of windings in the stationary outer framework that are connected to an electric power line to set up a rotating electromagnetic field. The inner core, supported on a shaft with bearings, carried conductors that are cut by the rotating field to induce currents whose fields react with the main field to produce a torque on the rotor, accelerating any load connected to the shaft. There is no contact between the stator and the rotor except through the bearings supporting the rotor shaft. When used to propel a moving vehicle the rotating induction motor must be fitted with a gear box and wheels that provide the necessary linear thrust and speed through friction with the roadbed.

The linear induction motor (LIM) is essentially the motor illustrated in Figure 3.5a split along a radius and laid out flat as shown in Figure 3.5b. Now when power is applied to the windings, the electromagnetic field is made to 'sweep' the conductors in the passive reaction element. Just as in the rotating configuration, there is a relative thrust generated between the active and the passive elements. When used to propel a vehicle the active element can be carried onboard so the relative thrust is

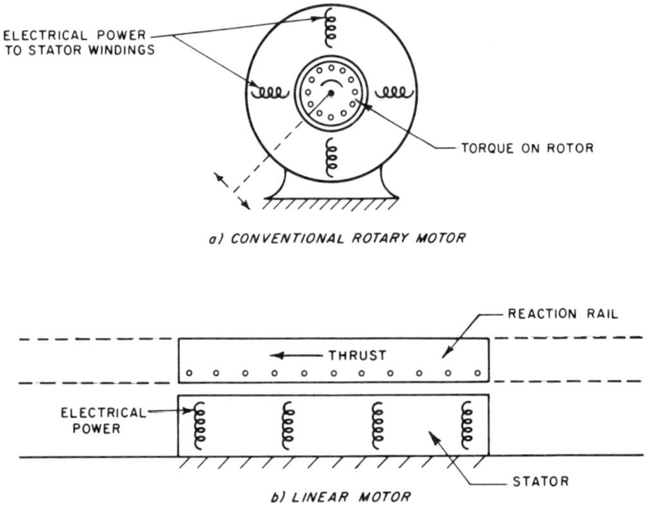

Figure 3.5 Concept of the linear induction motor.

generated directly against a passive stationary conducting element (the rail). Thus the need for gear boxes and wheels and the dependence on friction for tractive forces are completely eliminated.

The technology of the basic LIM has today reached the stage that the unit is in quantity production in standard sizes. The state of the technology is largely due to the efforts of Professor Eric R. Laithwaite, now of the Imperial College, London, and a Director of Linear Motors, Limited, who, since the early 1950s, has pioneered the development of practical machines of advanced designs.

The basic LIM using magnetic material in the reaction rail is, of course, a form of electromagnet. Therefore it not only provides thrust in the direction shown, but also provides a force of attraction between the active and the passive elements. This force can be as much as ten times the thrust of the motor and, with the appropriate controls for maintaining a stable air gap, can be used for levitation of the vehicle. The merits of this approach to levitation are to be explored in the design of the Jetrail modifications and compared with those of the transverse flux motor (TFM) described in the next section.

With an aluminium (nonmagnetic) reaction rail, the force between the active and passive elements becomes a repulsive one, which can also be used for levitation of the vehicle. However, the TFM has been developed to optimise the characteristics of repulsive electromagnetic levitation coupled with electromagnetic guidance and propulsion. Therefore the TFM is preferred in this mode of levitation.

The transverse flux motor

The transverse flux motor, invented and developed by Professor Laithwaite and his associates, is a linear induction motor at its highest level of evolution to date. It provides linear thrust in a way basically similar to the original LIM, by the action of a sweeping magnetic field along a conducting reaction rail. However, in the original

LIM the flux paths are loops in a vertical plane parallel to the line of thrust, as illustrated in Figure 3.6a. Since the maximum propulsion velocity is a function of the span of these loops, the original LIM becomes magnetically inefficient for high-speed applications. The TFM avoids this limitation by generating the flux loop in a plane perpendicular to the line of thrust (Figure 3.6b): thus the loops can be small for any velocity, simply the distance around the smallest dimensions of the linear motor.

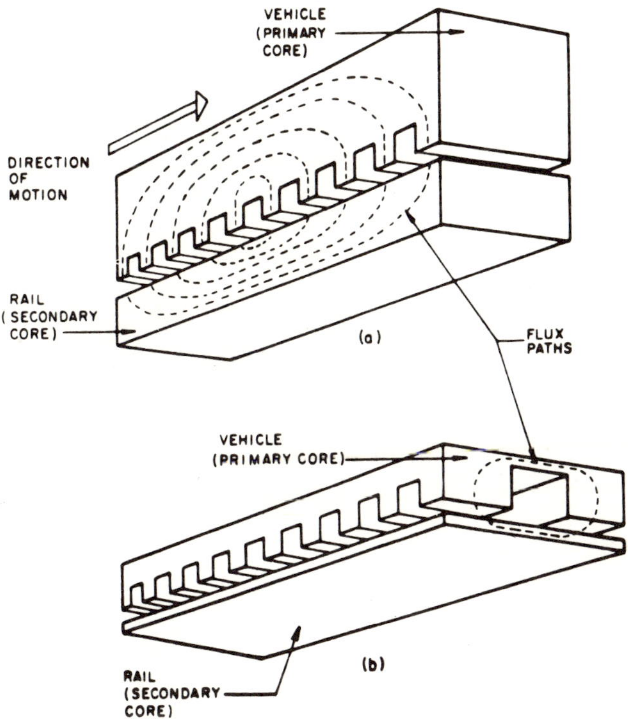

Figure 3.6 Comparison of (a) axial flux LIM and (b) transverse flux LIM, showing flux and relative thickness of motor and rail.

This arrangement is also more efficient for levitation by repulsion because there is less leakage of magnetic flux. Recognising the potential for levitation and propulsion with the TFM, Professor Laithwaite and his associates concentrated on refining it still further to provide guidance as well as levitation and propulsion in the same unit. We now have the capability for designing transverse flux motors with all three characteristics for any vehicle at any speed up to 300 m.p.h. In fact, the preliminary design of the TFMs for the Jetrail project have already been made and declared successful. The unique propulsion system powered by the linear induction motor (LIM) is another innovation of Astroglide. Use of the LIM (Figure 3.7) transfers all power into the overhead guideway for swift, silent, frictionless transportation. The motor reacts directly against the steel I-beam secondary which is also the major structural support. The LIM provides enough attractive force to virtually support the vehicle; thus it is a simplified form of MAGLEV in which the wheels provide the stabilising force but most of the vehicle weight is supported magnetically. There are NO moving parts in the system. Linear induction propulsion is a revolutionary advancement over the conventional rotary induction motor. The thrust and vehicle speed can be varied by producing thrust in the opposite direction when a braking effect is obtained. Because the propulsion drive and main braking forces are not applied through wheels or gears, the noise that accompanies all conventional forms of transportation is eliminated.

Figure 3.7 Linear motor and steel I beam reaction plate.

General characteristics of the Astroglide system

To connect suburbs with cities, in busy downtown shopping centres, between airline terminals or to parking lots, or connecting university buildings, wherever busy people need to move within major activity centres, Astroglide AGT (advanced ground transport) systems will move them quickly, safely and silently. Without interference from automobile or pedestrian traffic, the city-bound traveller, the air passenger, shopper or student, along with baggage and books, rides comfortably in automatically-controlled vehicles to his preselected destination.

The use of small cars (holding 10 to 25 passengers) makes the operation of the system (Figure 3.8) very flexible. In addition, the individual units are less expensive than larger ones would be, are more easily maintained and require less support. The number of vehicles in a given system can be preprogrammed to meet the existing demand and added to as requirements increase. Waiting time is minimised if there is an adequate number of cars standing by. When a boarding station has a vacancy, a car is automatically dispatched to that point whether or not there is a passenger demand at the moment.

Figure 3.8 Astroglide guideway showing switch and line turn off.

If a car must be removed from the system for cleaning, checking, routine maintenance or repair, the system continues to function with a minimum of delay. Should a car stall for any reason, exterior warning lights automatically come on to signal the monitor, and the next car on the line is moved up to push the stalled car to the station. The cars may be driven manually, if necessary, and all rails have stand-by power. The car doors have a rubber guarded safety edge and will automatically retract if contacted by a foreign object. The doors are locked shut when the vehicle is in motion and an escape panel is provided in the event of danger.

Each car is suspended on bogies which are magnetically levitated. The car is driven by linear induction motors sized to meet the propulsion requirements and levitation of the car and provide redundancy in the drive system. Each car is equipped with both torsion and automatically operated manual brakes to assure safe stopping, and these are applied instantly should an unsafe condition occur making *car collision impossible* (Figure 3.9).

Figure 3.9 View of electromagnetic back-up fail-safe friction brake.

Operation of the system

The system, for the purpose of illustration, could have ten passenger cabs and two baggage cars with three stations having two major and one secondary stops. The cycle could include two levels of loading capacity, designated 'low' and 'high' level. At the start of the cycle, six units are located at Station B, the terminal loading lobby and four units at Station A, which is the parking lot or remote building.

Low level loading

As an example of low level loading, a passenger arrives at the parking lot (Station A), checks his car and bags, takes an escalator to the second level, enters the waiting car and pushes a button. The system will initiate a count of 20 seconds, during which time a sign may count down numbers from 20 to 0 to advise when the cab is ready to go, or a voice command may be used to inform passengers of the cabs readiness to travel. Following the count, the door closes and the cab departs, carrying the passenger toward Station B, the terminal loading lobby.

High level loading

When there is a large group of returning passengers, perhaps 100 to 350 ready to travel from the arrival lobby (Station B) to the parking lot building (StationA), the high level cycle will be initiated, which is accomplished manually by means of a dispatching selector switch. This cycle begins with three cab doors open and ready to accept passengers.

The most forward cab will move first, approximately 20 seconds after the cycle is initiated. The second cab will close its doors and depart approximately 10 seconds later, and the third cab will follow in approximately another 10 seconds. The three remaining cabs on the storage spur at Station B will then move to the forward position and open their doors to accept approximately 60 more passengers. The first cab in this group will begin travelling toward Station A about one minute later with the remaining two following at 10 second intervals.

Simultaneously, three cabs will depart from Station A to Station B to initiate a third three-cab flow to Station A in the same sequence as described. The total time elapsing between the departure of the first cab and the time the third set of cabs begins to load passengers is approximately 3.5 minutes. Thus, approximately 180 passengers are transported in a continuous cycle of three units operating in three individual loading cycles.

Baggage handling

After the passenger checks his bag, it is placed in a baggage car pod along with others and the baggage car is automatically dispatched over the same rail system to the baggage loading area adjacent to the aircraft by an automatic switching arrangement.

The total concept is that the passenger, the bag, and the car are separated at one location and following the return flight, are brought back together again at the same location with a minimum of walking for the passenger, together with complete climate control for his comfort.

Control system

The control system consists of three sub-systems – vehicle dispatching, the safety sub-system, and propulsion control.

Vehicle dispatching sub-system

The dispatching sub-system determines when loaded vehicles should begin movement toward selected destinations and when empty vehicles should begin movement to maintain readiness for subsequent transportation requirements. The dispatching sub-system also controls the operation of station and vehicle doors. Dispatching functions according to either a 'high flow' or 'low flow' transportation pattern, which is controlled manually by means of a two-position selector switch located in a convenient position external to the vehicles. (The ASTROGLIDE dispatching system is a proprietary concept and method belonging to PRT Systems Corporation.)

Safety sub-system

The safety sub-system will prevent any event in the operation of the system that might be unsafe to passengers or property. All other sub-systems are designed and connected so that they may be overridden by the safety sub-system. The safety sub-system interfaces with both the dispatching and propulsion control sub-systems. The interface, however, is so designed that the safety sub-system will maintain its surveillance regardless of events that might occur in the other two sub-systems, including failures and malfunctions. Failure of any element or combination of elements will cause all portions of the system to assume an operating status that is known to be safe.

Routes are divided into portions having different authorised speeds by means of a 'safety block' system comparable to that used in the monorail system at Love Field, Dallas, Texas, Braniff International. The safety sub-system will keep moving vehicles separated by at least one and one-half times stopping distance by means of the safety blocks. The safety sub-system enforces safety in three ways – by speed control, door control and fail-safe breaking.

Speed control Speed control is accomplished by combined functioning of the safety sub-system and the propulsion control sub-system. The latter will provide for voltage to be applied to produce torque for moving at assigned speed.

The safety sub-system transmits to the vehicle in any route segment, intelligence which the propulsion control system will recognise and implement. As the propulsion control section explains, the sub-system will not allow any forward tractive effort unless and until it receives speed control intelligence. The safety sub-system will interrupt and open-circuit the supply of propulsion power if all conditions of safety are not satisfied. Thus, propulsion power is transferred to vehicles only by means of the safety sub-system throughout all portions of the route and for all speeds from zero to maximum. The safety sub-system prevents reverse direction drive of vehicles.

Door control The safety sub-system monitors closure of vehicle doors and causes vehicle brakes to remain set if doors are not firmly closed. The safety sub-system also provides devices that sense docking accuracy of vehicles at loading and unloading locations and prevents vehicle doors from opening unless the vehicle is docked within the prescribed centre lines of the docking doors.

Dynamic and regenerative braking Two additional drive system characteristics can be built in with proper control system design. The first is dynamic braking. This characteristic is due to the ability of the LIM to generate a negative value of thrust to 'brake' the vehicle as long as it continues to move. This is quite easily done in an induction motor control system by switching to a steady magnetising current and letting the eddy currents induced in the moving element generate the negative torque (or drag) passively.

The second characteristic can be provided with more sophisticated electronic control circuits. Regenerative braking is due to the ability of the LIM to act as an electrical generator when commanded to slow down or stop. Now the generated electrical power can either be fed back into the main power lines or stored in a battery carried on the vehicle. In either case, power is recovered every time the vehicle decelerates or goes down-grade. Therefore, the net power consumed is only that due to losses in the system.

Propulsion control sub-system

All functioning of the propulsion control sub-system must be in permissive agreement with the safety sub-system. Any propulsion control sub-system functions that might jeopardise operational safety are prevented by the safety sub-system. The propulsion control sub-system is interlocked with vehicle door operating mechanisms as provided by the safety sub-system. The failure mode of the propulsion control sub-system will de-energise the propulsion system, which in turn, will reduce forward propelling torque to zero and increase braking effort to its maximum rated level.

The dispatching programmes (high or low level loading) are part of the vehicle dispatching sub-system, and the propulsion control sub-system will function in accordance with the dispatching programme selected. Propulsion control functions are initiated by intelligence from the dispatching sub-system. After such functions have begun, they will continue according to the selected pattern without further intelligence from the dispatching sub-system, providing that safety requirements are satisfied. (This is a proprietary concept and method belonging to PRT Systems Corporation.)

Drive system considerations In the proposed personal rapid transit system, there may be more than one vehicle operating over the same area of track. This means that the main propulsion power will have to be common to all vehicles. This dictates that a motor speed control system must be carried on each vehicle.

Since the vehicle is to operate with different loads, up and down grades and maintain the optimum spacing, speed must be controlled precisely and on command from centrally-generated computer signals. And for passenger comfort and safety as well as long vehicle life, automatic limiting of acceleration and jerk is also required. To include all these capabilities in the vehicle drive system, a control loop is provided.

The LIM, coupled with the vehicle, is fitted with the means for sensing acceleration and speed. The measure of speed is fed back and compared with the command signal representing desired speed. If actual speed is too low, the difference signal is a command to apply more thrust. This signal is compared with the feedback signal of actual vehicle acceleration and allowable acceleration and jerk limits to generate a limited thrust command signal which will not over-accelerate the vehicle. The signal is amplified and fed to the LIM to generate the thrust necessary to accelerate to command speed; always within the allowable limits.

Conclusion

We are convinced that the advanced Astroglide system is the type of public transportation attractive enough to persuade city-bound motorists to leave their cars at home. We know it will release metropolitan areas from the twin strangleholds of pollution and congestion.

Reference

[1] Laithwaite, E. R. (1973) 'Linear motors for high-speed vehicles', *New Scientist*, 28 June 1973.

4

The Canadian MAGLEV Project

G. R. SLEMON

Passenger vehicles travelling at 500 kilometres per hour (300 m.p.h.), suspended and propelled by magnetic fields, have emerged from the realm of science fiction and are now considered to be both technically feasible and economically viable. Research, design and experimental studies on such systems have been in progress for several years in England, Germany, Japan, USA and Canada. The article which follows is a description of the Canadian MAGLEV Project, a design study undertaken by three universities – McGill, Queen's and Toronto – and funded by the Canadian Transportation Development Agency.

Canadian transportation needs

Almost half of Canada's population is concentrated along a relatively narrow strip between Windsor and Quebec City, including the large metropolitan cities of Toronto and Montreal, as shown in Figure 4.1. At present, inter-urban passenger transportation along this corridor is mainly by automobile and aeroplane. A multi-track rail system exists, but is used mainly for freight.

A considerable expansion in passenger travel along this corridor is expected. It is the industrial heart of the country, attracting more than two-thirds of Canada's substantial number of immigrants. Several options for passenger transportation are being considered. Aeroplane services can be extended but only with airports located further from city centres, increasing the total travel time. In any case, there is strong public opposition to the allocation of more land for airports. Also, air fares continue to rise disproportionally because of fuel costs. Serious consideration is being given to high-speed passenger trains. Any substantial improvement in rail service would require not only high-speed trains but also a new dedicated track, built with no grade crossings, a large radius of curvature and a very rugged roadbed. Experience with the Tokyo-Osaka high-speed rail line in Japan indicates that such a system can be effective and economic. Track maintenance is, however, a major problem with high-speed trains and promises to become more so with increased labour rates.

A third option is based on the new technologies of magnetic levitation and propulsion. A design and experimental study of this option is being undertaken to develop a sound technical and economic basis for a decision on which system should be adopted for the next half century.

The transportation corridor shown in Figure 4.1 contains only about 10 million peopl The traffic density is therefore low in comparison with similar corridors in Europe and Japan. To provide the reasonably frequent service to which passengers have become

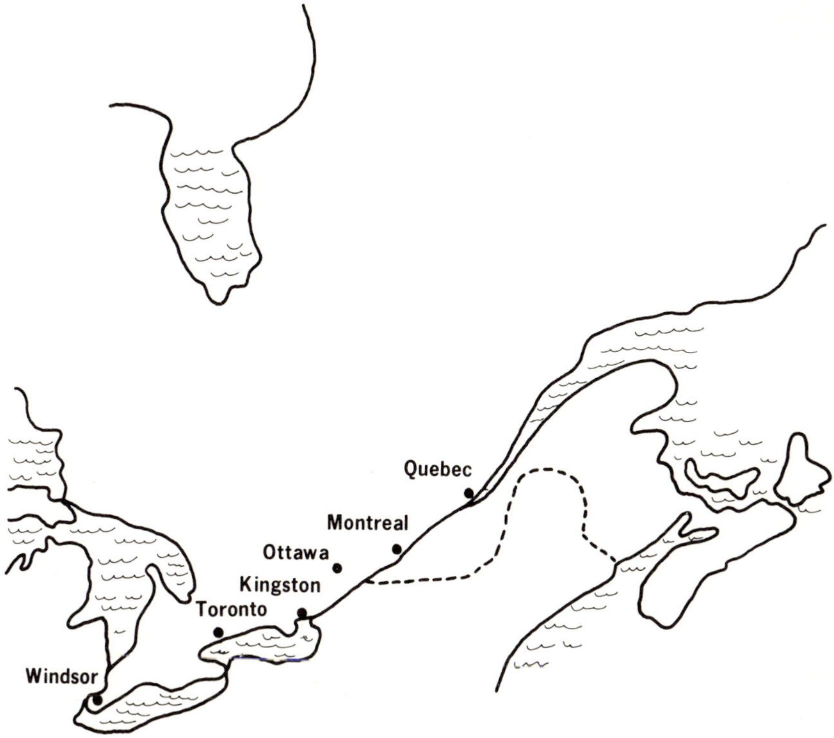

Figure 4.1 A major transportation corridor in Canada.

conditioned by the airlines, the Canadian MAGLEV study has concentrated on using single 100 passenger vehicles rather than trains. A 100 passenger vehicle would be similar in size and shape to the body of an aeroplane of equivalent capacity – 30 m long (98 ft) and about 3.2 m (10 ft) in diameter. Its mass is expected to be about 30 tonnes.The maximum speed of the magnetically-levitated vehicle is to be in the 400–500 km h^{-1} (250–300 m.p.h.) range. At this speed, the total trip time is expected to be lower than that provided by aircraft up to distances of at least 800 km (500 miles). In contrast, steel-wheeled vehicles on steel rails are limited to about 250 km h^{-1} (155 m.p.h.) by problems of track maintenance and wear. In a conventional rail vehicle, all the weight and impact of the vehicle is concentrated at a few small areas of contact between the wheels and the rails. The support structure must be designed to accept these highly-concentrated loads. If an alternative system can distribute this load over a wide area, the supporting track or guideway can be substantially simplified and reduced in cost. If, in addition, the support can be obtained without physical contact, the wear problem can be eliminated. These are features which the magnetic suspension sets out to achieve.

Electromagnetic suspension

A contact-free electromagnetic suspension derives from the phenomenon that oppositely-directed electric currents in two parallel conductors repel each other

(Figure 4.2). This phenomenon has not been previously exploited for vehicle suspension because a very large current is required to produce an appreciable force. For example, two conductors, spaced 30 cm (12 in) apart and each carrying 100 000 A would be able to support about 680 kg per metre of conductor length (450 lb ft^{-1}). A vehicle having a weight of 1360 kg per metre of length could be supported by an onboard conductor loop passing down one side and back the other at floor level, carrying 100 000 A and interacting with a similar, 100 000 A conductor loop on the guideway surface.

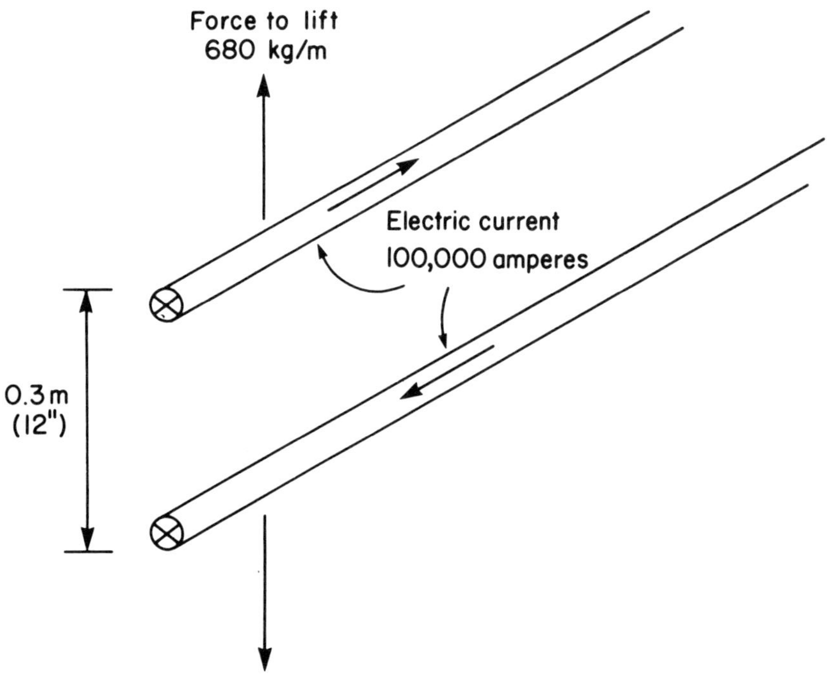

Figure 4.2 Force between two current carrying conductors can be used to levitate.

If ordinary electrical conductors such as copper and aluminium were used to carry the onboard current, a large mass of conductor (typically 400 kg) would be required and the power loss would be high (typically 16 kW), for each metre of vehicle length. These problems may be avoided by the use of superconducting conductors, such as niobium tin or niobium titanium, which have no resistance to the flow of electrical current provided their temperature is maintained near absolute zero. This is most conveniently accomplished by making a coil of superconductor and placing it in a vacuum-walled dewar which is then filled with liquid helium. The boiling point of liquid helium is 4.2 K (−268.8°C). At this temperature, a superconductor of a given cross-sectional area can carry more than 100 times as much current as a similar copper conductor, and can do so without electrical losses.

In principle, the guideway could also be fitted with a superconducting current loop to interact with the vehicle coil. This approach would be prohibitively expensive.

An alternative approach which has been adopted by most design groups is to use the onboard current loops or magnets and the speed of the vehicle to induce appropriate currents in sheets of aluminium placed on the guideway surface under the vehicle as shown in Figure 4.3. Ideally, the sheet acts as an electrical mirror, producing the same effect as an image coil with oppositely-directed current located at an equal distance on the other side of the sheet.

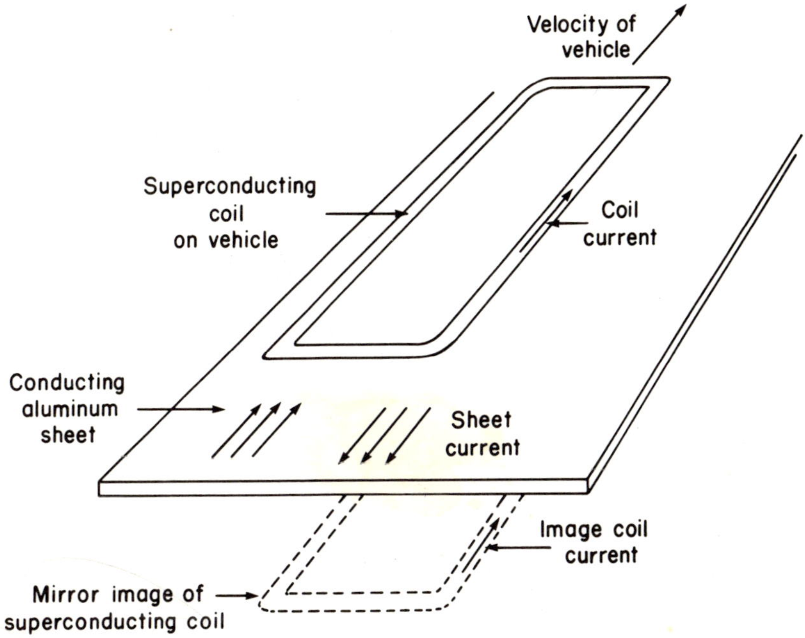

Figure 4.3 A superconducting magnet moving over a conducting plate induces a current in the plate and produces lift.

The suspension system proposed for the Canadian vehicle is shown in Figure 4.4. Two superconducting magnets are located at each of the four corners of the vehicle. Each magnet is 0.3 m wide and 1.0 m long (12 by 40 in) and carries a total current of about 400 000 A. The conductor required for this current has a cross-section of only about 20 cm^2 (3 in^2). Flat aluminium sheets, 80 cm wide and 1 cm thick (32 in x 0.4 in) have been placed on the guideway surface.

A flat-topped guideway has been chosen to avoid accumulation of ice, snow and debris. Tests conducted at the National Research Council, Ottawa, have shown that trough-shaped or inverted-T guideways used by other design teams would not be suitable in Canadian winter conditions because of snow accumulation.

The lift force provided by a magnet depends on the speed of the vehicle and also on the height of the magnet above the conducting sheet. It has been decided that, normally, a 15 cm clearance (6 in) should be maintained between the vehicle and the guideway. This clearance allows for quite large imperfections in the guideway surface and also for a reasonable amount of ice and snow. The corresponding

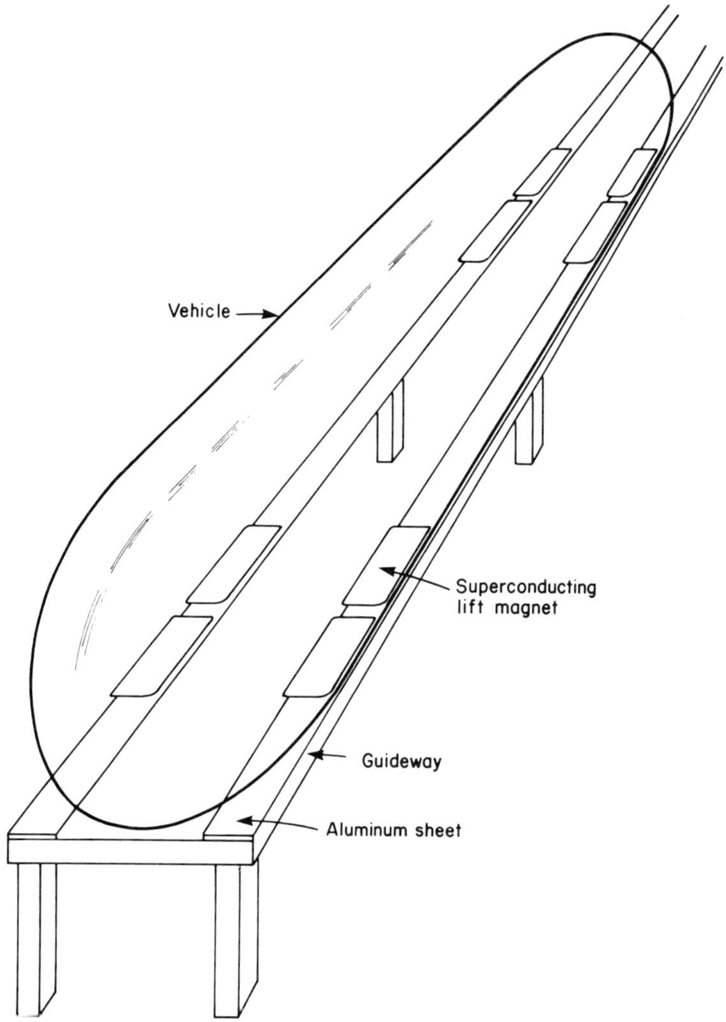

Figure 4.4 A view of the vehicle and its guideway.

separation distance between the magnet and the sheet has to be about 22 cm to provide for the dewar thickness. At this separation the lift force increases with vehicle velocity, as shown in Figure 4.5. Each magnet will support its share of the vehicle weight (estimated at 30 tonnes) at the 22 cm separation for a speed of about 300 km h^{-1}. For higher speeds, the separation distance will increase very slightly. At 100 km h^{-1}, the separation distance will decrease to about 20 cm. For low speeds up to about 50 km h^{-1} the vehicle will be suspended and guided on conventional retractable rubber-tyred wheels as shown in Figure 4.6.

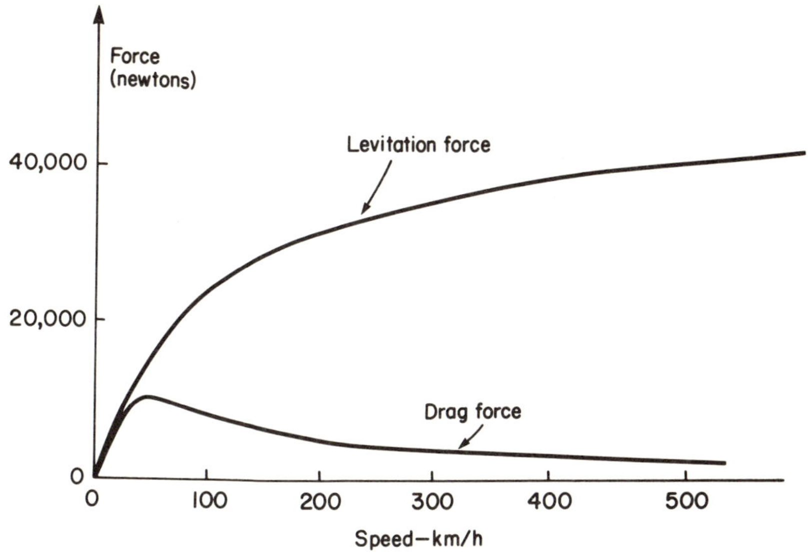

Figure 4.5 Lift and drag force for one magnet at a height of 22 cm above a conducting sheet.

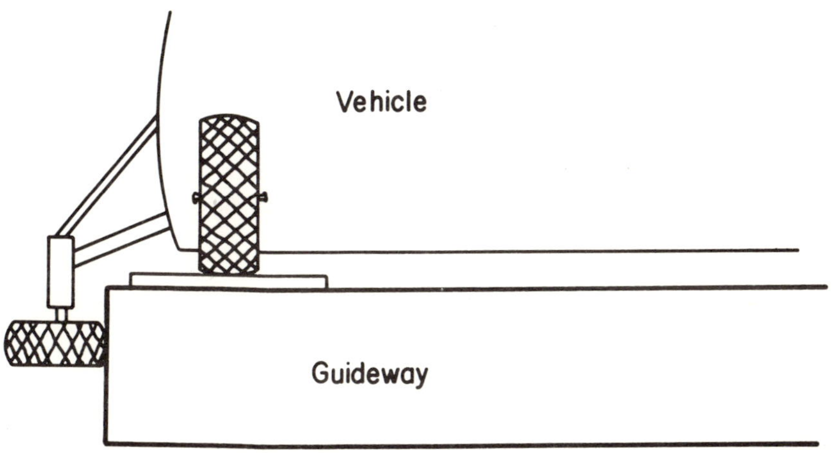

Figure 4.6 Landing gear for speeds up to 50 km h^{-1}.

Propulsion

The propulsion system must provide a thrust adequate to overcome the aerodynamic drag of the vehicle and also to provide adequate acceleration and grade climbing ability. In addition, it must provide thrust to overcome a magnetic drag caused by the levitation magnets moving over the levitation strips. The existence of this drag force is to be expected because the power loss of the currents in the levitation strips must eventually be supplied from the vehicle. As shown in Figure 4.5 this drag force reaches a maximum value at a speed of about 30 km h^{-1}. At higher speed, the drag force reduces as the speed increases so that the product of drag force and speed is equal to the roughly-constant power loss in the levitation strips.

The aerodynamic drag force on the vehicle is proportional to the square of the vehicle velocity. The proportionality constant depends on the skin friction and the streamlining of the vehicle. At the maximum speed of 500 km h^{-1}, this drag force is expected to be about 30 000 N (6700 lb). Figure 4.7 shows the major force components which must be produced by the propulsion system as it accelerates from a station.

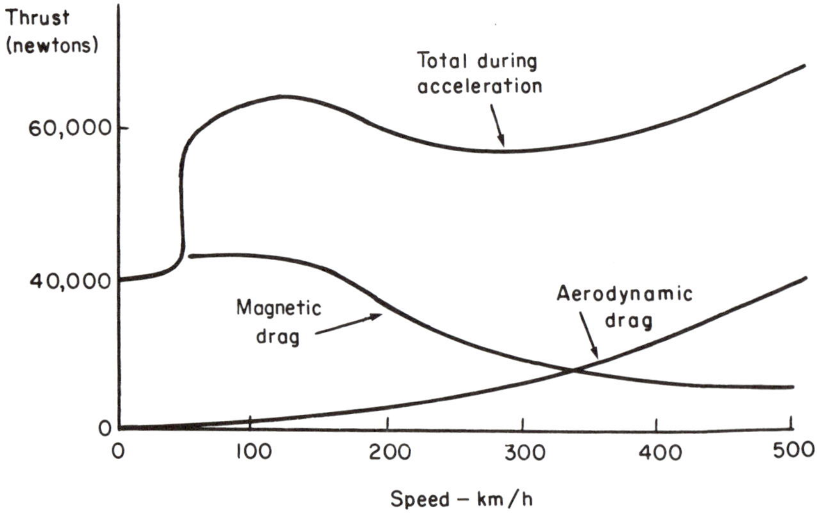

Figure 4.7 Thrust required for acceleration and cruising

Below a speed of about 50 km h^{-1} the vehicle is suspended on wheels and a force of about 30 000 N is required to produce the desired value of acceleration of 1 m s^{-2} (2.25 miles per hour per second). The levitation strips are omitted for the 100 m section of guideway necessary to reach this take-off speed of 50 km h^{-1}.

To limit the drag force at low velocity the thickness of the levitation strip is increased initially to about 3 cm and then progressively reduced to 1 cm over the first kilometre of guideway out of the station. In this way the magnetic drag force is maintained constant at about 30 000 N up to a speed of 150 km h^{-1} and then drops off with increase in speed. When added to the aerodynamic drag force, the total thrust required for cruising at constant speed varies only slightly in the range 35 000–40 000 N over the whole speed range. For full acceleration, the thrust required is relatively constant at about 60 000–65 000 N. This thrust is required for about 2.5 minutes.

At maximum speed, the cruising power required from the propulsion system is equal to the thrust (newtons) multiplied by the speed (metres per second), i.e. about 5.6 MW. This is approximately twice the power rating of an electric locomotive but is substantially lower than the power rating of a 100 passenger aeroplane.

Linear synchronous motor

Since the vehicle is magnetically suspended, its propulsion thrust must be provided without physical contact. This can best be done by use of a linear synchronous motor. This motor makes use of the same phenomenon employed for levitation, i.e. the force between parallel current-carrying conductors. As shown in Figure 4.8, it consists of a number of rectangular superconducting magnets, equally spaced and of alternating polarity, arranged along the underside of the vehicle. These interact with current-carrying cables embedded in the upper surface of the guideway between the levitation strips.

In contrast with most electric motors, this linear synchronous motor uses no iron either on the vehicle or on the guideway. Other groups have developed linear induction motors for high speed magnetically-levitated vehicles. A major advantage of the synchronous motor is that it can operate efficiently with a vehicle clearance of 15–20 cm whereas the induction motor must usually have an air gap less than 3 cm.

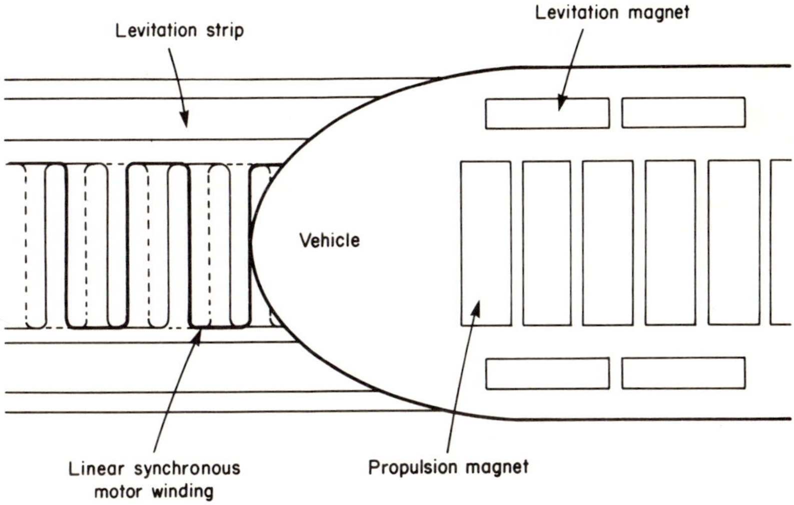

Figure 4.8 Linear synchronous motor.

The guideway winding consists of six stranded aluminium cables each about 1 cm in diameter. The cables are spaced about 9 cm apart, are covered with standard electrical insulation and are imbedded in asphalt. The winding is laid in sections of about 5 km length and each section is supplied with electrical power from a trackside power conditioner which converts power from the 50–60 Hz power grid to the appropriate current and frequency required by a vehicle while on that section of guideway.

The principle of operation of the motor is shown in Figure 4.9 where the magnet and guideway conductors are shown in cross-section. Currents away from the observer are indicated by a cross and toward the observer by a dot. Similarly-directed currents attract each other while oppositely-directed currents repel. The net thrust on the vehicle in Figure 4.9 is forward. As the vehicle moves forward the currents in the six guideway cables must be adjusted so that the cable carrying maximum current is always positioned approximately on the central axis of the magnet coil above it. This is accomplished by using a 3-phase alternating current supply to the cable windings and adjusting the frequency of this supply to precisely the value required for the vehicle velocity. The magnets in Figure 4.9 are spaced 57 cm centre to centre. The vehicle moves forward by two magnet spacings per cycle of the supply. Thus the frequency of the supply must be 122 Hz for a speed of 139 m s^{-1} or 500 km h^{-1}.

The motor designed for this vehicle includes 46 magnets each about 50 by 150 cm. Each magnet is superconducting and carries a total current of about 500 000 A. The required cruising thrust of 40 000 N can be achieved by use of a current in each aluminium guideway cable of only 245 A.

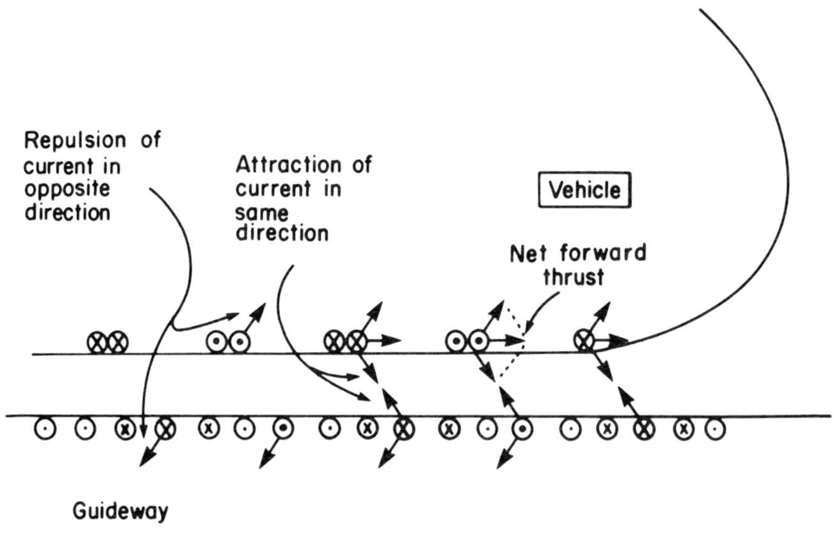

Figure 4.9 Principle of linear synchronous motor.

It might be expected that this motor would be very inefficient in view of the fact that only about 30 metres of a 5 kilometre energised track section is being used at any instant to produce thrust. However, 75% of the power supplied by the power conditioner to the track winding is converted into mechanical power. This efficiency can be increased, if desired, by increasing the size of the cables or by reducing the track section length. The values chosen for this design were obtained from an optimisation study which balanced the costs of cable, power conditioning apparatus and electrical energy to obtain a minimum overall annual cost.

The values obtained reflect the low traffic density which is anticipated in the early stages of operation of this system, i.e. about 150 vehicles per day per direction. As the traffic builds up, energy costs become more significant. When this happens, it is proposed to double the number of power conditioning stations and halve the length of each guideway conductor section (i.e. to 2.5 km).

At first glance, it might seem that the cost of this continuous motor winding in the guideway would be prohibitive. It should be noted that the amount of aluminium in the motor winding cables is only about 15% of that required for the two aluminium levitation plates.

A major advantage of the linear synchronous motor is that it requires no onboard power for propulsion. All the power is supplied from the guideway. Other motors which require onboard power must either carry fuel with the pollution problem this entails or must collect electric power from contact with trackside conductors. At 500 km h^{-1} it is difficult, if not impossible, to keep sliding contacts from bouncing off the trackside conductors to an unacceptable extent. In any case, it has been estimated that a power collection system would require considerably more trackside conductor material than is required for the linear synchronous motor winding.

The motor can provide a retarding thrust to the vehicle to decelerate it. All that is required is that the timing or phase of the guideway currents be adjusted so that currents away from and toward the observer in Figure 4.9 are reversed. The magnitude of the deceleration may then be adjusted by adjusting the magnitude of the current from the power conditioner. The motor is designed to provide normal deceleration of 3.6 km h^{-1} s^{-1} and an emergency deceleration of 10.8 km h^{-1} s^{-1}. At this emergency rate, the vehicle can be brought from full speed to a stop in a distance of about 3.2 km and a time of about 45 seconds.

Restarting the vehicle following an emergency stop on open guideway is more difficult than starting from a station because of the drag force of the 1 cm levitation strip (see Figure 4.5). This drag force of about 75 000 N can be supplied by paralleling the two power conditioners which normally supply adjacent track sections and using them to supply double current to the required section. This arrangement would be required for only about 30 seconds.

Guidance system

The vehicle will experience side thrusts caused by winds, cornering and guideway irregularities. It must remain on the guideway providing a safe and comfortable ride in spite of the maximum anticipated side thrusts. By itself, the simple levitation system of Figure 4.3 does not provide adequate guidance. But the complete levitation and propulsion system of Figure 4.8 can provide adequate guidance forces. When the vehicle moves to one side, the edges of the propulsion magnets induce currents in the edges of the levitation strips. These currents tend to repel the magnets and return the vehicle to its desired central position. It has been estimated that the maximum side force obtainable in this way is equivalent to the total weight of the vehicle and is therefore greatly in excess of that required even in severe gale winds.

While the interaction between propulsion magnets and levitation strips can provide adequate stabilising forces, there may be undesirable oscillations leading to poor ride quality. It is proposed to damp these out by use of guidance loops placed on the vehicle just below some of the superconducting magnets as shown in Figure 4.10. These guidance loops are made in the form of a figure-eight and are supplied with current from a damper control system. Current in the guidance loop strengthens the magnetic field one side and weakens it on the other. Onboard accelerometers can measure the sideways acceleration of the vehicle and control the guidance loop current to damp out oscillations. It is estimated that the guidance loop current will have to be only about 5% of the propulsion magnet current to provide adequate damping.

There are other degrees of freedom in which the vehicle may have undesirable

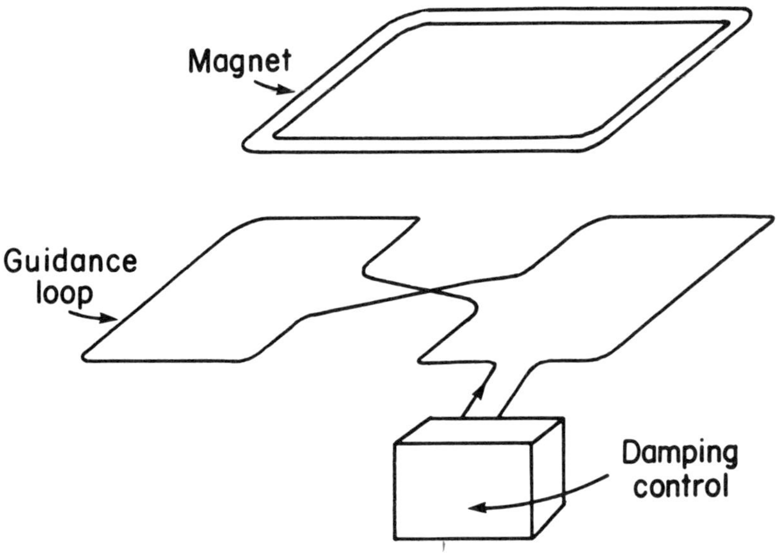

Figure 4.10 Damping control system.

oscillations requiring damping. Vertical or heave oscillations are expected to be troublesome because they are of relatively low frequency and are inherently undamped. It is proposed that the required vertical damping forces be provided by the linear synchronous motor. From Figure 4.9 it can be seen that, if the oppositely directed currents are directly under each other, a lift force is produced. The required lift force to damp out a vertical oscillation can therefore be produced by dynamically adjusting the phase of the cable currents either slightly ahead or slightly behind the positional relationship shown in Figure 4.9.

Pitch damping of the vehicle can be provided by dynamic control of the motor thrust, since the thrust is applied below the centre of mass of the vehicle.

Control system

Control of the thrust on a particular vehicle requires control of the magnitude, frequency and phase of the guideway cable currents in the track section occupied by the vehicle. The power conditioner which has been designed for this purpose takes electric power from the nearest convenient point on the transmission grid and first converts it to a direct current of the magnitude required by the motor winding. This current is then switched sequentially through the cables of the motor winding using solid-state power thyristors. The switching operations are done six times per cycle of the supply.

Control of the phase of the motor current is necessary in order to maintain the appropriate relative positions of vehicle and guideway winding currents, as shown in Figure 4.9. This is accomplished by measuring the voltage which is induced in the guideway winding by the motion of the vehicle magnets. The phase of this voltage is directly related to the instantaneous position of the vehicle magnets relative to the winding. The controller causes the power conditioner to inject a current into the winding at the desired time delay or advance with respect to the induced voltage, depending on whether forward or reverse thrust is required. By setting the phase in this way, the frequency of the power conditioner is automatically adjusted to the required value, six times per cycle.

The exact number of winding wavelengths between station-stops will be known from the initial design. These may be counted as cycles of the supply as the vehicle proceeds. It is therefore possible to programme a platform stop, even after hundreds of kilometres of travel to a precision of one sixth of a wavelength or about 19 cm. This precision is dictated by the six switching intervals per cycle of the power conditioner. The vehicle will come to rest at this pre-programmed position to within an error of less than one centimetre. The motor current will be maintained constant at zero frequency to hold the vehicle in position while stopped. The vehicle will be restarted by central control initiating the increase of supply frequency from the appropriate power conditioner to provide controlled acceleration.

Only one vehicle may occupy a guideway section at any time. Since it is necessary to ensure that the next guideway section is ready to receive the vehicle, only every second guideway section may be occupied simultaneously. With 5 km guideway sections and a speed of 500 km h^{-1}, the minimum headway is 10 km or 72 seconds. The maximum capacity of the guideway with 100 passenger single vehicles is 5000 passengers per hour in each direction. If greater capacity is required, the section lengths can be reduced. Alternatively larger vehicles or trains could be used.

Figure 4.11 shows the control system.

Onboard power

The vehicle has been designed to be inherently passive with all of its propulsion power provided by energising the guideway winding. Electric power will however be required onboard to provide lighting, air-conditioning and communications, and also to supply the refrigerators needed to cool the helium for the magnets. It has been estimated that the total of these requirements will be about 100 kW.

It is proposed that this power be generated using ram air turbines located in the nose of the vehicle with exhaust ducts on the sides of the vehicle. Two turbines would be used for reliability. Each will drive an alternating-current generator. Storage batteries with a capacity of about 25 kilowatt hours will be used to provide electric power in the station and when the vehicle speed is low.

Guideway

The guideway will be elevated above ground surface at an average height of about 7 m (23 ft) to provide adequate clearance over highways and rail lines. In some areas, the guideway may be built in tunnels. In any case, extreme care must be taken to prevent hazardous objects from falling on the guideway. Since vehicles pass each other at a relative speed up to 1000 km h^{-1}, separate guideways are required for the two directions of travel and they must be separated by several vehicle diameters.

Preliminary designs of the guideway have made use of a prestressed concrete box beam supported at intervals of 25–35 m. The distributed nature of the vehicle load makes the design considerably simpler than for a wheeled vehicle of similar weight

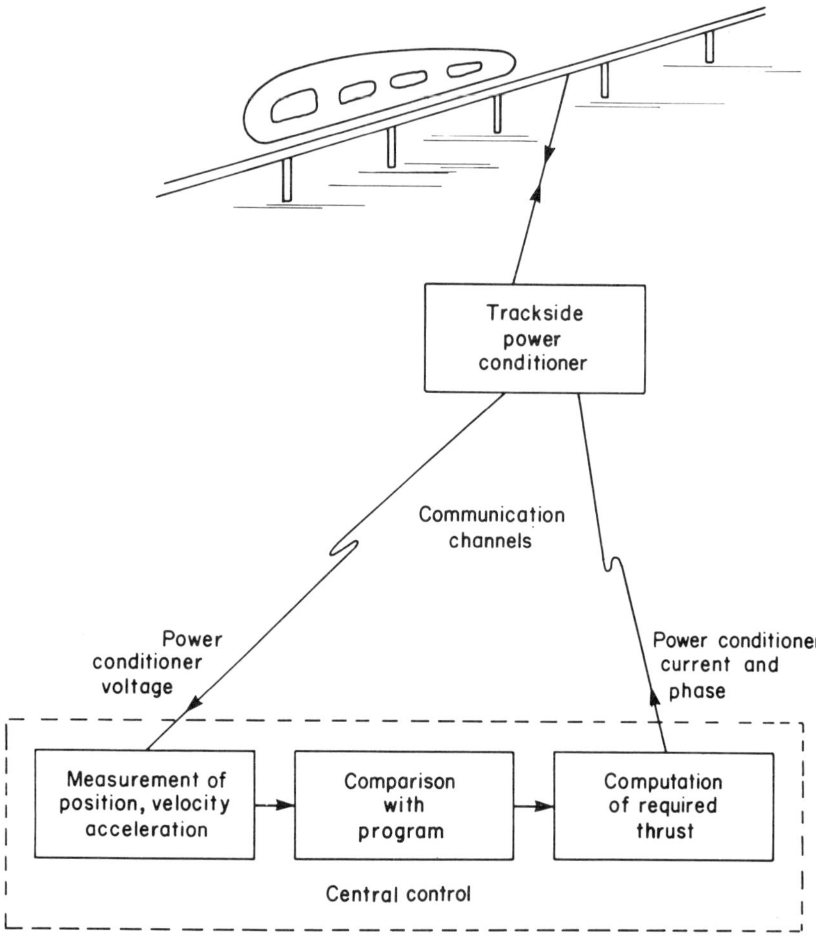

Figure 4.11 Control system.

and speed. A major difference in the design is that the use of iron and other conducting materials is somewhat restricted by the power losses in these structural materials which arise from the moving magnetic field of the vehicle.

The guideway will have to be relatively straight to limit the side thrust experienced by passengers to about 10% of the force of gravity. At the full speed of 500 $km\,h^{-1}$, and banking the guideway to a 5 degree angle, the radius of curvature must have a minimum value of about 10 km.

Stations along the route will be provided with a siding guideway for discharge and loading of passengers. Switching to a siding will occur about 1 km from the station at a vehicle speed of about 150 $km\,h^{-1}$. This kilometre of guideway will be provided with levitation strips of increasing thickness to limit the magnetic drag peak. Switching of the vehicle can be performed by replacing the levitation strips by conducting coils in the switching region and using solid-state thyristors to close the circuit on only those coils in the desired guideway routing.

Overall assessment

A passenger transportation system can be judged on the basis of its service, its reliability, its safety and its cost. This system has been designed to produce a travel time which is comparable and probably superior to that provided by aircraft for most journeys. The maximum speed is somewhat lower than for aircraft, but the stations can be conveniently located in downtown or suburban centres. The frequency of service exceeds that of aircraft because of the relatively small vehicles which are used. It is also feasible to locate stations at much smaller intervals than would be reasonable for airports, e.g. 50 km.

The reliability of a transportation system based on new technology is difficult to assess because of the lack of operating experience. The proposed system should be essentially independent of weather conditions because of its physical design and its automatic operation. Its potential reliability of service should therefore be much better than that of aircraft and also better than that associated with conventional rail systems.

The reliability of the vehicles can potentially be made very high because the major suspension and propulsion systems have no moving parts. The vehicle will be able to operate safely with the loss of one of the two levitation magnets at any corner. Several refrigerators will be provided onboard so that the loss of any one of them will not affect the operation of the vehicle.

The reliability of the system is, of course, dependent on the availability of electric power for the power conditioning stations. The loss of power to a single guideway section does not seriously affect operation, since a vehicle entering the unenergised section at full speed can drift through its 5 km length in less than 45 seconds and enter the next section at about two-thirds of maximum speed. If two sequential sections are unenergised, the vehicle can be brought to a safe stop on its wheels and can be started again when power is available.

The communication and control system must be made reliable by provision of high reliability components and by appropriate redundancy. The proposed system avoids the most troublesome aspects of presently automated transportation systems since no information need be transmitted to and from the vehicles. The only required communication links are between central control and the power conditioning stations.

The capital cost of this system will be dominated by the cost of the guideway structure, its supports, right-of-way and tunnels. The costs of levitation strips, guideway cables and power conditioning stations are expected to represent a relatively small proportion of the total capital cost. The cost of vehicles is also expected to be relatively small in view of the simplicity of the onboard system.

It is hoped that the relatively light guideway structure required for these vehicles can be constructed at a cost which is comparable with, if not lower than, the cost of a new reserved rail track for conventional high-speed passenger trains. If this is so, the proposed magnetically levitated system should be preferable because of its potential advantages of improved service and reliability.

At the present time, the major new technology components for this system have been laboratory tested. Full scale field testing is in progress or is planned in several countries. Much work remains to be done. A minimum of a decade of research, design, construction and testing must be envisaged before a system of this type can be placed in revenue operation.

5

A Japanese MAGLEV Project Using a D.C. Linear Motor

K. MATSUI T. SAIJO T. UMEMORI

5.1 Introduction

With increased urbanisation, the cost of travelling time becomes of increasing importance and the time would seem ripe for the birth of a new transport system.

Conventional railways began with the use of reciprocating engines and developed to use rotational prime movers. Perhaps now that the limit of adhesion has been reached, the time has come to make the reverse move from rotational movement to linear movement. With the development of the jet engine, the linear induction motor and the technology of air cushion suspension, it was easy to conceive a super-speed railway that was independent of adhesion and, initially, its realisation was considered not very difficult. For instance, the French 'Aerotrain', combining an aircraft engine with an air cushion attained a maximum speed of 420 km h^{-1}.

In these days, however, when pollution has become a significant factor and human rights are calling for due attention, the super-speed railway of tomorrow must be one that is pollution-free and fully automatic. One method of meeting these requirements, is to combine the linear induction motor with some form of magnetic levitation. There are two types of magnetic levitation: the induction type using a superconducting magnet, in which a levitation force proportional to speed can be obtained, and the feedback control type in which a constant levitation force can be developed by attraction between the electromagnet and the iron rail. In the former, the levitation performance at high speeds is good but large braking forces are developed at low speeds. In the latter, the levitation in the low-speed range is good but its application to the high-speed range introduces difficulties. With regard to propulsion by a linear induction motor, the end effect due to the finite length of the field causes the magnetic flux at the leading edge to decrease with increased speed, resulting in a reduced propulsive force. These problems may be solved by appropriate design techniques but, as viewed from the power supply system, the motor is liable to impose a low power factor on the circuit as a whole. The alternative of generating a variable voltage and frequency supply onboard using a large capacity inverter also gives rise to problems where a low power factor is involved.

In the case of a thyristor-commutated d.c. linear motor such as described in this chapter, levitation and propulsion take place simultaneously through electromagnetic forces developed between the vehicle field and a coreless, wound armature coil on the ground, avoiding the usual commutation problem of the d.c. motor by thyristor switching. Full exploitation of this system is possible, whereby the thyristor converter, installed alongside the d.c. substation, provides a constant current to the track conductors which are divided into separate feeder sections that are energised only during passage of a vehicle. Under this arrangement, the power loss in the system as a whole is minimised and reactance is not a serious problem. The system lends itself to fully automatic operation, control to a preset programme taking place at trackside.

Development of the thyristor-commutated type d.c. linear motor is an entirely new idea in wheel-less levitated transport. Feasibility investigations, both experimental and theoretical have been carried out. A summary of results now follows.

5.2 Design and operation of d.c. linear motors

5.2.1 *Construction and operating principles*

The ground armature coil system is shown in Figure 5.1. The conductors are split into a number of groups and the coil groups are staggered, as shown in the diagram. Figure 5.2 shows the coils arranged upright on the ground with a magnet system

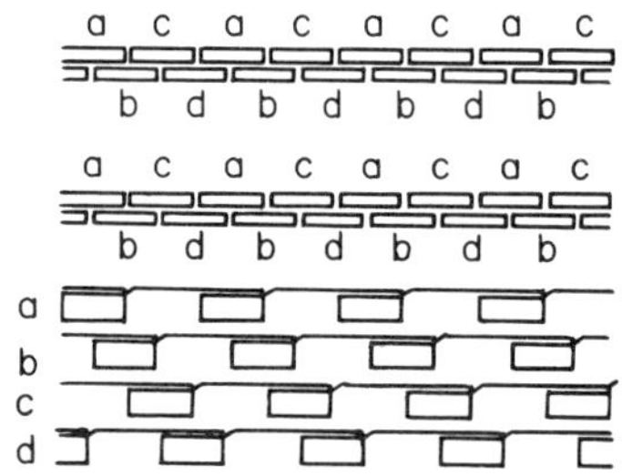

Figure 5.1 Splitting of the armature coil into groups.

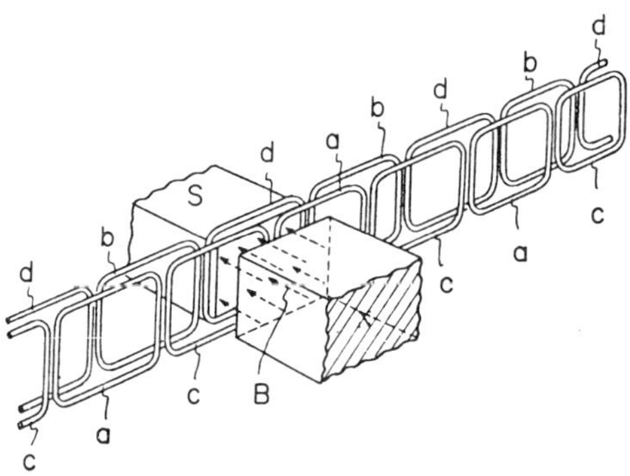

Figure 5.2 Relative position of field magnet and ground coils.

onboard the vehicle causing the magnetic flux to cross the lower and vertical parts of the coil. The vertical displacement between magnets and coils gives rise to levitation forces. The armature current is switched by the thyristors from one coil group to another in succession as the vehicle progresses.

It is essential that synchronism be maintained between the movement of the field and the armature current. If the field falls out of step, the system fails. For the purpose of ensuring this synchronism, two synchronising methods are possible: automatic switching, in which the current flow in the coil groups is changed by detecting the field position from the ground side, and external control, in which the switching of the coil groups takes place automatically according to a pre-determined programme that takes account of the load characteristic of the d.c. linear motor. An example of one arrangement is illustrated in Figure 5.3 in which the ground armature coils are split into four groups. In each of the coil groups a, b, c, d, the wound coils are series-connected in pairs and the coils in each group are series-connected in a ring through diodes. Meanwhile, the coils in a, b, c, d groups are integrally layered with a half-pitch stagger. The thyristor switches S_1–S_4, $S_1{}'$–$S_4{}'$ are connected in a 4-phase bridge, the part corresponding to the d.c. terminals being made the input terminal,

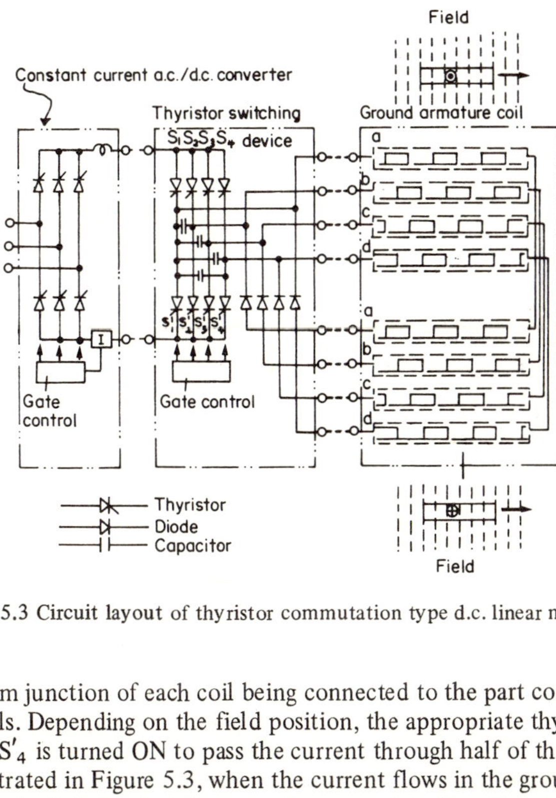

Figure 5.3 Circuit layout of thyristor commutation type d.c. linear motor.

and the ring-form junction of each coil being connected to the part corresponding to the a.c. terminals. Depending on the field position, the appropriate thyristor switch of S_1–S_4, S'_1–S'_4 is turned ON to pass the current through half of the coil groups. In the case illustrated in Figure 5.3, when the current flows in the groups a and b, the field magnets are levitated and at the same time moved to the right. As the thyristors are switched ON or OFF to make the current flow change from b to c, from c to d and from d to a in succession, the field magnets are continuously levitated and propelled.

In such a system, constant current supply and feeder sectioning of armature coil are essential for safety of operation.

(1) *Constant current supply* On starting, a speed-induced back e.m.f. is not yet developed, and connection to fixed voltage is likely to cause excessive current. This

can be suppressed by the insertion of a resistance but a supply of a steady current by a thyristor converter is obviously ideal.

(2) *Feeder-sectioning* As the vehicle progresses, the coil in a feeder section ahead of the vehicle may be energised before the vehicle moves into that section and the coil in a feeder section behind the vehicle may be de-energised as soon as the vehicle leaves it. Depending on the vehicle length, the feeding system and circuit arrangement may differ, but generally the longer the feeder section, the more advantageous the system from the vehicle point of view.

The feeding circuit may be arranged with the feeder line inserted between the constant current a.c./d.c. converter and the thyristor switching device or between the thyristor switching device and the ground coil, the former being a d.c. feeding system and the latter an a.c. one.

5.2.2 *Levitation and propulsion*

The levitation force (see Figure 5.4) can be calculated as follows.

Field magnet length, L
Magnetic flux density, $\mathbf{B}$
Coil pitch, l
Height of coil cross-section or height of field magnets, h
Number of turns in split coil, N
Current in split coil, I(A)
Number of split coil groups, $2n$

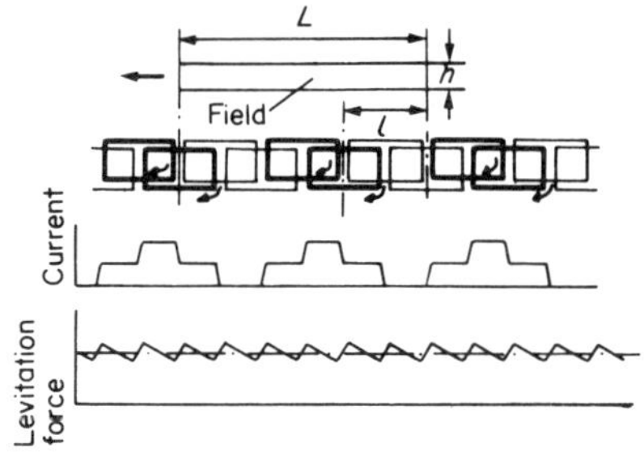

Figure 5.4 Levitation by thyristor commutation type d.c. linear motor.

The field magnet length will be given by $L = (2n + 1)l$, being an odd number of coil pitches.

Meanwhile, since only one half of the split coil group are successively energised, the number of energised, split coil groups in the field magnet length is given by:

$$\left[mn + \left(\frac{1}{n} + \frac{2}{n} + \ldots + \frac{n-1}{n}\right)\right] \sim \left[mn + \left(\frac{1}{n} + \frac{2}{n} \ldots + 1\right)\right]$$

$$= \left(mn + \frac{n-1}{2}\right) \sim \left(mn + \frac{n+1}{2}\right)$$

Thus the minimum and maximum levitation forces are expressed respectively by:

$$f_{\min} = \frac{\mathbf{B}}{9.8} NI \left(mn + \frac{n-1}{2}\right) \left(1 - \frac{h}{l}\right) L$$

$$f_{\max} = \frac{\mathbf{B}}{9.8} NI \left(mn + \frac{n+1}{2}\right) \left(1 - \frac{h}{l}\right) L \qquad (1)$$

where f is the levitating force in kilogrammes and $\left(1 - \frac{h}{l}\right)$ is the end factor of the coil.

Accordingly the average levitating force is given by:

$$f = \frac{\mathbf{B}}{9.8} NI \left(mn + \frac{n}{2}\right) \left(1 - \frac{h}{l}\right) L$$

$$= \frac{\mathbf{B}}{2 \times 9.8} nNI \left(1 - \frac{h}{l}\right) L \qquad (2)$$

The propulsion force can be calculated as follows:

First the effective height of field in the vertical part of the armature coil must be known. Suppose a field magnet is levitated a distance x, as illustrated in Figure 5.5.

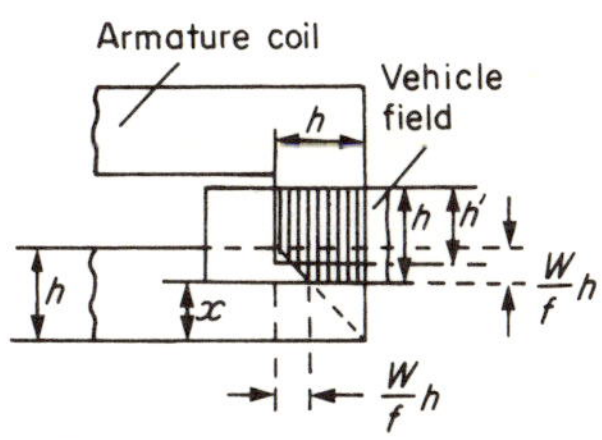

Figure 5.5 Effective height of field.

Then the height x is given by

$$x = \left(1 - \frac{W}{f}\right) h \qquad (3)$$

where W is the vehicle weight
h is the field height
f is the levitating force

The effective height h' of field is, from Figure 5.5:

$$h' = x + \frac{(h - x)(h + x)}{2h}$$

$$= x + \frac{h^2 - x^2}{2h} \qquad (4)$$

Substituting (3) into (4) yields:

$$h' = \left(1 - \frac{W^2}{2f^2}\right) h \qquad (5)$$

Thus the propulsion force F can be given by:

$$F = \frac{\mathrm{B}}{9.8} h' nNI \left(1 - \frac{h}{l}\right) L = \frac{\mathrm{B}}{9.8} h \left(1 - \frac{W^2}{2f^2}\right) \left(1 - \frac{h}{l}\right) nNIL \qquad (6)$$

5.3 Power converter for d.c. linear motor

The most important element in the system is the power converter. It consists of a constant current a.c./d.c. converter which converts the a.c. power of commercial frequency to a steady direct current and of a thyristor switching device which changes the d.c. thus obtained into an intermittent d.c. square wave form synchronised with the movement of a vehicle.

5.3.1 *Constant d.c. converter*

If the power source for the d.c. linear motor is a steady current the levitating force is constant regardless of propulsion. Power can be supplied and regenerated smoothly depending on the drive or brake conditions required and the system ensures that field and armature never get out of step. Figure 5.6 illustrates, for simplicity, the case of a single-phase thyristor bridge.

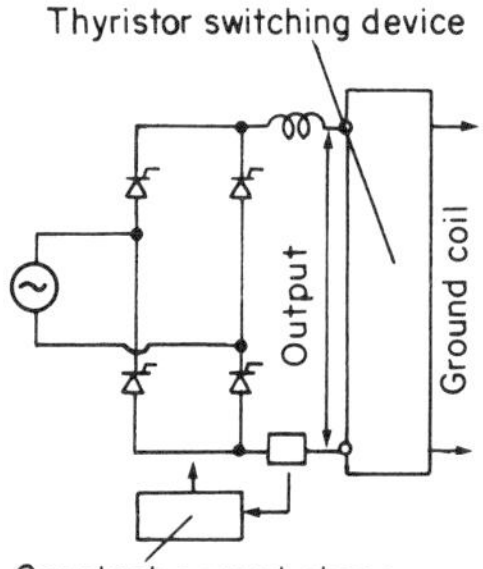

Figure 5.6 Example of a constant current a.c./d.c. converter.

When a thrust is being developed by the linear motor, a positive speed e.m.f. is generated on the load side of the bridge; conversely, when a braking force is being developed, a negative speed e.m.f. is generated (Figure 5.7). Since the thyristor bridge is so phase-controlled that the current is constant through the forward-reverse change, a so called 'forward' change occurs with the control angle in the range of $0 < \alpha < 90°$ when a thrust is being developed by the linear motor; and thereby power is supplied from the supply side to the linear motor. When a braking force is being developed, a so called 'reverse' change occurs with the control angle being in the range of $90° < \alpha < 180°$; and thereby power is regenerated on the supply side.

Thus the system can automatically make a forward change, make the power factor zero, and make a reverse change, depending on the acceleration, deceleration and coasting of the linear motor; and meeting these conditions, power can be supplied, or regenerated.

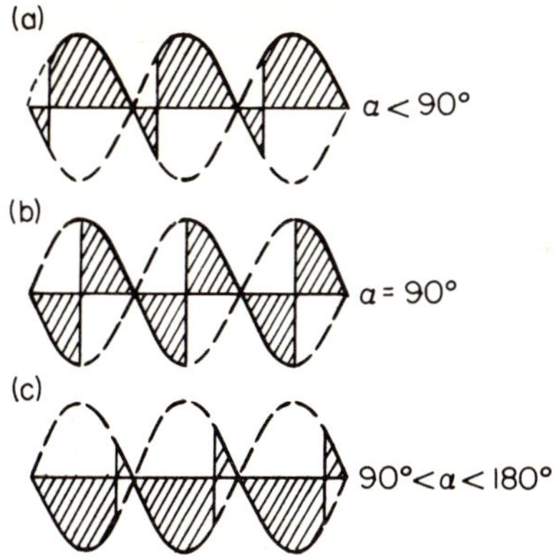

Figure 5.7 Output voltage waveform.

High-speed linear motors for which the capacity of the d.c. source reaches several tens of megawatts from a single converter unit are wound for 3, 6, 12 or 24 phases, minimising ripple on the d.c. side and higher harmonics on the a.c. side.

5.3.2 *Thyristor switching device*

There are two kinds of thyristor switching device: bridge type and flip-flop type. Figure 5.3 illustrates a bridge type, but in practice the flip-flop type is to be preferred to the bridge type in many respects. The authors have generally used the flip-flop type in their experimental design and for a number of applications.

The results of the analysis on the flip-flop thyristor switching device now follow.

Figure 5.8 shows an example of the circuit connections in a flip-flop thyristor switch when the ground armature coil is split into four groups.

Suppose that in a flip-flop circuit:

(1) S_A is switched OFF by making S_B ON. Then the equivalent circuit will be as shown in Figure 5.9 and i_1, i_2 will be calculated as follows:

$$i_1 = I - \left(\frac{C}{L_2}\right)^{\frac{1}{2}} V \sin \frac{t}{(L_2 C)^{\frac{1}{2}}}$$

$$i_2 = \left(\frac{C}{L_2}\right)^{\frac{1}{2}} V \sin \frac{t}{(L_2 C)^{\frac{1}{2}}} \qquad (7)$$

The terminal voltage of the capacitor V_c is given by:

$$V_c = -V \cos \frac{t}{(L_2 C)^{\frac{1}{2}}} \qquad (8)$$

If the time from S_B ON to S_A OFF is T_1, from $i_2 = 0$ the following expression is obtained:

$$T_1 = (L_2 C)^{\frac{1}{2}} \sin^{-1} \frac{I}{V} \left(\frac{L_2}{C}\right)^{\frac{1}{2}} \qquad (9)$$

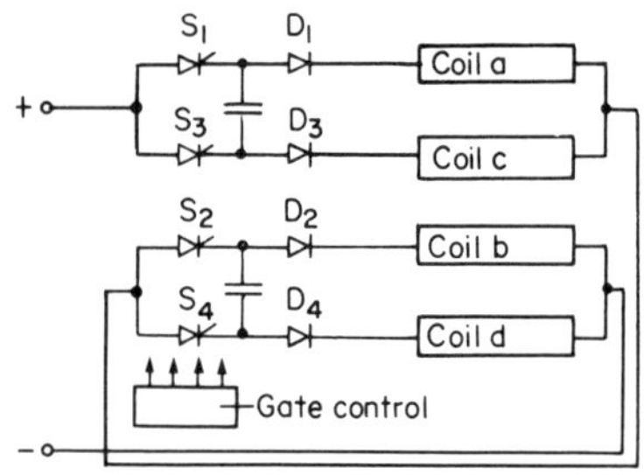

Figure 5.8 Flip-flop circuit.

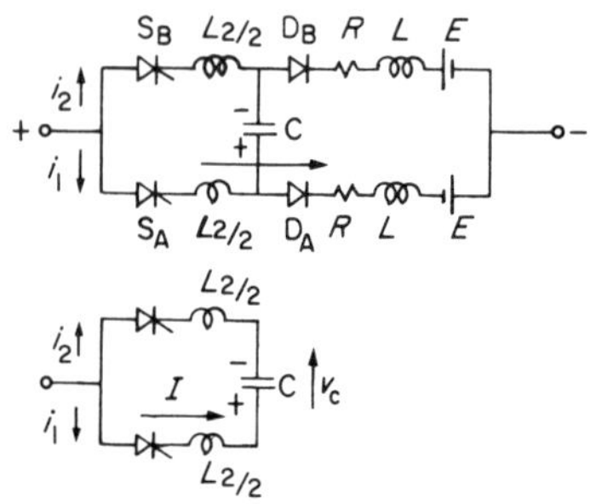

Figure 5.9 Equivalent circuit at S_B ON and S_A OFF.

Thereby the terminal voltage of the capacitor V_1 will be:

$$V_1 = V\cos\frac{T_1}{(L_2C)^{\frac{1}{2}}} = V\left(1-\frac{I^2L_2}{V^2C}\right)^{\frac{1}{2}} \tag{10}$$

(2) When the period in which the terminal voltage (V_C) of the capacitor C rises and the current flowing to the D_B side is considered, the equivalent circuit will be as shown in Figure 5.10 and the following will hold:

$$v_C = \frac{It}{C} - V_1 \tag{11}$$

If the time of current flow to D_B side is T_2 then using $2E + IR = \frac{I}{C}T_2 - V_1$ the following equation is obtained:

$$T_2 = (V_1 + 2E + IR)\frac{C}{I} \tag{12}$$

(3) Commutation from D_A side to D_B side.

In this case the equivalent circuit will be as shown in Figure 5.11 and i_1, i_2 will be calculated as follows:

$$i_1 = I\exp\left(-\frac{R}{2L}t\right)\left[\cos\alpha t - \frac{R}{2I\alpha}\sin\alpha t\right]$$

$$= I\left[1+\left(\frac{R}{2L\alpha}\right)^2\right]^{\frac{1}{2}}\exp\left(-\frac{R}{2L}t\right)\left[\cos(\alpha t+\beta)\right] \tag{13}$$

$$i_2 = I\left\{1 - \left[1 + \left(\frac{R}{2L\alpha}\right)^2\right]^{\frac{1}{2}} \exp\left(-\frac{R}{2L}t\right)\cos(\alpha t + \beta)\right\}$$

where,

$$\alpha = \left[\frac{1}{2LC} - \left(\frac{R}{2L}\right)^2\right]^{\frac{1}{2}}$$

$$\beta = \tan^{-1}\frac{R}{2L\alpha}$$

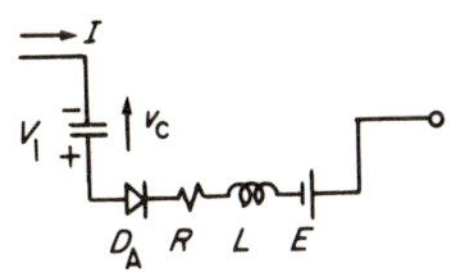

Figure 5.10 Equivalent circuit before the current switches to D_B.

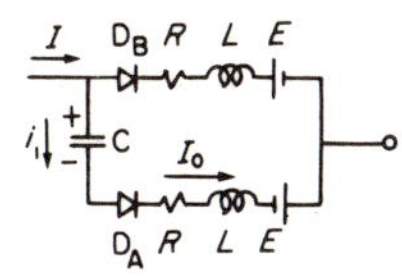

Figure 5.11 Equivalent circuit for movement from D_A to D_B.

Thereby the terminal voltage v_C of the capacitor will be:

$$v_C = (IR + 2E) + I\exp\left(-\frac{R}{2L}t\right)\left[\left(2L\alpha + \frac{R^2}{2L\alpha}\right)\sin\alpha t\right] \tag{14}$$

If the time to $i_1 = 0$ is T_3,

$$T_3 = \left(\frac{\pi}{2} - \beta\right)/\alpha \tag{15}$$

whereby the terminal voltage V of the capacitor will be:

$$V = (IR + 2E) + I\exp\left(-\frac{R}{2L}T_3\right)\left[\left(2L\alpha + \frac{R^2}{2L}\right)\sin\alpha T_3\right] \tag{16}$$

Thus the time T taken for commutation in a flip-flop circuit system will be

$$T = T_1 + T_2 + T_3 \tag{17}$$

An example of commutation is given in Figure 5.12.

For thyristor commutation at high speed to be completed within the specified time, a large surge voltage must be taken in commutation on account of an enormous self-inductance of the circuit. Thereby, with the surge voltage becoming higher than the speed e.m.f. it would be impossible to adopt a power source commutation or load commutation system using an ordinary inverter. Thus a super forcible commutation system in which the surge energy is received and delivered using a capacitor

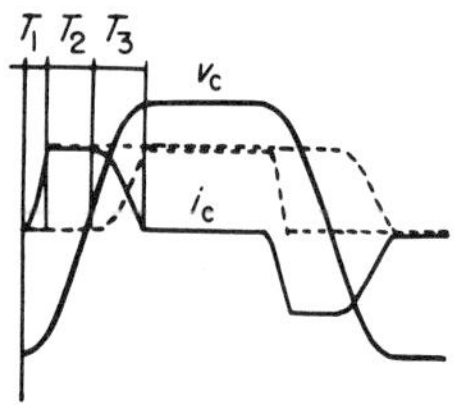

Figure 5.12 Commutation waveform under flip-flop circuit.

bank, parallel-connected, in the circuit must be employed. Increased surge voltage in commutation, however, would be undesirable because of installation considerations and accordingly it is necessary to minimise the surge energy.

Since the surge voltage in commutation depends on $L(di/dt)$, the surge energy may be minimised by any one of the three methods; by decreasing the self-inductance L of the armature coil, by decreasing the coil current, and by increasing the commutation period.

For the purpose of decreasing the self-inductance of the armature coil, the coil is split into many groups. Splitting into many groups without mutual coupling can decrease the self-inductance of the coil in inverse proportion to the square of the number of divisions; therefore, even if the commutation period is shortened in inverse proportion to the number of divisions, the surge voltage will be reduced in inverse proportion to the number of divisions.

For the purpose of decreasing the coil current, the field is strengthened; the ampere turns of the coil will be reduced to the extent of this field increase. There is, however, limitation to strengthening the field; it would be hard to obtain more than 0.9 T when a ferromagnetic core is used. Even in the case of a coreless superconductive field, it would be unlikely to be able to increase this larger value when the various restrictions of such a system are considered.

5.4 Vehicle field

Theoretically the force is generated through interaction between magnetic flux and current. Therefore the linear motor field can be generated normally or by a superconducting magnet. It is the former that can be manufactured easily at the present level of technology. The authors have investigated the normal-conducting field and the results are now discussed.

The magnetic flux density of the field can be found from the following formula:

$$\mathbf{B} = 1.257 \times 10^{-6} \frac{N_f I_f}{D} \xi \qquad (18)$$

where B is the magnetic flux density of field; N_f is the number of turns in field coil; I_f is the current in field coil; ξ is the effective magnetic flux factor; and D is the gap length in field.

The effective magnetic flux factor, which is significant, indicates the proportion of the magnetic flux produced by the field coil that is concentrated in the field gap.

Generally speaking, the linear motor field has a wide gap and naturally a heavy leakage of magnetic flux occurs. Therefore, if the selected profile and dimensions of the field are inadequate, it will not be possible to secure the necessary magnetic flux density on account of magnetic saturation, however much the ampere-turns for excitation may be increased. For this reason, numerous trials have been made in respect of the profile and dimensions of the field system.

Typical fields for a d.c. linear motor are as illustrated in Figure 5.13: external core type, internal core type and symmetrical type, each with the following features.

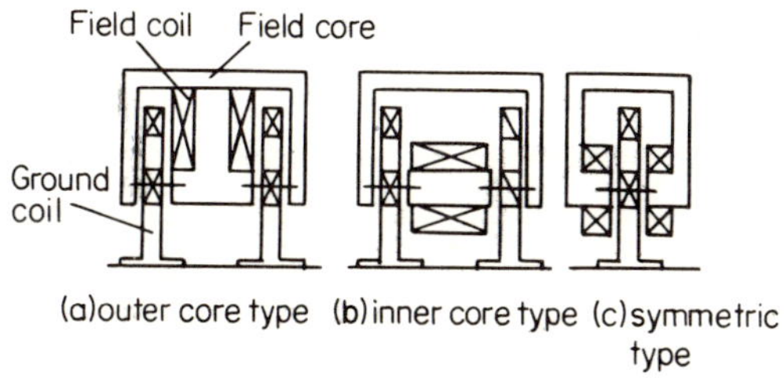

Figure 5.13 Field profiles.

1. At first sight, the external core type looks compact and ideal in profile; but the field gap being wide, the leakage of magnetic flux to the core bottom proves to be enormously heavy and even after dimensional modification in each part, the effective magnetic flux factor was only 15–23%, and the maximum magnetic flux density was only 0.56 T. See Figure 5.14.

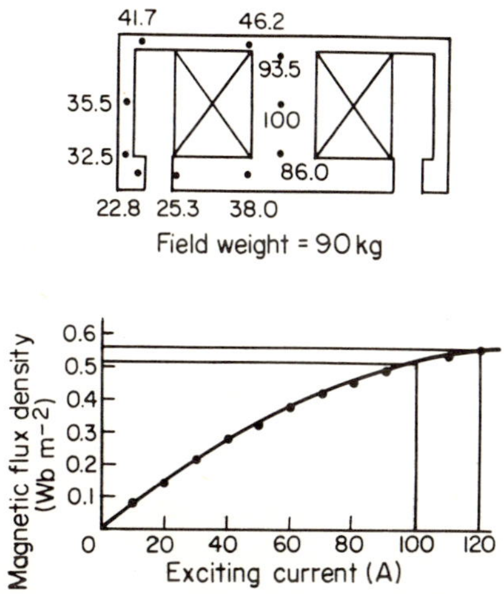

Figure 5.14 Characteristics of outer core type field.

2. The internal core type was easily saturated magnetically in the portion of the core where the field coil was wound; and only after this part was reinforced, could a maximum magnetic flux density of about 0.69 T be attained. See Figure 5.15.

3. The symmetrical type suffered the least leakage of magnetic flux; and after dimensional modification of each part, a magnetic flux density 0.8 T could be obtained. In the case of the symmetrical type particularly, in which the leakage flux was perfectly symmetrical, practically no attraction occurred between the field and the ground armature coil. Thus the symmetrical type may be the most suitable for the d.c. linear motor. See Figure 5.16.

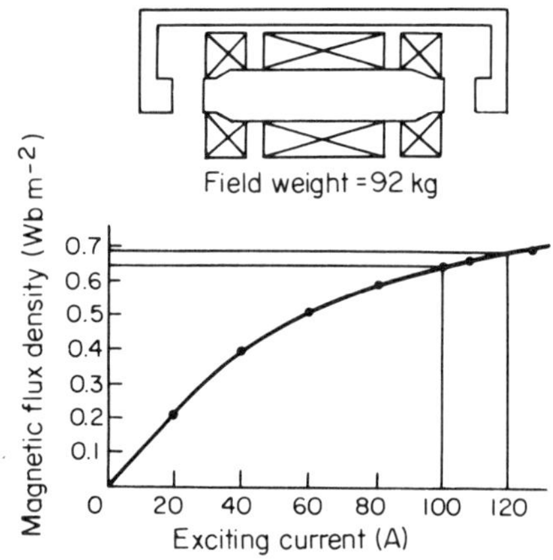

Figure 5.15 Characteristics of inner core type field.

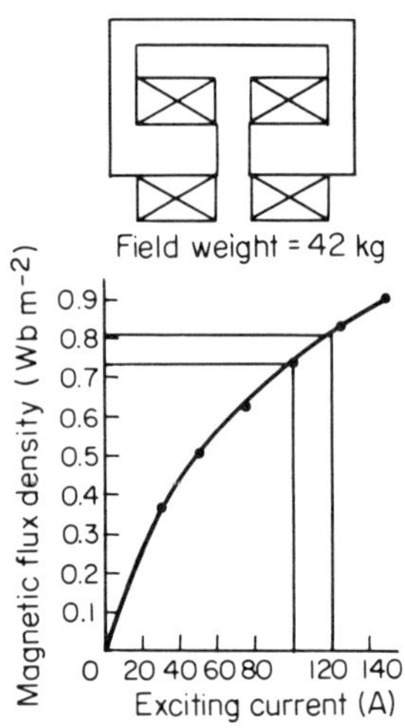

Figure 5.16 Characteristics of symmetric type field.

5.5 Electromagnetic rail

The authors define the ground armature coil of a d.c. linear motor, complete with the current collection and mechanical guidance devices as an electromagnetic rail.

Figure 5.17 illustrates an example of such an arrangement, for which various factors have to be taken into consideration.

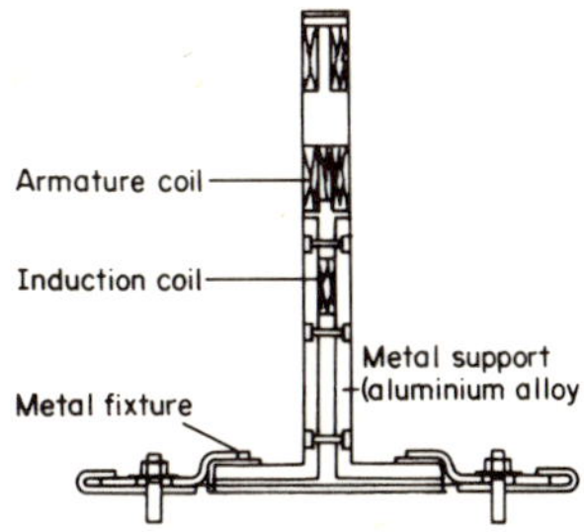

Figure 5.17 Sectional view of ground coil.

5.5.1 *General consideration of the ground coil*

The ground coil has to fulfil various requirements. It has to provide the forces for levitation, propulsion and guidance of the vehicle. The voltage impressed on the winding should be high for the purpose of minimising losses in the feeding system, thermal stress due to temperature rise as the result of a large current flowing instantaneously and to use only natural cooling. The structure must be weatherproof in view of long service outdoors. When combined with the requirement that the linear motor be compact and of high performance, technical difficulties arise and conventional techniques of winding and insulation for rotary machines and cables are not directly applicable.

To maintain mechanical strength of the coil, its outermost layer is sheathed in FRP and the outside of the coil is fixed on to the concrete guideway by means of a metal fixture.

The following problems must be solved:

– how to dissipate the heat generated in the winding
– how to absorb the thermal stress developed through temperature rise in the winding
– how to allow for thermal expansion and contraction of the fixture
– how to prevent the aging of the external skin
– how to connect the external skin with the metal fixture
– how to fasten the metal fixture to the guideway
– how to execute the work in the field (how to correct the alignment, level or distortion of the ground coil; and how to join the winding)

5.5.2 *Structure of ground coil in general*

Coils of length l in the travel direction are arranged with equal spacing (Figure 5.18). With every traverse of the vehicle over the distance l, the current in the winding is disrupted or reversed once and the frequency of the current, f, in the winding can be expressed as:

$$f = \frac{V}{2l} \tag{19}$$

Target values are:
$V \doteqdot 140\ \mathrm{m\,s^{-1}}$ for a vehicle running at $500\ \mathrm{km\,h^{-1}}$,
$f \doteqdot 70$ Hz at $l = 1$ m
$f \doteqdot 35$ Hz at $l = 2$ m and
$f \doteqdot 17.5$ Hz at $l = 4$ m.

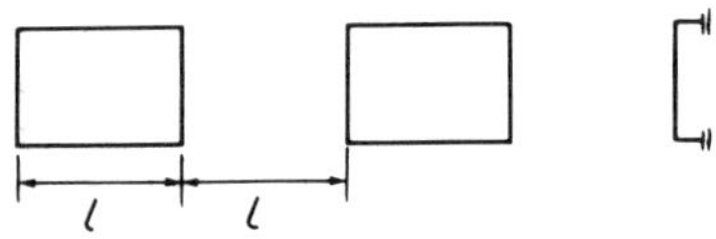

Figure 5.18 Coil length and coil gap.

In the thyristor d.c. linear motor and the linear synchronous motor, unlike the induction motor, the period of disruption or reversal of the winding current must be in perfect register with the speed of the vehicle. Hence the power supply frequency must be controllable from zero to the maximum value mentioned above. It should be borne in mind that for this reason a transformer cannot be used for such a power supply.

Generally speaking, when the coil length l is made large, the supply frequency f can be made small for a specified maximum speed V, resulting in a small reactive drop in the feeding wire, which is advantageous for power supply. In that case, however, the vehicle field has to be made large to develop the necessary thrust, which conflicts with weight requirements. Conversely, when the coil length l is made small, the vehicle field needed to develop the necessary thrust may be made small, but the frequency of supply to the ground coil will increase, resulting in an increased reactance drop in the feeding wire.

For the thyristor d.c. linear motor a length of 2 metres was chosen for high speeds.

The current density of the armature coil of a conventional rotary motor is of the order of 200–400 $\mathrm{A\,cm^{-2}}$. In the case of a linear motor with ground-mounted primary, the current flows only for a short time and accordingly a larger current density can be used. This helps make the coil compact and reduces the weight of the field coils. Moreover the ground coil, which constitutes the major part of the ground facilities can be made less expensive, but the power loss in the armature coil is increased, resulting in a decreased efficiency of the linear motor and a temperature rise in the winding, which complicates the cooling system and adds to the cost of insulating materials. Allowing for all these factors, the current density in the thyristor linear motor is estimated to be about 1000 $\mathrm{A\,cm^{-2}}$.

5.5.3 *Design of polyphase ground coil*

In a high-speed, high thrust linear motor the coil current and the inductance become large. To commutate such a large current in such a large inductance efficiently within a specified time is the key factor in the design. It is believed that the means of achieving this lies in the technique of a forced-commutation thyristor inverter for polyphase.

Suppose a single-phase ground coil carries a current I and has inductance L. The power supply to the coil, in addition to the drive power and the resistance loss should be such that at the current ON, energy equivalent to $\frac{1}{2}LI^2$ is supplied, while at the current OFF, $\frac{1}{2}LI^2$ is released (Figure 5.19); the reactive power thereby is exchanged between the power source and the ground coil. Now consider the same coil split into n phases. The thrust remains the same as it was before splitting but provision is made

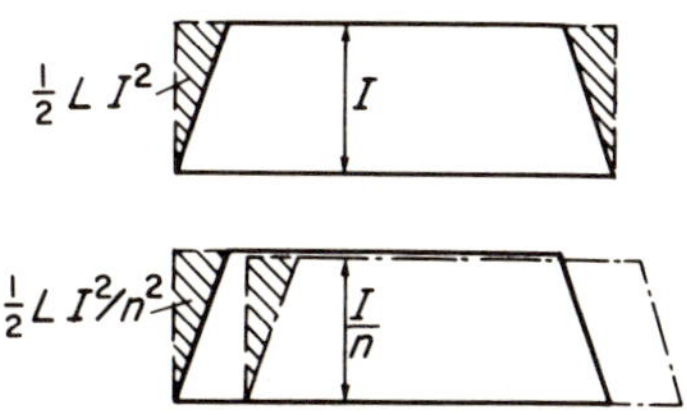

Figure 5.19 Polyphase design of ground coil and magnetic energy.

such that at current switching the mutual inductance between the phases may have no influence. The energy per phase at current switching is $\frac{1}{2}\ LI^2/n^2$ which is repeated for each of n phases and finally, the total energy transfer at switching is:

$$(\tfrac{1}{2}LI^2/n^2)\,n = \tfrac{1}{2}LI^2/n \qquad (20)$$

The energy to be disposed of is seen to decrease by $1/n$ for an n-phase design. The thyristor, commutating condenser and commutating reactor in the thyristor switching device will generally become less expensive therefore in a polyphase design. Though the feeder may be enlarged a polyphase design will have an overall advantage in a specific range corresponding to a required thrust. The number of phases, which should be chosen, considering the coil ampere-turns and the length of a section, would appear to be of the order of 10 or so for a railway of 500 km h^{-1} and 4–6 for a transport system of 200 km h^{-1}.

5.6 Self-control and external control

For the operation of a thyristor commutated d.c. linear motor, two central methods are available: one in which the field phase is detected and correspondingly the ground coil current is switched; and the other in which the coil current is switched in accordance with a preset pattern and the field is synchronised therewith for propulsion. The former is provisionally named the self-control type and the latter the external control type.

In Figure 5.20 (a) (b) (c), when arrival of the field tip at x_1 is detected and the first, second, third . . . nth phase coils are energised and when arrival at x_2 is detected and the second, third, fourth . . . nth, $(n + 1)$th phase coils are energised, the thrust F_1 will be given by:

$$F_1 = \mathbf{B}(I/n)\,nl = \mathbf{B}Il \qquad (21)$$

where I/n is the ampere-turns for one phase of coil, l is the length of coil side

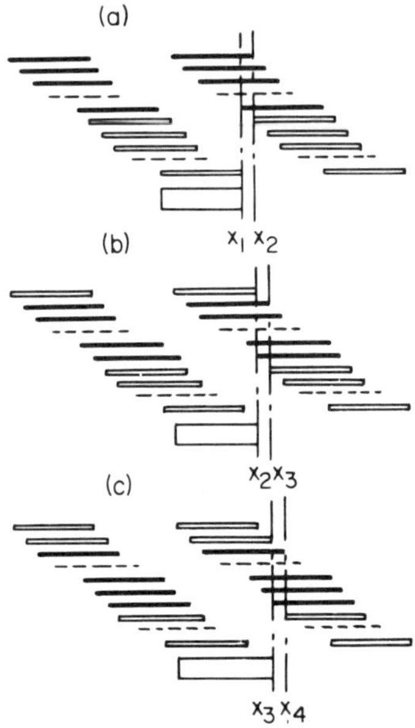

Figure 5.20 Energising phase of ground coil and field position, driving.

corresponding to the coil and F_1 is the maximum value of thrust obtainable.

When with the current-switch timing is delayed by one pitch, the first, second, third . . . nth phase coils energised upon arrival of the field tip at x_2, and the second, third, fourth . . . nth, $(n + 1)$th coils energised upon arrival at x_3, then one phase of the coil will have its leading end, instead of its trailing end, in register with the field. Thus the thrust F_1 will reduce to:

$$F'_1 = B(I/n)(n-2)l = \frac{n-2}{n} F_1 \qquad (22)$$

Further, when the first, second, third . . . nth coils are energised upon arrival of the pole tip at x_3 and the second, third, fourth . . . nth, $(n + 1)$th coils energised upon arrival at x_4, then the thrust F'' will reduce still more to:

$$F'' = \frac{n-4}{n} F_1 \qquad (23)$$

and so on.

Similar conditions will be true for braking, as illustrated in Figure 5.21 (a) (b) (c). If it is arranged such that the first, second, third . . . nth phase coils may be energised just when the field tip reaches x'_1 which is a position one coil short of x_1 and the second, third, fourth . . . nth $(n + 1)$th phase coils are energised just when the tip reaches x'_2 which is one pitch short of it, the maximum braking force will be attained.

On the contrary, if it is arranged that the same coil phases as above are energised upon arrival of the tip at a position short of x'_1, the braking force will be reduced.

By gradually controlling the ground coil switching by detection of the position of the pole tip in the stated range (the field positions giving maximum thrust and maximum

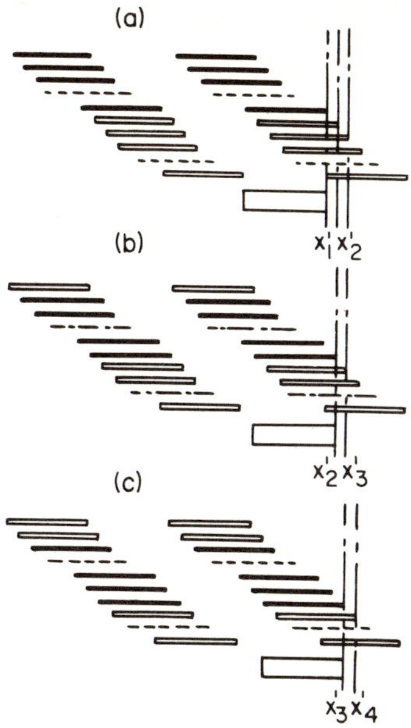

Figure 5.21 Energising phase of ground coil and field position, **braking.**

braking force deviate from each other by one coil length), the thrust can be continuously controlled from maximum drive to maximum brake with the levitation force remaining unchanged.

Writing E_a as the speed e.m.f. generated per one phase of a coil, as shown in Figure 5.20, where the maximum thrust is developed, the sum speed e.m.f.s generated in the energised coils will be equal to nE_a as shown in Figure 5.22 (a); and in the condition shown in Figure 5.7, where the maximum braking force is developed it will be equal to $-nE_a$ as indicated in Figure 5.22 (e). Change of phase in the ground coil will take place arbitrarily within this range, that is, as indicated in Figure 5.22 (b) (c) (d). In the case of (b) the thrust will be 1/2 of that in (a), and the mean speed e.m.f. will be $\frac{1}{2}nE_a$, i.e., also 1/2 of that in (a). By phase control of the energised coils, the sum of the speed e.m.f.s can be made variable in the range of $+nE_a$ to $-nE_a$ and correspondingly the drive power becomes variable in the range of maximum thrust to maximum braking force. By supplying the drive energy from a constant current source, power can be supplied or regenerated automatically.

The method of automatic control by detecting the field position and correspondingly by changing the coil-energising phase resembles that of thyristor motor control in rotary machines. The drive and brake control characteristics up to a circumferential velocity 500 km h^{-1} have been confirmed using a disc-type thyristor linear motor testing device using a combination of a light signal, a photo transistor and a perforated circular light shield.

Further, as described later, a self-control using the cross-induced radio for detection of vehicle position is being utilised in practice in a 1/3 scale power control system model for a loop track of about 30 m diameter consisting of a track of about 100 m

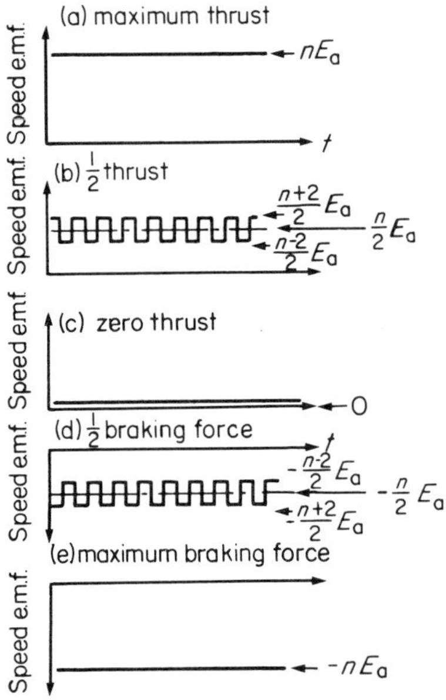

Figure 5.22 Thrust, braking force and speed e.m.f. waveforms.

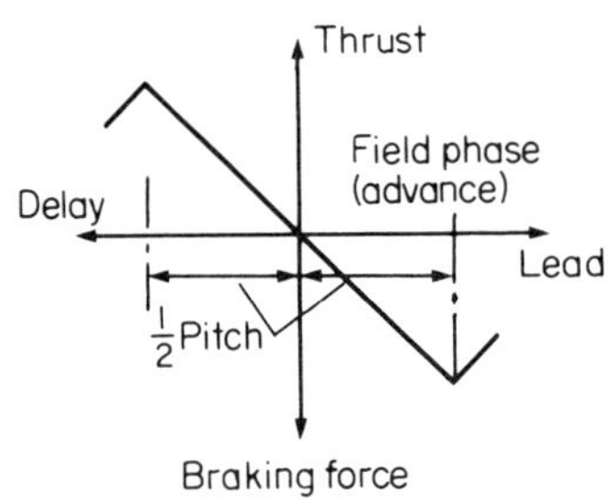

Figure 5.23 Field phase and thrust.

divided into 24 feeder-sections; and in a track of about 350 m length on which a test vehicle weighing about 3.5 tonnes can be accelerated to a maximum speed of 100 km h^{-1}.

The thyristor commutation type d.c. linear motor can be operated also by external control. In Figure 5.20 (a), suppose the coil current is not switched to the stage of (b) even when the field is displaced by the thrust, and as a result, the field tip passes x_2. Then between x_2 and x_3 the thrust decreases to $(n-2)/n$ times, in accordance with Equation (22). If the power supply to the coil is the same, even when the tip enters the interval between x_3 and x_4, the thrust will further decrease to $n-4/n$ times. If the field is further displaced so as to be located over half the length of the coil in the forward direction, an additional braking force will come into operation. Dynamically represented, the relation between the ground coil switching and the field displacement can be shown as in Figure 5.23. With maximum thrust and maximum braking force as the critical limits, automatic adjustment takes place such that, when the field displacement goes faster than the ground coil switching, the thrust decreases; and when the former goes slower than the latter, the thrust increases. As mentioned above, such an automatic adjustment takes place also in the braking mode.

As indicated in Figure 5.24, when the control is effected in a pulse pattern with a period inversely proportional to the speed, the field will move in accordance with the speed pattern. A notable feature is that there is no need for a feedback loop for speed control. During acceleration the speed e.m.f. is positive; under a deceleration pattern the speed e.m.f. becomes negative. Correspondingly, the constant current power source supplies and regenerates the power.

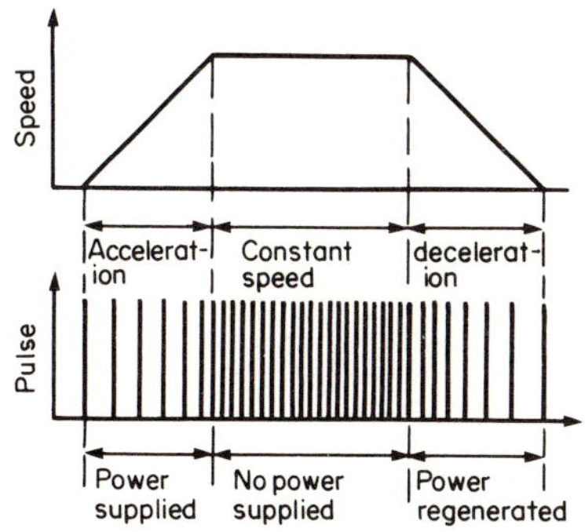

Figure 5.24 Example of pulse pattern by external control.

The possibility of operating a linear motor by external control has been confirmed in the low-speed range using a 1/4 scale linear motor with a 10 m long track which can levitate and run carrying one adult passenger and a 1/3 scale outdoor type linear motor with a 100 m long track which can carry a number of passengers. It is now planned to test the external control characteristic in the high-speed range using a rotary plate and linear motor. Future aspects to be studied include the hunting and out-of-step phenomena in which the field travel ceases to match the ground coil switching phase.

5.7 Power supply control system

As stated above, in a high-speed railway using a thyristor commutation d.c. linear motor in which the frequency control has to be made for each train, the ground coil will be electrically split at certain intervals and a thyristor switching device will be provided independently for each split division.

Power supply to such divisions of the ground coil and the resulting train control are now discussed.

5.7.1 *Feeding system and fed power*

Broadly there are two ways of feeding the thyristor commutation type d.c. linear motor: a d.c. feeding system of scattered converter type and a polyphase feeding system of concentrated converter type.

Figure 5.25 illustrates a typical example of the former, in which a constant d.c. is generated at the feeder substation and it is fed in parallel through two substations; this constant d.c. power is converted to a polyphase power by the thyristor switching device installed at each feeder sectioning post; and then it is supplied to the ground coil. Thereby it should be arranged that the ground coil currents in two adjacent feeder sections when a train runs across the feeder sections can remain equal to each other. One method available for this purpose is the double feeding system illustrated in Figure 5.25.

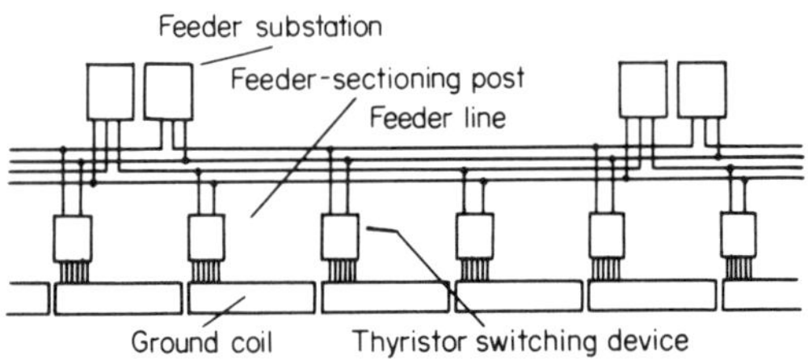

Figure 5.25 D.c. feeding system with converters scattered.

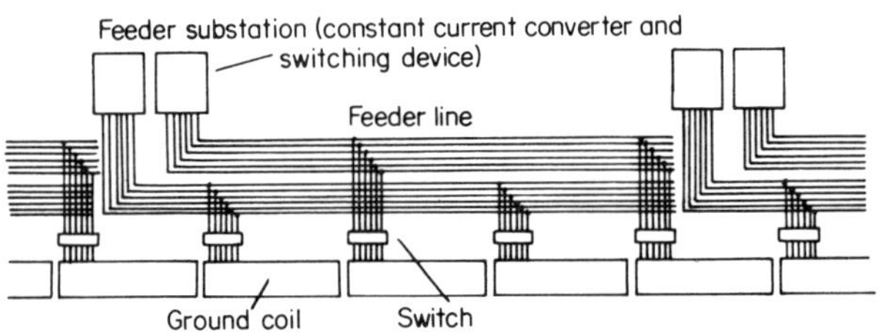

Figure 5.26 Multi-phase feeding system with converters scattered.

Figure 5.26 illustrates a specific example of the polyphase feeding system, in which the feeder substation is equipped with both a constant current a.c./d.c. converter and a thyristor switching device and polyphase power is generated and fed from the substation. The feeder sectioning post is equipped with only a changeover switch for a feeder section, and with progress of the train, only two feeder sections in which a train is present are energised at any one time. In this system, just as with the d.c. feed, a double-feeding system is provided when considering the passage of trains across different sections.

A major and characteristic difference between these two feeding system lies in whether or not the inductance of the feeder line has to be considered in the commutation of a ground coil phase.

For example, if the number of the ground coil divisions is 6 (groups), the length of the ground coil is 2 m, the maximum working voltage of ground coil is 10 000 V, the magnetic flux density of vehicle field is 0.8 Wb m^{-2}, the height of the vehicle field magnets is 0.15 m, the number of field magnets per vehicle is 4 and the number of vehicles in a train is 12, the maximum feedable power (P) for each system will be as follows:

1. d.c. feeding system (when the feeder section is longer than the train):

$$P = 1.74 \times 10^4 \frac{v}{Nl} \tag{24}$$

2. d.c. feeding system (when the feeder section is shorter than the train);

$$P = 5.73 \times 10^4 \frac{v}{N} \tag{25}$$

3. Polyphase feeding system (when the feeder section is shorter than the train):

$$P = \frac{6.94 \times 10^3 v}{1.2 \times 10^4 N + \frac{(N+1)^2}{N} v^2 lS} \tag{26}$$

where P is the maximum feedable power per vehicle (W), v is the train speed (m s^{-1}), N is the ratio of total speed e.m.f. to commutation surge voltage, l is the feeder section length (km) and S is the substation interval (km).

When a train is operated at high speed, a running resistance occurs. In a super high-speed range where the air resistance is predominant, however, the required power P(W) for one vehicle may be approximated by:

$$P = 2 v^3 \tag{27}$$

The relations between Equation (24) and Equation (27), between Equation (25) and Equation (27), and between Equation (26) and Equation (27) are graphically expressed in Figures 5.27–5.29 respectively.

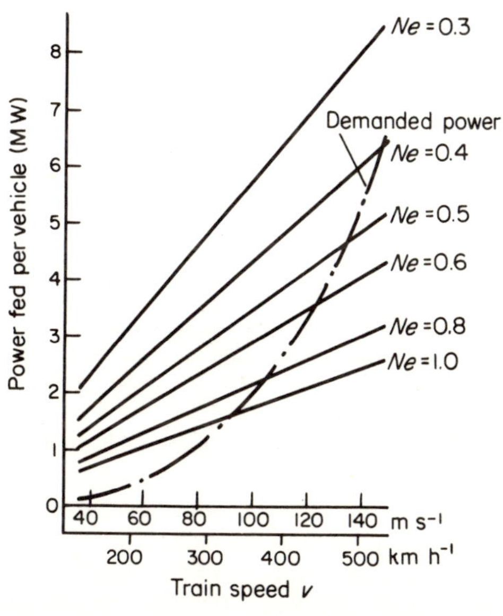

Figure 5.27 Power fed by d.c. feeding system (when the feeder sectioning is longer than the train.

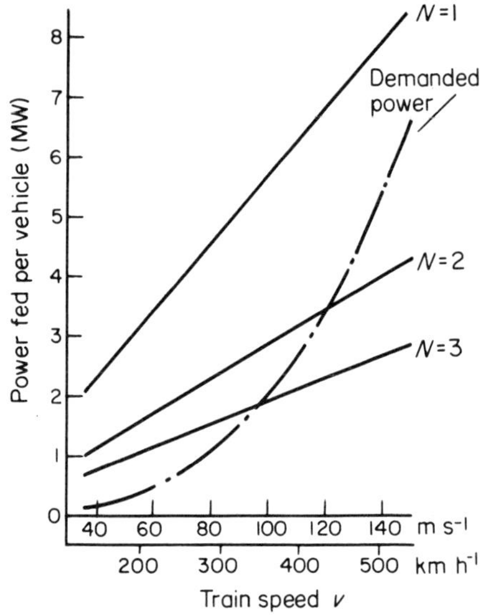

Figure 5.28 Power fed by d.c. feeding system (when the feeder-section is shorter than the train).

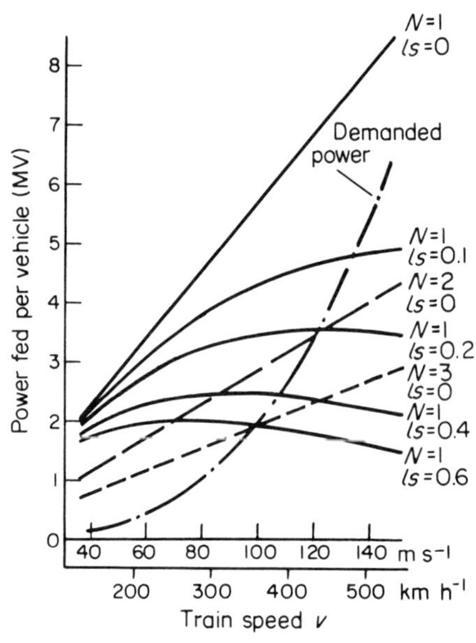

Figure 5.29 Power fed by polyphase feeding system.

Analysis of these relations leads to the following conclusions:

1. Each feeding system has a maximum feedable power which depends on the feeder circuit composition and on the working voltage of the coil. This maximum feedable power is one of the factors restricting the maximum operating speed of a train.

2. For a superspeed train which needs high power and high-speed commutation, the d.c. feeding system would be more suitable than the polyphase feeding system. As evident from Figure 5.27, if $Nl = 0.3$ (say, $N = 1$, $l = 0.3$ km), a power corresponding to the maximum speed of 600 km h^{-1} must be supplied.
3. The polyphase system is not suitable for large power feeding. It will exhibit high economic merit when it is applied to a transport system of medium capacity.

5.7.2 *Feeder section switching*

As mentioned in the above, the feeder sections are essential to the feeding system. How specifically to carry out the switching of feeder sections is therefore an important problem.

Taking a case of d.c. feeding, the feeder system for one set of constant current a.c./d.c. converters is illustrated in Figure 5.30. For the feeder section switches SW_A,

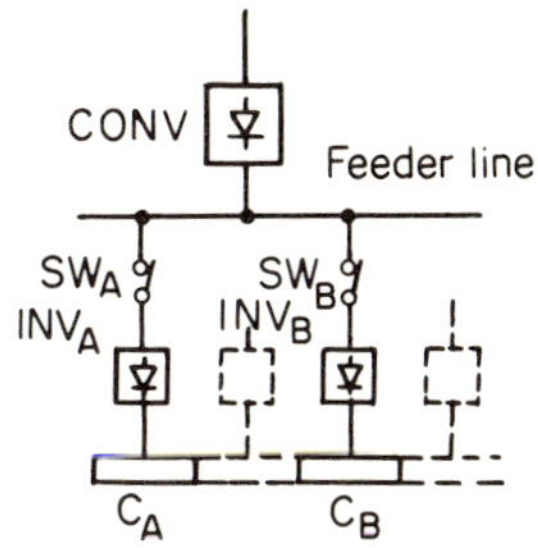

Figure 5.30 Feeder system for one set of constant current a.c./d.c. converters.

SW_B, etc., a thyristor switch and a mechanical switch are available; adoption of the former is highly controversial in respect of cost and reliability. A widely adopted practice with experimental devices and others at present is to use a mechanical switch for both SW_A and SW_B for the purpose of zero-current switching. The scheme of switching the feeder sections from the ground coil C_A to C_B is shown in Figure 5.31.

INV_A and INV_B are set in a fixed phase energised condition at the main circuit current ON and OFF for the following reasons: the charged state of the commutation capacitor of a flip-flop thyristor switching device depends on the coil current, and commutation under a sudden change of the coil current will make stable commutation impossible and risk inducing a thyristor element failure and similar troubles.

The above is a description of the d.c. feeding system, but precisely the same applies to the composition of a polyphase feeding system and already successful applications have been made for several model systems.

In the case of a commercial system, however, even zero-current switching with the use of a mechanical switch as described is not without its problems. First, the constant d.c. must be set ON or OFF each time the train passes the feeder sectioning point. In addition to the complexity of interlocked handling and the difficulty of maintaining the mechanical switches, the interference due to magnetic induction or the adverse effect of the reactive power variation on the power supply must be taken into account.

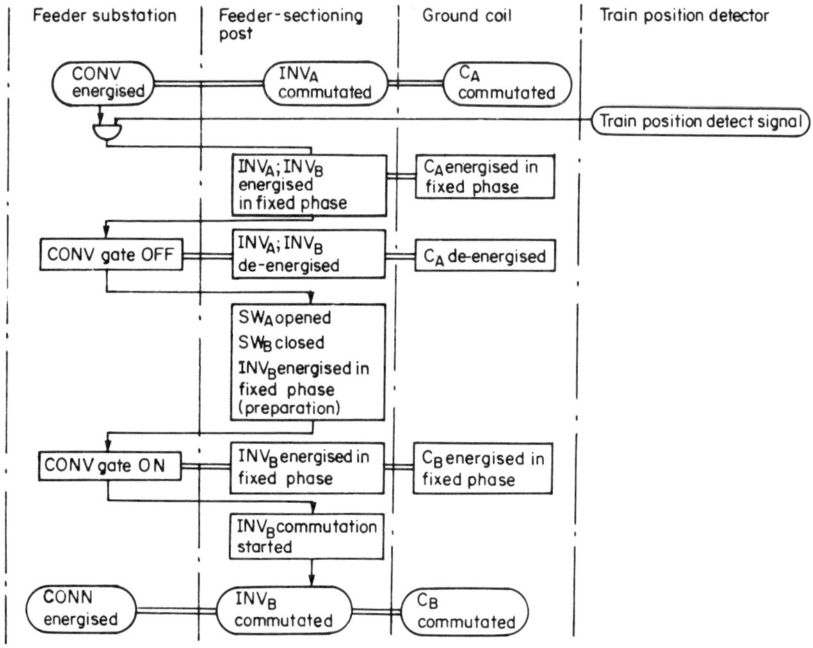

Figure 5.31 Interlocking for feeder-sectioning.

In view of all this, consideration is being given to the use of the flip-flop type thyristor switching devices located at the feeder-sectioning posts for feeder section switching as well as for commutation of the ground coil phase. A circuit illustrated in Figure 5.32 is an example; Figure 5.33 summarises the experimental data on this circuit.

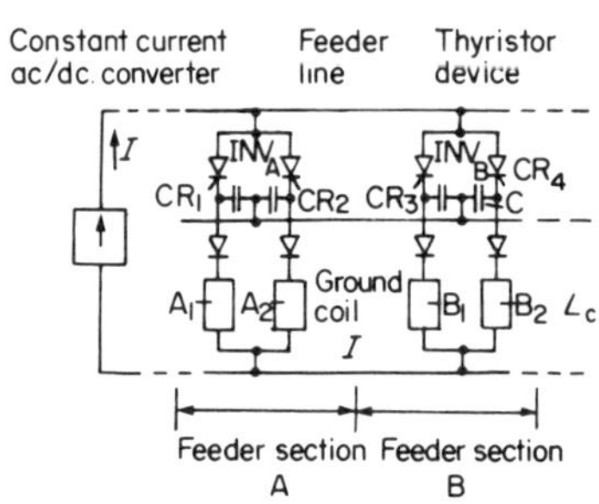

Figure 5.32 Switchless feeder-section system.

The conditions for the feeder section switching system to operate are as follows, where C is the charge voltage under phase commutation:

$$V \doteqdot \frac{I}{2}\left(\frac{L_c}{C}\right)^{\frac{1}{2}} \tag{28}$$

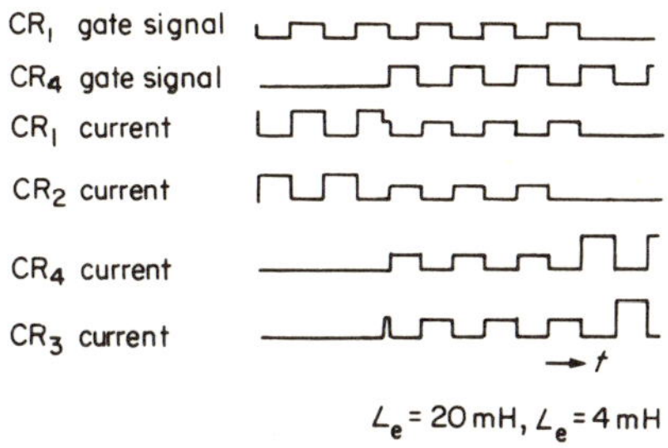

Figure 5.33 Test results of feeder-section switching system.

Current that can be passed in one C under section switching:

$$I_s \doteqdot \frac{V}{\left(\frac{2L_f}{C/2}\right)^{\frac{1}{2}}} = \frac{I}{4}\left(\frac{L_c}{L_f}\right)^{\frac{1}{2}} \tag{29}$$

(for notations in the above, refer to Figures 5.32 and 5.33).

Thus the feeder-section switching conditions may be approximated by:

$$I_s \geq \frac{I}{2}, \frac{L_c}{L_f} \geq 4 \tag{30}$$

This result has been verified experimentally. Further investigations into this system are planned.

5.7.3 *Power control system*

Figure 5.34 illustrates a general layout of the power control system. The ground coil split into groups at certain intervals (say, several hundred metres) detects the passage of a train, thereby closing or opening the feeder section switch. Substations equipped with a power converter are located at intervals (say, several kilometres) covering a certain number of feeder-sections, and thereto are attached local control posts.

In the case of a thyristor-commutation type d.c. linear motor, the feeding section of one substation, as illustrated in Figure 5.34, is involved with one block section of train operation so that one thyristor switching device may correspond to one train. Therefore if a local control post is provided with the means of controlling the commutation of the thyristor switching device at the relevant substation and, as shown in Figure 5.34, all-time information exchange is established between the adjacent local control post and the control centre, both the operation of a levitated train and the power supply for it can be controlled at the same time.

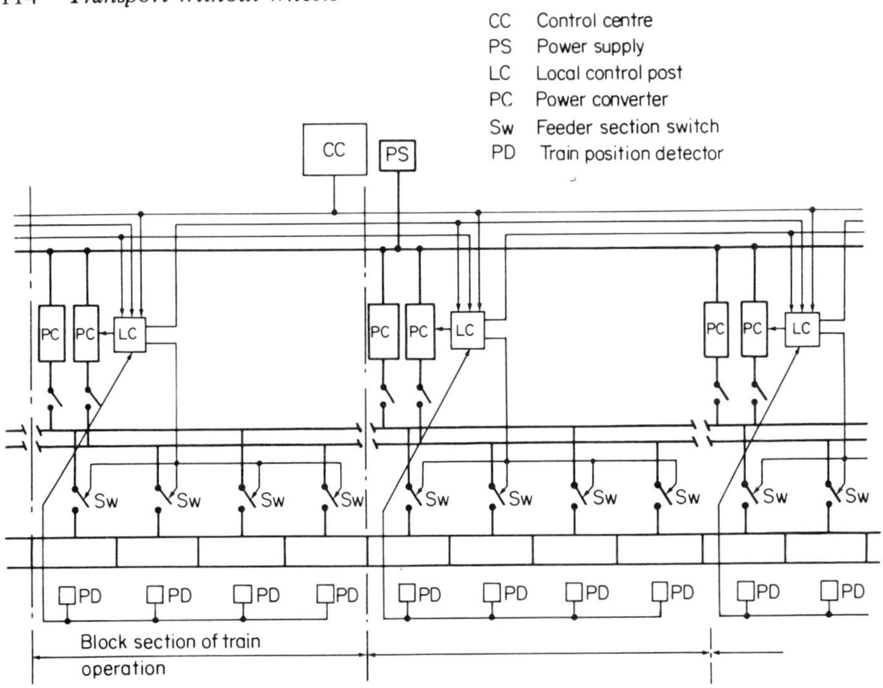

Figure 5.34 Layout of a complete system.

5.8 Experimental results and examples of applications

5.8.1 *High-speed characteristic test*

The thrust developed by a d.c. linear motor can basically be calculated as a Fleming force expressed by $\mathbf{B}\,I\,l$; but at low speed the following factors must be taken into consideration: (a) profile of ground coil, (b) extent of magnetic field flux and (c) armature reaction; at high speed, the following must be considered additionally: (d) eddy current loss in the conductor and core, (e) skin loss in the conductor and, (f) commutation loss.

To examine the high-speed characteristic of the d.c. linear motor, a test disc capable of obtaining a circumferential speed of 500 km h^{-1} has been developed with a ground coil fitted into the disc (see Figure 3.35 and Table 5.1). Figure 5.36 shows the shape of the coil.

Figure 5.36 summarises the measured data of thrust and power factor in the whole range of driving and braking up to a circumferential speed of 500 km h^{-1}, the dotted line representing the corresponding calculated results.

(1) *Thrust at low speed* The thrust at low speed can be calculated by the following formula:

$$F = \mathbf{B}\,I\,l\,k_1\ k_2/9.8\ \text{kg} \qquad (31)$$

where **B** is the field magnetic flux density (0.87; measured), I is the armature coil

Table 5.1

Dimensions and materials of the rotary disc

Material of disc	FRP
Diameter of disc	1730 mm
Thickness of disc	80 mm
Composition of 1 turn	four 2.7 mm x 1.6 mm Al PEW
No. of turns per coil	16 turn x 2 rows, symmetrical disposition
No. of phases	8
Coil current	130 A
Total ampere-turns	8320
Interphase insulation	0.5 mm thick FRP
Working voltage	800 (phase-to-phase)
Rating time	10 minutes

Figure 5.35 High-speed test rig.

Figure 5.36 Shape of coil inserted in the rotary disc (16 turns of 8-strand bundles of 2.7 x 1.6 mm square aluminium wire).

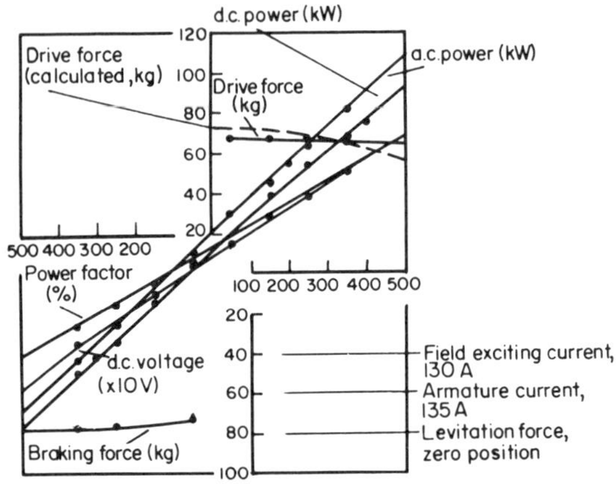

Figure 5.37 Test results – high-speed characteristics.

ampere-turns (8320), l is the path length of magnetic flux (0.111 m) [assuming a magnetic flux (gap length/4) to extend on both sides of the magnet pole, the conductor-crossing part being included by multiplying by a factor of 1/2.] k_1 is the correction factor for coil profile, $(744 - 55)/744 = 0.926$, k_2 is the factor for converting the circumferential thrust to a horizontal component

$$\frac{1}{\pi/6}\int_0^{\frac{\pi}{6}} \cos\theta \, d\theta = 0.955$$

Substitution of the above yields:

$$F = 72.5 \text{ kg(f)}$$

The measured thrust at zero speed is equal to about 68 kg, which agrees with the calculation to within 7.5%.

(2) *Thrust at high speed* Basically, the thrust at high speed can be calculated by the same formula as that for low speed, but at high speed the thrust is reduced by:

$$\frac{\text{(Commutation time)}}{\text{(Energising period of one phase)}}$$

because the effect of commutation time, as described in Section 5.3.2, cannot be neglected at high speed.

The dotted line in Figure 5.37 indicates the result of calculating the thrust using the constants of a rotary disc linear motor:

$$L = 0.46 \text{ mH}, C = 52.5\ \mu\text{F}.$$

The thrust reduction seems to be lower in the measurement than in the calculation. This is probably because the commutation signal phase is given slightly ahead.

5.8.2 *1/3 scale straight track test facility*

For the purpose of confirming various characteristics of a linear motor capable of developing simultaneously levitation and propulsion forces, a 1/3 scale straight test track as shown on Figure 5.38 with dimensions given in Table 5.2 has been built, and numerous tests have been carried out.

A ground coil of unit length equal to 7.2 m both physically and electrically was built and a track of about 110 m in total length was constructed, divided into 15 units; Figure 5.39 is a sectional view of the back. The whole structure was moulded with FRP and the bottom was fixed with an aluminium support.

Experimental operation was found to be successful with both d.c. and a.c. feeding, and both self-control and the external control were tested.

Figure 5.38 Experimental linear motor.

Table 5.2
Major technical data of 1/3 scale test device

Track length	about 110 m
Magnetic rail	7.2 m x 15 sections
Weight of running truck	about 1.2 tonne
Overall length of running truck	about 4 m
Levitation force	1.5 tonne(f)
Thrust	150 kg(f)
Height of levitation	30 mm
No. of vehicle field magnets	2
Magneto-motive force of field	88 kAT
Magneto-motive force of ground coil	15 kAT
No. of ground coil divisions	8
Equivalent pole pitch	0.4 m

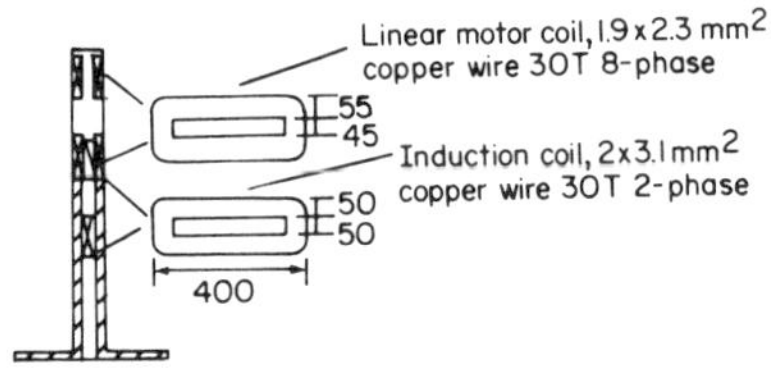

Figure 5.39 Composition of ground coil.

5.8.3 *Power control system model*

Of the subjects relating to the realisation of a railway system based on the linear motor, the power supply to the linear motor and the method of control are most important.

Figure 5.40 Whole power control system model.

Table 5.3

Major technical data of power control system model

Track length	about 100 m (circular loop)
Magnetic rail	4.32 m x 24 section
Weight of running truck	about 750 kg
Overall length of running truck	about 2 m
Thrust	70 kg(f)
Maximum running speed	25 km h^{-1}
No. of vehicle field magnets	2
Magneto-motive force of field	63 kAT
Magneto-motive force of ground coil	3.8 kAT
No. of ground coil divisions	6
Equivalent pole pitch	0.24 m
Power converter	d.c. flip-flop switching device and cyclo-converter

To test such systems a circular loop model with a track about 100 m long has been built; Figure 5.40 shows a general view and Table 5.3 lists its major technical data.

The ground coil (Figure 5.39) is designed with unit length equal to 4.32 m both physically and electrically and, combining 24 units of them, a track of about 100 m in total length was constructed. There are two running trucks. The operation of the running truck can be freely controlled by regulating the power supply from the power converter to the ground coil.

For power conversion, a cyclo-converter of continuously variable output frequency as well as the above-mentioned d.c. flip-flop inverter was available so that both the thyristor commutation type d.c. linear motor and the a.c. linear synchronous motor could be studied.

5.8.4 *Drive for current-collection test device*

In building a test device for analysis of the current-collection phenomena the idea of utilising the linear motor as the drive was implemented. Figures 5.41 is a photograph and Figure 5.42 is a diagram of the drive; Table 5.4 lists the major relevant technical data. To perform a running test of 100 km h^{-1} maximum over a distance of about 350 m, an enormous acceleration is needed, but the thrust calculated from the magnetic flux developed in the vehicle field and from the current in the ground coil indicates that this is feasible. Using a combination of an eddy current brake and a hydraulic brake the system is so designed that the running truck will never overrun.

Figure 5.41 Current collection test device.

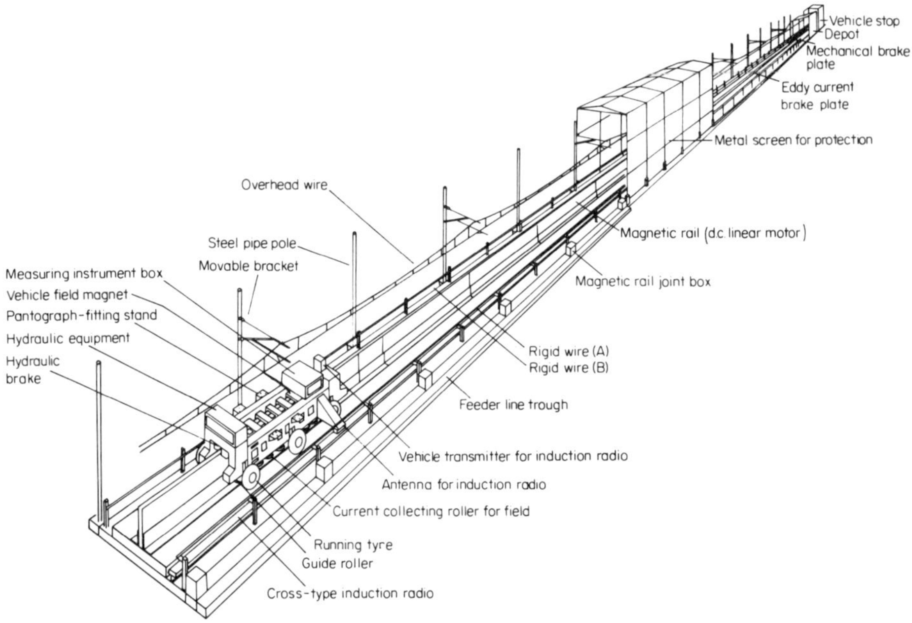

Figure 5.42 Complete current collection test rig.

Table 5.4

Major technical data of drive for current-collection test device

Track length	about 350 m
{ accelerating section	about 110 m
{ coasting section	about 90 m
{ braking section	about 150 m
Maximum running speed	100 km h^{-1}
Weight of running truck	3.5 tonne
Thrust	2.1 tonne(f)
No. of vehicle field magnets	8
No. of ground coil divisions	4
Magneto-motive force of field	82 kAT
Magneto-motive force of ground coil	24 kAT
Equivalent pole pitch	0.3 m

5.9 Concluding remarks

Investigations on the d.c. linear motor, begun in 1970, have yielded the results described in this chapter.

The d.c. linear motor and the synchronous linear motor develop thrust essentially by the same mechanism. The only linear motor so far developed in Japan that will be available for a superspeed railway is this synchronous linear motor which can provide the necessary thrust in both the high-speed and the low-speed range, with low end effects and the ability to develop a strong thrust even with a large air gap.

When a coreless armature coil is laid on the ground, the advantages are that there is no need for high-speed collection of large currents and the process of automating the train operation is relatively simple.

For the vehicle field magnets, both a superconducting and a normal conducting system are possible. It is even possible to design a practical normal conducting magnet with a gap of about 40 mm. MAGLEV railways may be said to have attained the stage where a practical system can be produced in the present state of the art.

The applications of the MAGLEV railway which is free from noise and air pollution, will cover both superspeed vehicles and new transport systems for the high-density traffic demands in urban areas at lower speeds.

6

Vibration Problems in Magnetic Levitation and Propulsion

F. C. MOON

6.1 Introduction

In the development of electromagnetic transportation, much of the research has centred on developing sufficient thrust in the case of linear electric motors, and high lift forces with minimum magnetic drag, in the case of levitation schemes, in order to demonstrate the feasibility of these new transportation systems to the skeptics. Questions concerning vehicle dynamics, stability, and ride quality while not entirely ignored, have received much less attention. The feasibility tests of electromagnetic transportation have been very successful. Linear induction motors have been built and tested to 375 km h^{-1} [1] and both attractive and repulsive magnetic systems have successfully levitated personnel carrying vehicles weighing many tons [2–4]. However, successful development of this technology into operating systems will require much more research into the dynamics of electromagnetic vehicles, especially those designed for speeds greater than 400 km h^{-1}.

A similar technical history was followed by the airplane. Wings with sufficient lift to carry humans were developed for gliders in the early nineteenth century and while Langley and Santos-Dumont built engine powered aircraft in 1896 and 1898, their vehicles had stability problems and went out of control. One of the contributions to the success of the Wright brothers in 1903 was the addition of the proper control or stabiliser surfaces to permit sustained flight. With further increases in speed, new dynamic problems, such as aeroelastic flutter, arose which were not properly understood until the 1930s and 1940s. Even in the sophisticated jet age, important dynamical instabilities continued to plague aviation with the engine-wing instability of the Electras, and the 'Pogo' instability and panel flutter experienced in the Saturn rockets in the US space programme. Unless magnetically propelled and levitated vehicles are placed in evacuated tubes, they will undoubtedly be subject to the whole range of aerodynamic instabilities coupled to the effects of electromagnetic forces.

Dynamical problems in magnetic levitation have been studied in applications other than transportation. A review of the dynamics of electric and magnetic suspension devices may be found in Geary [5]. Further research was made in connection with electromagnetic melting and levitation of metals using high frequency magnetic fields [6, 7]. Another set of problems stem from experiments in plasma physics in which superconducting magnets are levitated in a vacuum [8, 9]. Since a stable suspension using steady magnetic or electric fields is not possible according to Earnshaw's Theorem [10], these magnets were stabilised using feedback controlled magnetic fields.

This paper will review the dynamics of the repulsive magnetic levitation systems in which steady electric currents in superconducting magnets on the vehicle induce lift, drag, and guidance forces from eddy currents generated in a passive conducting guideway. In the attractive magnetic levitation systems, feedback control forces are required to stabilise the vehicles as a consequence of Earnshaw's theorem and a discussion of vehicle dynamics requires detailed knowledge of the control systems as well as the magnetic forces (see for example, Meisenholder and Wang [11]).

The vertical heave oscillations of a repulsively levitated train were discussed in the original Powell and Danby paper [12] proposing superconducting magnets for high-speed travel, The magnetic lift and drag forces versus speed are shown in Figure 6.1. As one can see, increase in velocity for high speeds brings a decrease in

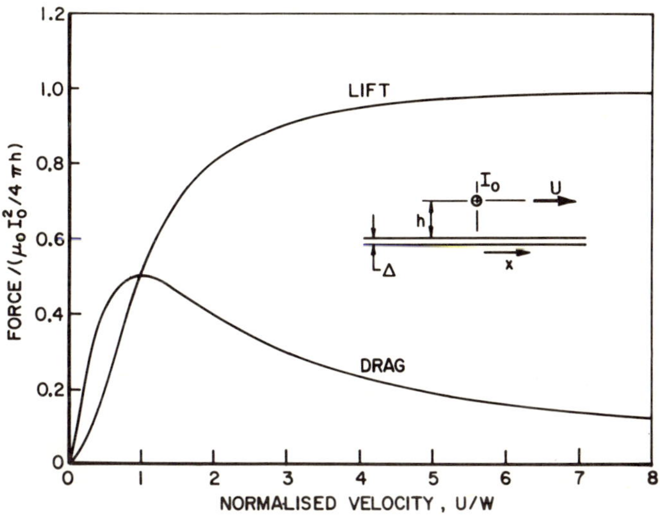

Figure 6.1 Lift and drag versus speed for a levitated wire moving over a thin conducting sheet.

magnetic drag. This was recognised early as one of the sources of potential instability in MAGLEV systems. Richards and Tinkham [13] studied this longitudinal instability for a number of propulsion-speed relations. Davis and Wilke [14] included the heave motion to analyse the coupled heave-longitudinal dynamics of a levitated wire. They also calculated the magnetic 'damping' force created by an oscillatory wire and concluded that *negative damping* occurs for velocities greater than the characteristic velocity based on the thin track theory. Iwamoto *et al.* [15] also made damping calculations of an oscillating current loop for a discrete loop track and claimed experimental evidence for this 'negative damping'.

Pitch-heave motion of a levitated wire pair, simulating a wide rectangular magnet, was analysed by Fink and Hobrecht [16]. Another analysis for a four magnet supported vehicle was given by Wilke [17] who also studied the lateral-roll vibrations of a vehicle in an inverted 'V' guideway. Finally the complete rigid body dynamics of a magnetically levitated body have been incorporated in a computer code by Coffey *et al.* [18] in conjunction with actual dynamic tests of a 2 metre long vehicle.

While several personnel carrying MAGLEV research vehicles have been tested, little hard data is available in the literature on their dynamic performance. Reports of visitors to these sites record a bumpy ride [19] due either to the flexible rail in the case of the attractive system or due to the discrete loops in the Japanese repulsive MAGLEV vehicle. Model tests performed for the US Department of Transportation at the Stanford Research Institute have recorded the effects of guideway irregularities on vehicle dynamics and no major instabilities have been reported [18]. Films of the linear synchronous machine levitated model at MIT Cambridge, Mass., show however, a yawing or corkscrew type motion of large amplitude [20]. In addition, the author has reported observation of longitudinal and yaw-roll instabilities in a small model test of a levitated model on a rotating wheel [21]. Iwasa [22] has developed an experimental technique based on an energy criterion to predict static stability properties of a magnet-guideway system but cannot predict dynamic instabilities such as flutter.

In less related papers Ohno *et al.* [23] have studied the pulsating lift forces in a linear synchronous motor. Such forces may not only drive heave and pitch oscillations of the vehicle but may have potential parametric instabilities associated with a time varying vertical stiffness. Also Thornton has concluded in a general analysis of MAGLEV vehicles that efficiency is directly proportional to the suspension stiffness [24].

While quite a gamut of questions relating to MAGLEV dynamics have been posed and analysed, it is the author's opinion that much more remains to be done. Certainly the combined effects of aerodynamic and magnetic forces must be incorporated in the analytic models [25] as well as the interaction of the propulsion and levitation forces.

6.2 Natural frequencies of levitated vehicles

Two methods for calculating natural frequencies of levitated vehicles are reviewed here. In the first method, the differential forces acting in the vehicle, arising from small motions about an equilibrium position, are used in a linear vibration analysis. In the second method, the variation of magnetic energy with the position variables of the vehicle is used in Lagrange's equations to obtain the equations of motion. In many cases the magnetic energy can be calculated using the inductance between the magnet and the guideway. Iwasa [22] has used this technique to examine experimentally the static stability of various magnet-guideway configurations. This method, while attractive in its avoidance of vectors and forces, can only treat the conservative parts of the magnetic forces, such as lift and guidance, and cannot include the non-conservative effects such as drag.

In the force perturbation method distinction must be made between quasi-steady and frequency-dependent forces. In general the magnetic force on a vehicle, moving over a guideway with velocity V_0 at a height h, will depend on the frequency of the perturbing motions i.e. for motions $h = h_0 + u_0 \exp(j\omega t)$ the force has the form

$$\mathbf{F} = \mathbf{F}_0(V_0, h_0) + \mathbf{F}_1(V_0, h_0, u_0, \omega) \exp(j\omega t) + \mathbf{F}_2(V_0, h_0, u_0, \omega) \exp(j2\omega t) \quad (1)$$

The frequency doubling effect is a consequence of the dependance of the forces on the square of the magnetic fields and currents. In linear vibration studies this term is usually ignored. Further, the function $\mathbf{F}_1$ is expanded in the perturbation variable u_0 i.e.:

$$\mathbf{F} = \mathbf{F}_0 + \mathbf{F}_1(u_0 = 0)\exp(\mathrm{j}\omega t) + \frac{\partial \mathbf{F}_1}{\partial h_0}(u_0 = 0)\exp(\mathrm{j}\omega t)\,[u_0] \tag{2}$$

The second term in Equation (2) is present only in guideways with periodic irregularities, such as a linear synchronous system, and represents a forcing term.

For homogeneous guideways (e.g. sheet conductors) $\mathbf{F}_1(u_0 = 0) = 0$. The third term is usually a complicated function of frequency and is usually simplified by expanding this term in the frequency parameter i.e.:

$$\mathbf{F} = \mathbf{F}_0 + \left[\frac{\partial \mathbf{F}_1}{\partial h_0}\Bigg|_{\substack{u_0 = 0\\ \omega = 0}} + \frac{\partial^2 \mathbf{F}_1}{\partial h_0 \partial \omega}\Bigg|_{\substack{u_0 = 0\\ \omega = 0}} \omega\right] u_0 \exp(\mathrm{j}\omega t)$$

or,

$$\mathbf{F} = \mathbf{F}_0 + \frac{\partial \mathbf{F}_1}{\partial h_0} u(t) + \frac{1}{\mathrm{j}}\frac{\partial^2 \mathbf{F}_1}{\partial h_0 \partial \omega}\frac{\mathrm{d}u}{\mathrm{d}t} \tag{3}$$

where $u(t) = u_0 \exp(\mathrm{j}\omega t)$. A quasi-static vibration analysis contains only the first two terms, which can be obtained from the force versus height characteristics under steady motion. The last term is the 'damping' term and has only been included in a few studies (see for example [14]). Section 6.3 will examine the effect of magnetic 'damping' while Section 6.4 on coupled dynamics will treat only the quasi-steady forces. In general the effective damping and quasi-steady approximations are only good in a low frequency limit. Studies of transient vehicle motion incorporating the complete frequency effects on forces have not been found in the literature. Also non-linear analyses of MAGLEV dynamics have not been presented, except those incorporated in numerical computer codes (see for example [22]).

6.2.1 *Linear perturbation methods*

In the analysis of vibration of levitated vehicles the starting assumption is of an equilibrium state about which the vibration and stability of perturbed motions are studied. From this point of view, this study falls in the purview of classical linear vibration theory and in particular the literature of aircraft dynamics[26, 27]. The difference, of course, is that in magnetic levitation the proximity of the guideway creates restoring forces and moments proportional to displacements, whereas velocity-dependent forces are paramount in aeronautical problems. A growing body of literature also exists on the dynamics of tracked air cushion vehicles some of which may be relevant to magnetic levitation.

In the direct method for calculating natural frequencies of levitated bodies, one determines the magnetic field and induced currents from Maxwell's equations and in turn determines the magnetic forces **F** as a function of the relative position of the magnet with respect to the guideway. The forces are then expanded in a Taylor series in the perturbed displacement variables about the equilibrium position.

For heave motion, the height h, of the magnets above the conducting track guideway is perturbed i.e.:

$$\mathbf{F}(h) = \mathbf{F}(h_0) + \left.\frac{\partial \mathbf{F}}{\partial h}\right|_{h = h_0} (h - h_0) \tag{4}$$

For magnets, the force is proportional the square of the total ampere-turns in the magnet i.e.:

$$\mathbf{F} = \frac{\mu_0 I^2}{4\pi}\, \mathbf{f}\left(\frac{D}{h}\right) \tag{5}$$

where D is some characteristic dimension of the magnet. At equilibrium, the lift should balance the gravitational force, g or:

$$\frac{\mu_0 I^2}{4\pi}\, \mathbf{f}\left(\frac{D}{h_0}\right) = mg$$

where D is a characteristic dimension of the magnet.

The equation of motion for heave or vertical motion is:

$$m\ddot{h} - \frac{\partial F}{\partial h}(h - h_0) = 0$$

or,

$$\ddot{u}_z - \frac{g}{f(D/h_0)}\frac{\partial f}{\partial h}\, u_z = 0 \tag{6}$$

$$\text{where } u_z = h - h_0$$

For oscillatory motion exp ($j\omega t$), the frequency is given by:

$$\omega^2 = -g\, \frac{\partial \ln f}{\partial h} \tag{7}$$

One can see that the *vertical frequency is independent of the mass of the vehicle* and depends directly on the levitation height and velocity.

As an example, consider the case of a levitated wire of length L, moving with a velocity V_0 normal to itself[28]. The lift and drag forces are given by (Figure 6.1).

$$F_z = \frac{\mu_0 I^2}{4\pi}\, \frac{{V_0}^2}{V_0^2 + w^2}\left(\frac{L}{h}\right) \tag{8}$$

$$F_x = -\frac{w}{V_0} F_z$$

where w is a characteristic velocity inversely proportional to the track thickness Δ, and electrical conductivity σ.

$$w = 2/\mu_0 \sigma \Delta$$

The expressions in Equation (8) are good only for the thin track approximation in which the skin depth is greater than the track thickness Δ.

For this case, the frequency of vertical oscillation is simply:

$$\omega = (g/h)^{\frac{1}{2}}$$

which is the frequency of a pendulum of length h. If the current is held constant, the levitation height varies as a function of speed, as does ω. The vertical frequency as a function of velocity is shown as a function of velocity in Figure 6.2 for constant current. This relation is also valid for a closed planar wire filament such as a circular magnet, so long as $h \ll D$, the magnet diameter.

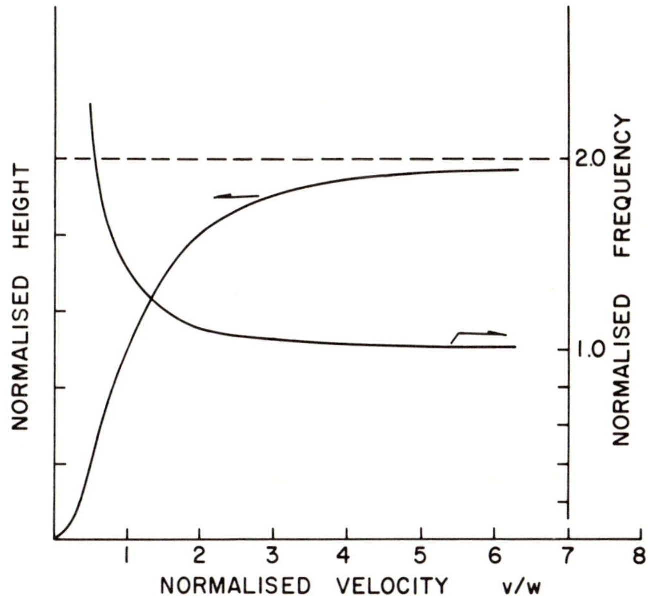

Figure 6.2 Levitation height and heave frequency versus speed for a wire moving over a thin conducting sheet for constant current.

For arbitrary shaped magnets numerical methods must be used to determine the lift/height relation. One example, using numerical integration of the Biot-Savart law for a rectangular superconducting magnet is shown in Figure 6.3 [29]. One can see that for a given value of h, the lift/height law can be written approximately as:

$$f = \left(\frac{D}{h}\right)^{\beta}$$

so that the frequency is given by:

$$\omega^2 = \beta g/h \tag{9}$$

For persistant current operation of the superconducting magnet, the flux is held constant which results in a higher value of β and a higher material frequency than the constant current case (Figure 6.4).

6.2.2 *Energy method*

Let a vehicle have a potential energy function V with generalised coordinates U_k. These coordinates may represent, for example, the heave amplitude of a vehicle or

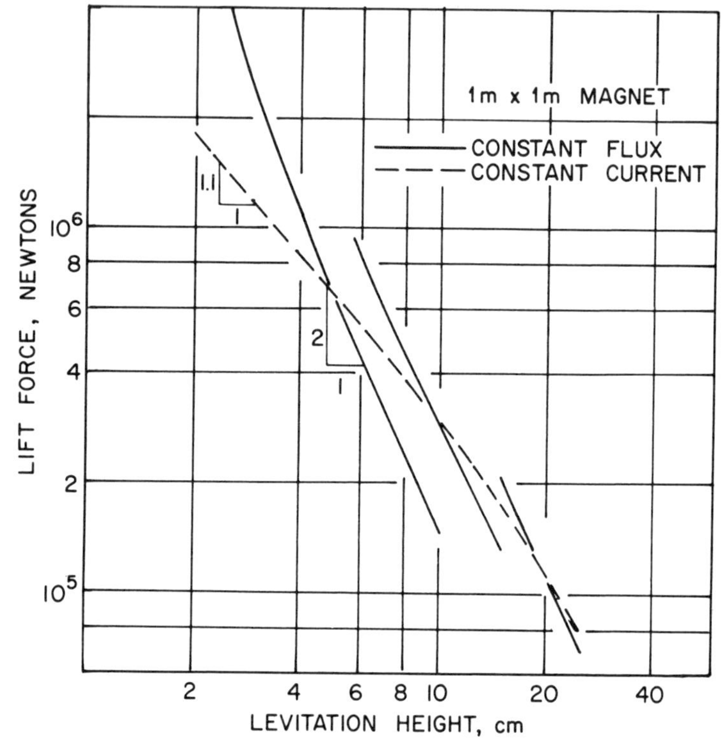

Figure 6.3 High speed lift force versus height for a rectangular coil for constant current and constant flux conditions.

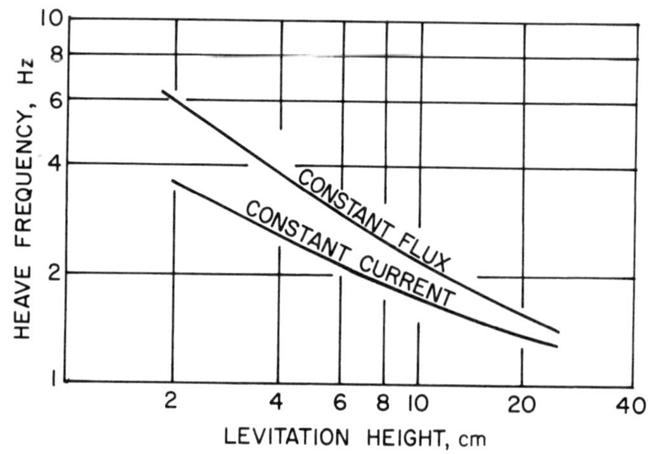

Figure 6.4 Natural heave frequency for rectangular levitation magnets versus height for constant current and constant flux conditions.

the change in pitch angle. Whatever the interpretation, the potential energy is assumed to be a known function of the U_k.

Now imagine that electric generators pump M independent currents I_k through the magnets on the vehicle and that the voltage at each generator is e_k. Thus the total work done in a small time dt is:

$$\Sigma I_k e_k \mathrm{d}t$$

From Maxwell's equations the rate of change of flux ϕ_k through the kth circuit is given by:

$$\frac{\partial \phi_k}{\partial t} = e_k$$

where the circuit is assumed to be non-resistive.

The stored magnetic energy in the circuit is assumed to be a known function of the fluxes and displacements, i.e. $W = W(\phi_n, U_n)$.

Then the variational principle states that the virtual changes in potential and magnetic energies are equal to the virtual work i.e.

$$\Sigma I_k \mathrm{d}\phi_k = \Sigma \frac{\partial W}{\partial \phi_k} \mathrm{d}\phi_k + \frac{\partial W}{\partial U_k} \mathrm{d}U_k + \Sigma \frac{\partial V}{\partial U_k} \mathrm{d}U_k$$

Since the dU_k and $\partial\phi_k$ are independent we have:

$$I_k = \frac{\partial W}{\partial \phi_k} \tag{10}$$

$$\frac{\partial W}{\partial U_k}(\phi_n, U_n) = -\frac{\partial V}{\partial U_k}$$

Often, however, the magnetic energy is known as a function of the currents. We can make a Legendre transformation by defining a coenergy [30] :

$$W^*(I_k) = \Sigma I_k \phi_k - W(\phi_i)$$

which leads to the equations:

$$\phi_k = \frac{\partial W^*}{\partial I_k} \tag{11}$$

$$\frac{\partial W^*}{\partial U_k}(I_n U_n) = \frac{\partial V}{\partial U_k}$$

Thus if $Q_k^e = -\dfrac{\partial V}{\partial U_k}$ is called the generalised mechanical force, then the equilibrium between mechanical and magnetic forces is:

$$\frac{\partial W^*}{\partial U_k} + Q_k^e = 0$$

or the generalised magnetic force is given by:

$$Q_k^m + Q_k^e = 0$$

$$Q_k^m = \left.\frac{\partial W^*}{\partial U_k}\right|_{I\text{ (constant)}} = -\left.\frac{\partial W}{\partial U_k}\right|_{\phi\text{ constant}} \tag{12}$$

For a single linear magnetic current $W = W^* = LI^2/2$ and $\phi = LI$ where L is the inductance and is assumed to be dependent on the deformation U_k. Then the magnetic force is:

$$Q^m = \frac{I^2}{2}\frac{\partial L}{\partial U_k} = -\frac{\phi^2}{2}\frac{\partial(1/L)}{\partial U_k}$$

Energy method-dynamic effects To model the mechanics of vehicles under magnetic forces we incorporate the magnetic energy of the currents, W, into Hamilton's principle (see for example, Crandall *et al.* [30]). For a device with N current paths and N currents I_k:

$$W = \tfrac{1}{2}\Sigma\Sigma\, L_{ij}I_iI_j \tag{13}$$

Suppose then we can represent the displacements of a vehicle by amplitudes $U_n(t)$. Using generalised masses and stiffnesses, the kinetic and potential energies, T and V, can be written in terms of the modal amplitudes $U_n(t)$. Likewise, the magnetic energy can also be written in terms of the variables I_k, U_k. Thus,

$$T = \tfrac{1}{2}\Sigma\, m_n \dot{U}_n^2(t),\ V = \tfrac{1}{2}\Sigma\, k_n U_n^2$$

$$W = W(I_k, U_k)$$

For general rigid body motion, there will be six generalised coordinates. For a flexible vehicle and/or guideway there could be an infinite number of degrees of freedom. If we are satisfied with a finite number of modes as an approximation, a Lagrangian can be constructed:

$$L = T - V + W$$

The independent variables are U_n, q_n where $I_k = \mathrm{d}q_n/\mathrm{d}t$.

The usual variational method leads to the set of equations:

$$\begin{aligned} \ddot{U}_n(t) + \Omega_n^2 U_n - \frac{1}{m_n}\frac{\partial W}{\partial U_n} &= \frac{1}{m_n}F_n \\ \frac{\mathrm{d}}{\mathrm{d}t}\frac{\partial W}{\partial I_k} + R_k I_k &= E_k \end{aligned} \tag{14}$$

where F_n are generalised external forces or non-conservative forces on the structure such as damping, and Ω_n are the natural frequencies. Also R_k are circuit resistances and E_k are circuit voltages which maintain the currents.

For superconducting devices operating in the persistant mode, $E_k = 0, R_k = 0$. Thus the magnetic flux is conserved,

$$\phi_k = \frac{\partial W}{\partial I_k} = \text{constant}$$

For linear circuits the magnetic flux (see Equation (13)) is given by:

$$\phi_k = \Sigma\, L_{kj}I_j \tag{15}$$

For a single magnet and its image, the magnetic energy is:

$$W = \tfrac{1}{2}(L_{11} - L_{12})I^2$$

where the image magnet has currents of opposite sense and L_{12} is the mutual inductance between the two magnets. The total flux through the vehicle magnet is:

$$\phi = I(L_{11} - L_{12})$$

For one degree of freedom – such as pitch or heave,

$$\ddot{U} + \Omega^2 U + \frac{I^2}{2m}\frac{\partial L_{12}}{\partial U} = \frac{F}{m} \tag{16}$$

Note that if the levitated magnet moves an amount U, then the change in the magnet-image system is $2U$. An alternate form of Equation (16), if the flux is given instead of I, is,

$$\ddot{U} + \Omega^2 U + \frac{\phi^2}{2m(L_{11} - L_{12})^2}\frac{\partial L_{12}}{\partial U} = \frac{F}{m} \tag{17}$$

Consider the case of small oscillations about an equilibrium state where U is the generalised perturbation coordinate from equilibrium, then we expand L_{12} in a Taylor series in U, i.e.

$$L_{12} = L_0 + L'U + \tfrac{1}{2}L''U^2$$

If the mechanical force is balanced by the magnetic forces at equilibrium, then,

$$F = I^2 L'/2$$

and the linear equation for vibration becomes:

$$\ddot{U} + (\Omega^2 + \frac{I^2}{2m}L'')U = 0.$$

In the case of constant flux through the levitation magnet:

$$\ddot{U} + \left\{\Omega^2 + \frac{\phi^2}{2m(L_{11} - L_0)^2}\left[L'' + \frac{2(L')^2}{(L_{11} - L_0)}\right]\right\} U = 0$$

For gravity forces acting alone, $\Omega^2 = 0$, and the frequency of oscillation is given by:

$$\omega_0 = \left(\frac{I_0^2 L''}{2m}\right)^{\frac{1}{2}} \text{; constant current} \tag{18}$$

or

$$\omega_1 = \omega_0 \left[1 + \frac{2(L')^2}{L''(L_{11} - L_0)}\right]^{\frac{1}{2}} \text{; constant flux}$$

Thus when $L_{11} > L_0$, the constant flux case results in a higher natural frequency.

As an example, consider a circular magnet of radius R levitating a total mass M at a height h above a conducting sheet in the high speed limit. The mutual inductance of the magnet with its image for $h \ll R$ is [10]:

$$L_{12} = \mu_0 R \left(\ln\frac{4R}{h} - 2\right)$$

For a circular cross-section of the magnet arc of radius r the self inductance is

$$L_{11} = \mu_0 R \left(\ln \frac{8R}{r} - \frac{7}{4}\right)$$

Since $h > r$, one can show that $L_{11} > L_{12}$ i.e., the self flux linkages of the magnet are greater than those from its image,

$$L_{11} - L_{12} = \mu_0 R \left(\ln \frac{2h}{r} + \frac{1}{4}\right) > 0$$

Then substituting $L' = -\mu_0 R/h, L'' = \mu_0 R/h^2$, in Equation (18), the constant current and constant flux frequencies are given by

$$\omega_0 = (g/h)^{\frac{1}{2}}$$
$$\omega_1 = \omega_0 \left[1 + \frac{2}{\left(\ln \frac{2h}{r} + \frac{1}{4}\right)}\right]^{\frac{1}{2}} \tag{19}$$

The increased frequency of a constant flux magnet over a constant current one at the same levitation height is illustrated in Figure 6.4 for a rectangular coil.

6.3 Negative magnetic damping in levitated magnets

The quasi-steady dependence of the magnetic lift and drag forces on train velocity and magnetic coil geometry have been studied in a number of papers (e.g., Guderjahn *et al.* [31], Reitz and Davis [32]). In those papers, the motion of the current filament or coil was steady. In this section the effect of vertical oscillations of the wire is investigated.

It is shown that the vertical oscillations of the moving current filament can produce a positive or negative damping force depending on the speed. Negative damping forces for a track of discrete loops has been predicted by Iwamoto *et al.* [15] who also claim experimental evidence for this effect. The first study of this problem, however, was given by Davis and Wilke [14], using an image-wake method for an oscillatory wire moving over a thin conducting plate. In this section a Fourier transform method is used to obtain the same low frequency results as in [14].

6.3.1 *Theoretical analysis*

Levitating magnets on actual vehicles are planar superconducting coils, usually rectangular in shape. In this section only the current filaments transverse to the motion are examined (AB, CD in Figure 6.5). This will permit a one-dimensional approximation of the induced currents in the track and a closed form solution for the damping force and plate stresses.

Consider a long current filament moving normal to itself over a flat conducting plate (Figure 6.5). We assume a steady velocity in the direction parallel to the plate while the motion of the filament normal to the plate is assumed to be a prescribed sinusoidal motion, small compared with the mean height above the plate, h_0. The moving magnetic field will induce currents in the conducting plate. It is assumed that the horizontal velocity of motion is small enough so that the distribution of

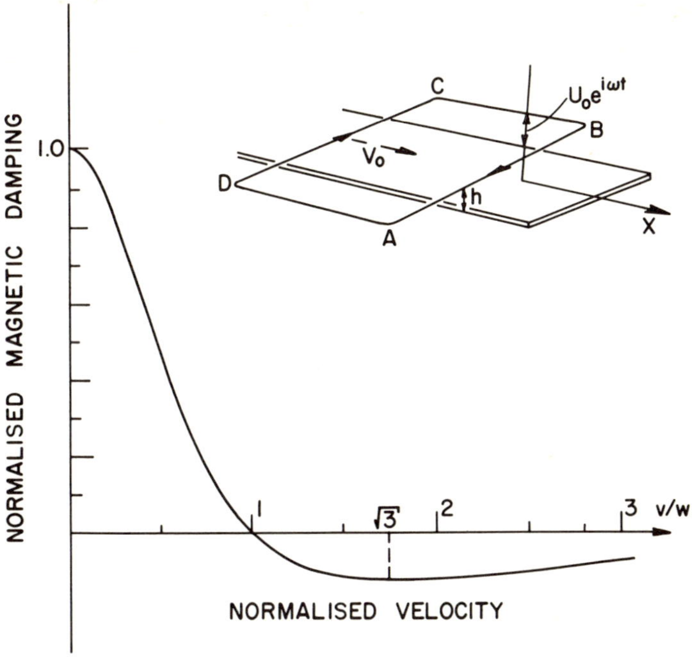

Figure 6.5 Magnetic damping of a levitated wire versus speed.

current across the plate thickness is uniform. Return current paths which must exist in a finite plate are also neglected. Thus the plate is considered to be very wide with respect to the height h_0 and its thickness, Δ. The coil is also assumed to be operated in the constant current mode.

Under these assumptions the induced current density vector in the plate **J** will be parallel to the moving filament. Since the current density is uniform across the thickness $K \equiv \Delta J$ A m^{-1} is defined.

For convenience, the solution is obtained relative to a coordinate system that moves with the horizontal velocity of the filament v. Ohm's law for a moving conductor in an electric field **E** and magnetic field **B** becomes:

$$K = \sigma\Delta(E_z - B_y v) \tag{20}$$

where σ is the plate conductivity. Maxwell's equation for the induced electric field E_z is:

$$\frac{\partial E_z}{\partial x} = \frac{\partial B_y}{\partial t}$$

or using Equation (20):

$$\frac{1}{\sigma\Delta}\frac{\partial K}{\partial x} = \frac{\partial B_y}{\partial t} - v\frac{\partial B_y}{\partial x} \tag{21}$$

The magnetic field **B** is the sum of two parts; one part generated by the wire itself B_0; and another $\mathbf{B}^1$ produced by the induced current in the plate. For low frequencies we assume that $\mathbf{B}^0$ is the quasi-steady field produced by the filament in its instantaneous position i.e.

$$B_y^0 = \frac{\mu_0 I_0 x}{2\pi[h^2(t) + x^2]}; \quad B_x^0 = \frac{\mu_0 I h(t)}{2\pi[h^2(t) + x^2]} \tag{22}$$

where $h(t)$ is the instantaneous height above the plate. For small vertical (normal) motions i.e.

$$h(t) = h_0 + u_0 \exp(j\omega t), u_0 \ll h_0 \tag{23}$$

we can write $\mathbf{B}^0(t)$ as:

$$\mathbf{B}^0(t) = \mathbf{B}_0(x) + \mathbf{b}_0(x) \exp(j\omega t) \tag{24}$$

and where $\mathbf{b}_0$ is linearly proportional to u_0.

For a single filament,

$$B_{oy} = \frac{\mu_0 I_0 x}{2\pi(h_0^2 + x^2)}, b_{oy}(x) = -\frac{\mu_0 I_0 x u_0 h_0}{\pi(h_0^2 + x^2)^2} \tag{25}$$

Similarly one can emulate Equation 24 and write:

$$K(x, t) = K_0(x) + K_1(x) \exp(j\omega t) \tag{26}$$

where K_1 is linear in u_0.

Also, the induced magnetic field can be written as:

$$B_y^1(x, t) = B_{1y}(x) + b_{1y} \exp(j\omega t) \tag{27}$$

where the total field is given by $\mathbf{B} = \mathbf{B}^0 + \mathbf{B}^1$.

At a position x' in the plate, the differential current filament $K_0(x')\mathrm{d}x'$ contributes to the magnetic field at $x(y = 0)$ according to $\mathrm{d}B_{1y} = \mu_0 K_0 \mathrm{d}x' / [2\pi(x - x')]$. Thus for B_{1y}, and by analogy b_{1y}, the expressions are:

$$B_{1y} = \frac{\mu_0}{2\pi} \int_{-\infty}^{\infty} \frac{K_0(x')\mathrm{d}x'}{(x - x')}, b_{1y} = \frac{\mu_0}{2\pi} \int_{-\infty}^{\infty} \frac{K_1(x')\mathrm{d}x'}{(x - x')} \tag{28}$$

Equations (21) to (28) are sufficient to determine the mean and oscillatory current distributions in the plate.

Once the current in the track is determined, the force on the wire can be found from the integrals.

$$\begin{aligned} F_x &= \int_{-\infty}^{\infty} K B_y^0 \mathrm{d}x = \frac{1}{2\pi} \int_{-\infty}^{\infty} T^*[K] T[B_y^0]\, \mathrm{d}k \\ F_y &= -\int_{-\infty}^{\infty} K B_x^0 \mathrm{d}x = -\frac{1}{2\pi} \int_{-\infty}^{\infty} T^*[K] T[B_x^0]\, \mathrm{d}k \end{aligned} \tag{29}$$

where the real parts of K and $\mathbf{B}$ are used and $T[K]$, $T[\mathbf{B}]$ are Fourier transforms on

x of K and **B** respectively, and T^* denotes the complex conjugate of T, where:

$$T[K] \equiv \int_{-\infty}^{\infty} K(x) \exp(jkx) dx$$

The steady part of **F** is obtained from integrals of $K_0(x)$, $\mathbf{B}_0(x)$ and the dynamic part of the force denoted by $\mathbf{F}^1$, is given to 1st order in u_0 by:

$$F_x^1 = \int_{-\infty}^{\infty} [K_0 b_{oy} + K_1 B_{oy}] dx \exp(j\omega t)$$

$$F_y^1 = -\int_{-\infty}^{\infty} [K_0 b_{ox} + K_1 B_{ox}] dx \exp(j\omega t) \tag{30}$$

Using the Equations (21), (27) and (28), the induced oscillatory current is given by:

$$\frac{1}{\sigma\Delta}\frac{\partial K_1}{\partial x} = \frac{\mu_0}{2\pi}\int_{-\infty}^{\infty} \frac{\left[j\omega K_1 - v\dfrac{\partial K_1}{\partial x'}\right]}{(x - x')} dx' + j\omega b_0 - \frac{v \partial b_0}{\partial x} \tag{31}$$

In seeking a particular solution to this equation using the Fourier transform of $K_1(x)$, i.e. $T[K]$, the following identities are useful:

$$T[K'_1] = -jkT[K_1]$$

$$T\left[\int_{-\infty}^{\infty} \frac{K_1}{x - x'} dx^1\right] = T[K_1]T\left[\frac{1}{x}\right]$$

$$T[1/x] = j\pi, k > 0$$
$$= -j\pi, k < 0$$

$$T[1/x] \; T^*[1/x] = +\pi^2$$

Using these identities, the transform of K_1 is given by $T[K_1] = -(\omega/k + v)\,\sigma\Delta T[b_{oy}]/\{1 + (v + \omega/k)T[1/x]/\pi w\}$, where $w = 2/\mu_0\sigma\Delta$ is a characteristic velocity. The ratio v/w is sometime called a magnetic Reynold's number. This expression can be simplified by multiplying numerator and denominator by $\{1 - (v + \omega/k)T[1/x]/\pi w\}$. The resulting expression for the transform of $K_1(x)$ is given by:

$$T[K_1] = -\frac{2}{\mu_0 w}\frac{(v + \omega/k)T[b_{oy}]\;\{1 - (v + \omega/k)T[1/x]/\pi w\}}{[1 + (v + \omega/k)^2/w^2]} \tag{32}$$

If one considers 'ω' as a Fourier transform parameter on the time domain, then the oscillatory current as a function of time is given by a Fourier inversion integral on the variable ω. This will result in a complicated convolution integral, in time, of the perturbed magnetic field $\mathbf{b}_{oy}$. To obtain an approximation to this time function $T[K_1]$ is expanded in powers of ω.

$$T[K_1] \simeq T[K_1]\Big|_{\omega=0} + \omega \frac{\mathrm{d}T[K_1]}{\mathrm{d}\omega}\Big|_{\omega=0}$$

To find the vertical damping force we use Equation (29) and observe that only the integral of $K_1(x)B_{0x}(x)$ has a linear term in ωu_0. The coefficient of $\mathrm{j}\omega u_0$ will give the classical damping constant as a function of v. Since $B_{0x}(x)$ is symmetric, only the symmetric part of $K_1(x)$ or $T[K_1](k)$ will yield a non-zero force. Thus the damping force is proportional to:

$$\text{(even part)}\ \frac{\mathrm{d}T[K_1]}{\mathrm{d}\omega}\Big|_{\omega=0} = \sigma\Delta \frac{[1-(v/w)^2]\ T[b_{0y}]/k}{[1+(v/w)^2]^2} \tag{33}$$

One can immediately see that the damping coefficient will change sign at $v = w$. It is shown below that this leads to a negative eddy current damping force for $v > w$.

With no oscillation of the wire, a quasi-steady current distribution $K_0(x)$ is induced in the track by the moving wire. It is easy to show that $K_0(x)$ must satisfy Equation (30) for the case $\omega = 0$, and b_{0y} replaced by B_{0y}. This current distribution is given by

$$K_0(x) = -\frac{2}{\mu_0}\frac{v}{w}\frac{B_{0y} + B_{0x}\,v/w}{1 + v^2/w^2}$$

where B_{0y} and B_{0x} are evaluated at $y = 0$.

6.3.1 *Dynamic levitation forces*

To calculate the force on the wire or the force distribution on the track, one uses $\mathbf{K} \times \mathbf{B}^0$, where $\mathbf{K}$ is directed normal to the x, y plane. Only the field of the wire $\mathbf{B}^0(x, t)$ is used to calculate the forces since $\mathbf{K} \times \mathbf{B}^1$ only contributes to the self stresses in the track which are equilibrating and produce no net force on the track.

As discussed in a previous section, the total force on the wire has a steady term, an oscillatory part proportional to u_0 exp $(\mathrm{j}\omega t)$ and a harmonic proportional to u_0^2 exp $(\mathrm{j}2\omega t)$ which is neglected in this paper.

The linear oscillatory force on the wire F_y^1 is given by the integral of Equation (29) where the oscillatory track currents are given by the Fourier inversion of Equation (31) and the perturbed field b_{0y} by Equation (25). The term $K_0\, b_{0y}$ contributes a force dependent on the displacement u_0 exp $(\mathrm{j}\omega t)$ with no other dependence on the frequency ω. The term $K_1 B_{0x}$ contributes a term proportional to both u_0 exp $(\mathrm{j}\omega t)$ and a damping term, $\mathrm{j}\omega\, u_0$ exp $(\mathrm{j}\omega t)$, when K_1 is expanded in Taylor series in ω. As discussed above, the damping term is proportional to $(w^2 - v^2)$ and hence changes sign for $v > w \equiv 2/\mu_0 \sigma\Delta$, (see [14]).

Without belabouring the details of integration, which are straightforward, the total oscillatory force on the wire is given to order ω, by:

$$F_y^1 = -\frac{\mu_0 I_0^2}{4\pi h_0}\left(\frac{v^2}{(v^2+w^2)}\frac{u}{h_0} + \frac{(w^2-v^2)}{(v^2+w^2)^2}\frac{w\mathrm{d}u}{\mathrm{d}t}\right) \tag{34}$$

where $u = u_0$ exp $(\mathrm{j}\omega t)$.

A negative damping force has been calculated for a loop of current moving over discrete loop track by Iwamoto *et al.* [15]. They have also performed experiments which seem to confirm their prediction of negative damping for velocities beyond a characteristic speed. The damping force as a function of speed is shown in Figure 6.5.

6.3.2 *Experiments on magnetic damping*

The results of analysis of a vibrating wire above a moving plate show that the damping coefficient decreases with speed and in fact becomes negative for speeds greater than a characteristic velocity. Experiments on this effect were presented by Iwamoto *et al.* [15], in which they use a permanent magnet on a pendulum. The magnet then vibrates normal to a moving conducting plate attached to a rotating wheel. Their results show a decrease in total damping with speed. However, it was not clear whether this decreased damping was due to the negative damping effect or simply to an increase in magnet-plate gap as lift developed on the pendulum.

Thus to check the results of Iwamoto *et al.* [15], and because the idea of negative magnetic damping was intriguing, the author constructed a similar experiment. The magnet was mounted to a stiff cantilevered beam (Figure 6.6) in order to keep the gap from increasing too much as magnetic lift developed.

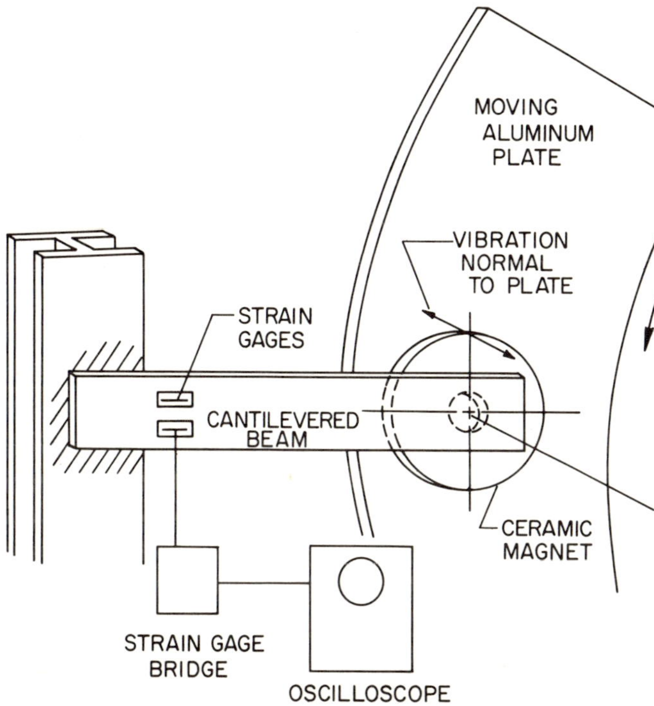

Figure 6.6 Experimental set-up to measure magnetic damping versus speed.

The magnet was a flat, circular, ceramic permanent magnet 5.3 cm in diameter. The polarisation was oriented normal to the magnet face and had reversed polarisation on each half of the magnet. Along a line parallel to the magnet, the normal magnetic field reversed sign as is the case for a long wire (Figure 6.7). The moving plate then saw a normal magnet geometrically similar to the moving wire for which the theoretical results were derived. The peak magnetic field was about 300 G.

The natural frequency of the beam was 22 Hz and did not change much with

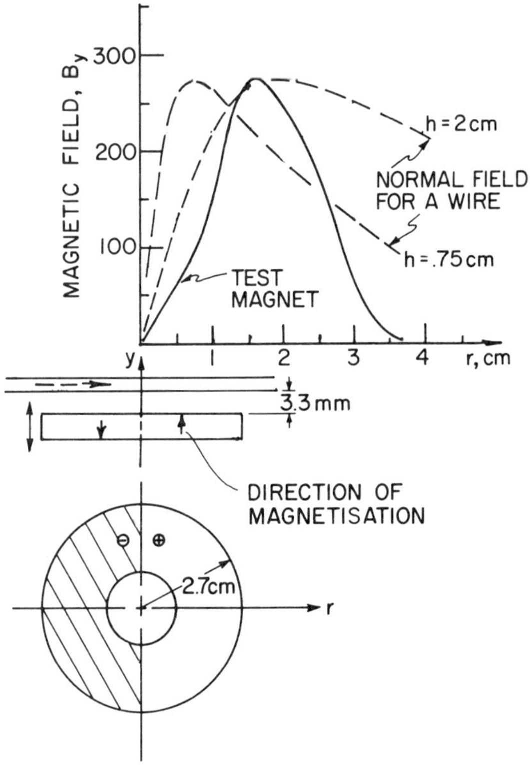

Figure 6.7 Normal magnetic field along the test magnet used in magnetic damping tests.

speed. The magnet-plate gap was 3.3 mm and increased with speed to 4.3 mm. The change in total damping versus gap for zero speed is shown in Figure 6.8. The decrease in damping due to increased gap cannot account for the larger change measured for increased speed. The plate was a 3.2 mm thick aluminum annulus clamped to a rotating wheel machine. The magnet was placed at a radius of 66.7 cm and the plate rotated up to 600 r.p.m. The calculated characteristic velocity was found to correspond to around 210 r.p.m.

The cantilevered beam was about 15 cm in length and had resistance strain gauges to measure both the static and vibratory motions. The cantilever was plucked at various speeds and the damped motion recorded on film from an oscilloscope.

The results of these tests are shown in Figure 6.9. One can see that the magnetic damping at zero field more than doubles the total damping. The damping constant shown plotted is the inverse time constant β in exp $(-\beta t)$. As one observes, the total damping decreases with speed and approaches a constant at 600 r.p.m.

It was suspected that, in addition to the initial structural damping, significant aerodynamic damping was occurring between the magnet-plate gap. Therefore a plastic dummy 'magnet', geometrically identical to the test magnet, was run at various speeds and attached to the same cantilever beam. Lead weights were added to match the natural frequency of the dummy with the test magnet. The resulting

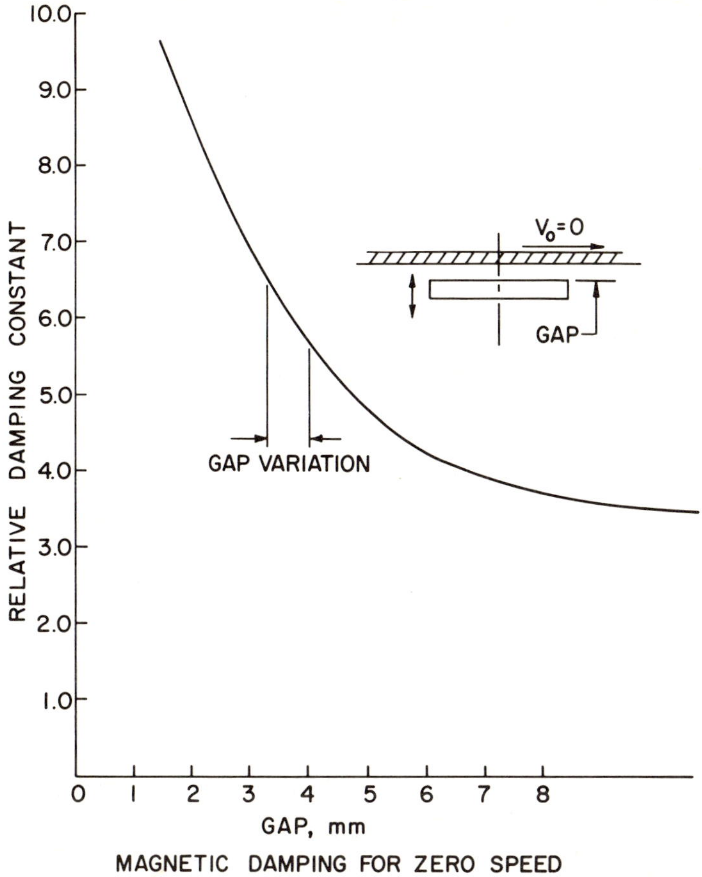

Figure 6.8 Magnetic damping versus gap for zero speed.

aerodynamic and structural damping is shown as curve 2 in Figure 6.9 and shows a linear increase with speed.

The final results are shown in Figure 6.9 as curve 3 and are the result of subtracting curve 2 from the total damping, (curve 1). The magnetic damping shows a zero crossing around 325 r.p.m. and continues to increasing negative values with speed. The zero crossing r.p.m. was higher than the calculated value of 210 r.p.m. Further tests could be done to determine the difference between theory and experiment, but the results clearly show the predicted decrease and negative damping with speed when aerodynamic and structural damping are subtracted. Instead, effort is being directed toward tests of magnetic damping using superconducting magnet coils.

While this effect is not expected to lead to instabilities in actual magnetically levitated vehicles, due to the presence of aerodynamic damping, one cannot rely on the lift magnets alone for passive damping of vehicle motions as this effect is likely to decrease with speed.

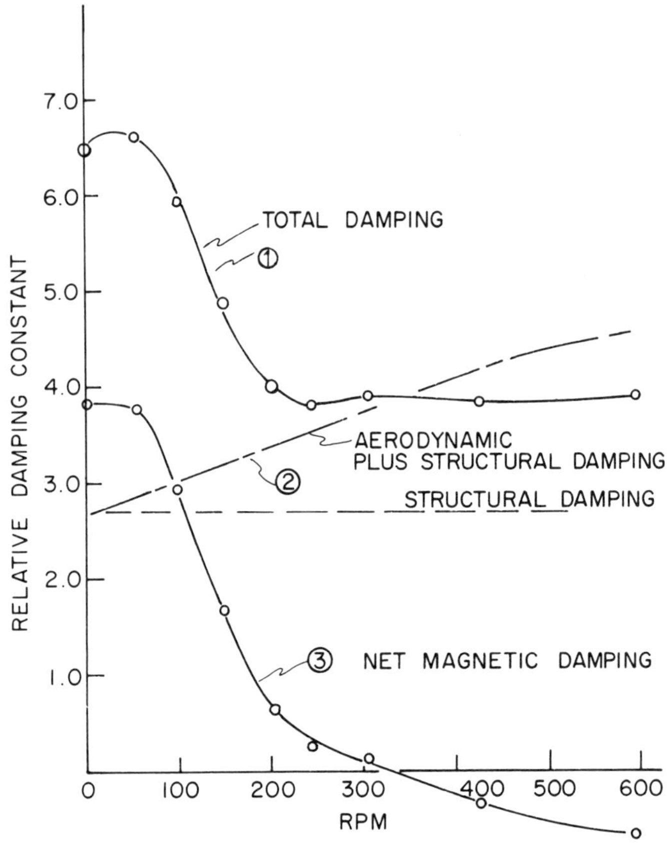

Figure 6.9 Experimental data on magnetic damping versus speed; 1; total damping curve; 2. aerodynamic and structural damping curve; 3. magnetic damping curve.

Finally one might suspect that similar negative magnetic damping might be present in the motion between a linear induction motor primary and reaction rail. This problem should receive further study.

6.4 Dynamic instabilities in levitated bodies

Since ride quality is one of the requirements a levitated vehicle system must satisfy, it is of primary importance to identify the sources of dynamic motion. The dynamic forces on a vehicle are numerous and include:

1. Magnetic lift and guidance forces
2. Magnetic drag and 'damping' forces
3. Aerodynamic pressures due to both mean flow and turbulence
4. Aerodynamic viscous forces (e.g. skin friction)
5. Power plant thrust forces
6. Power pickup (e.g. from catenary-pantograph dynamics) – sliding contact forces

Dynamic motions of vehicles can be classified as rigid-body or elastic motions. In this section we will treat only rigid-body dynamics. Thus the total force and first moment of the forces on the vehicle are important. For long flexible vehicles the elastic vibrations may become important. An analogy can be made here with aircraft where the rigid-body motion can be stable but the wings can experience violent aeroelastic vibrations or *flutter*.

Another useful classification of dynamic motions is the separation of forced motion from self-excited motions. Forced motions result from forces which are independent of the vehicle position or motion such as guideway irregularities, wind gusts, or propulsion thrust transients. Self-excited motions result from forces which are proportional to the motion of the vehicle and act in such a way as to increase the motion. One example is the negative magnetic damping forces treated in Section 6.3. Most designers anticipate the forced motions and a number of studies have investigated methods for damping out these transients. Self-excited motions on the other hand cannot often be tolerated since they lead to violent oscillations. Thus a vehicle has to be designed from the outset to avoid the conditions which might produce self-excited motions.

In this section only self excited motions due to magnetic forces alone are discussed. Of course, aerodynamic forces might, in fact, be a greater source of such instabilities in actual full scale vehicles. However, the motive here is to anticipate problems by presenting a few models of potential, self-excited instabilities, namely the roll-yaw instability (or Dutch roll in aeronautics) and the pitch-heave vibrations or 'porpoising' motion.

These analytical models have been motivated by experiments in the laboratory on magnetically levitated models[21] and from films of the MIT model shown by Thornton at the IEE meeting in London on Linear Electric Machines[20]. The author has relied on analogies with aircraft instabilities and the reader is encouraged to review the texts in this field (for example, Seckel[26], Dickinson[27]) for further details.

6.4.1 *Lateral yaw-roll dynamics*

Consider the vehicle shown in Figure 6.10. For simplicity the 'V' guideway is considered, but the analysis can be extended to cover other guideway shapes such as the split channel. Were one to consider only the magnetic lift and guidance forces, the dynamics would be a planar coupling of roll and lateral motions. However, lateral motion to starboard, for example, can produce greater magnetic drag on the starboard magnets, hence creating a yawing moment about the vertical axis. Thus, in general all three motions are coupled. It is worthwhile at this point however to discuss conservative and non-conservative forces.

Conservative magnetic force Conservative forces are those which can be derived from a potential. Thus magnetic lift and guidance forces can be derived from the magnetic energy function W (Section 6.2) and in the case of high magnetic Reynolds number, W can be calculated from the inductance between the magnet and its image below the conducting sheet. If h is the height above the sheet, then $W = W(h)$ and the magnetic force is $\mathrm{d}W/\mathrm{d}h$.

On the other hand, the magnetic energy is not a function of the position along the track and hence the drag cannot be related to it via the gradient of W along the track.

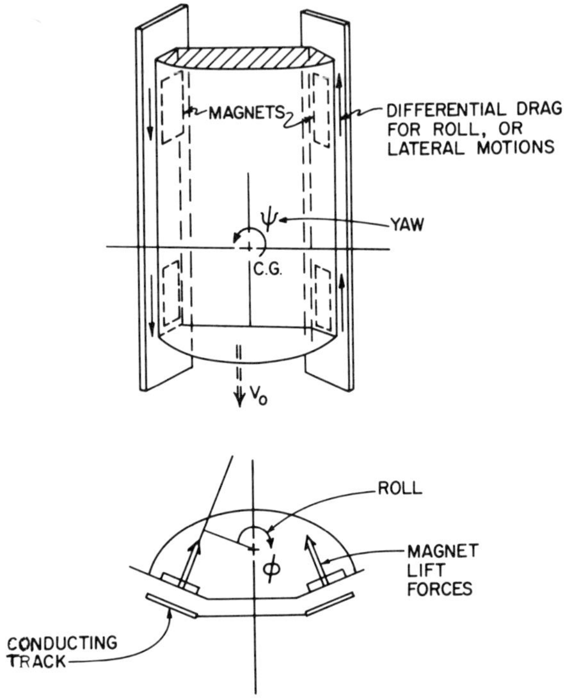

Figure 6.10 Sketch of a levitated vehicle in a 'V' guideway showing lateral, yaw and roll motions.

The next point to note is that conservative magnetic forces produce either neutrally stable motions (i.e. undamped vibrations) or divergent motions (i.e. those proportional to exp $(-\alpha t)$). Conservative forces produce reciprocal forces in the coupled equations of motions which action does not occur for some drag-related forces. To show this, the magnets and guideway in the levitated vehicle system in Figure 6.10 are considered to have a magnetic energy function $W(u_y, \phi, \psi)$ in which u_y is the lateral displacement, and ϕ the roll about the longitudinal axis parallel to the guideway, and ψ the yaw angle about the vertical axis. For linear vibrations, W is quadratic in the generalised coordinates $U_1 = u_y$, $U_2 = \phi$, $U_3 = \psi$,

$$W = \tfrac{1}{2}\Sigma K_{ij} U_i U_j + W_0$$

where $K_{ij} = K_{ji}$, and Lagrange's equations lead to the following eigenvalue problem for harmonic motion exp $(j\omega t)$

$$\begin{bmatrix} (K_{11} - m\omega^2) & K_{12} & K_{13} \\ K_{12} & (K_{22} - I_R\omega^2) & K_{23} \\ K_{13} & K_{23} & (K_{33} - I_Y\omega^2) \end{bmatrix} \begin{bmatrix} \bar{u}_y \\ \bar{\phi} \\ \bar{\psi} \end{bmatrix} = 0 \qquad (35)$$

where the bar indicates the amplitude of the motion. (It should be noted that when two rotations of a rigid body about different axes are present, one should use Euler's equations for a rigid body. The added terms are proportional to the nonlinear term $(I_R - I_Y)\dot{\phi}\dot{\psi}$, (see for example, [18]), and can produce pitching. This is a gyroscopic effect and might be important in a nonlinear analysis of dynamic motions.) If the matrix is symmetric, only real values of ω^2 can satisfy this equation. If $\omega^2 > 0$, then undamped oscillatory motion results. If $\omega^2 < 0$ the motion diverges with no oscillations. Further we can write these equations in the form:

$$\mathbf{MU}\omega^2 = \mathbf{KU} \tag{36}$$

where **M** is a diagonal mass matrix and **K** a 'stiffness' matrix. If both **M** and **K** are real symmetric and positive definite, that is the perturbed magnetic and kinetic energies are positive, the eigenvalues ω^2 must be real and positive and the motion is oscillatory.

Of course we have neglected coupling with the heave or vertical motion. If it is important, then gravity will enter the Lagrangian and the total stiffness matrix **K**, due to both magnetic and gravitational forces, can become negative definite. Such would be the case if we tried to levitate an inverted pendulum.

(2) *Non-conservative forces – yaw-roll* So that the reader may follow the arguments without recourse to numerical calculations, the case where the lateral motion is small and the roll and yaw motions are coupled is considered. This can occur if the centre of gravity is not symmetric between forward and aft magnets. For example, roll might produce a greater change in the lift force on the forward magnets than on those aft (see Figure 6.10) thereby creating a yaw moment. By the same mechanism yaw can create a roll moment due to the difference in lift between fore and aft magnets. Such effects can be represented by conservative forces and lead to the following eigenvalue problem:

$$\begin{bmatrix} (K_{22} - \omega^2 I_R) & K_{23} \\ K_{23} & (K_{33} - \omega^2 I_Y) \end{bmatrix} \begin{bmatrix} \bar{\phi} \\ \bar{\psi} \end{bmatrix} = 0$$

To include the magnetic drag, it is to be noted that when the vehicle rolls, increased drag on one side and decreased drag on the other side produce a yaw moment, (Figure 6.10). However, the reciprocal effect, i.e. yaw-induced roll moment due to magnetic drag, cannot occur since the longitudinal drag cannot create a moment about the roll axis – hence the effect is non-conservative. Inclusion of this effect into the dynamic equations results in a modified eigenvalue problem:

$$\begin{bmatrix} (K_{22} - \omega^2 I_R) & K_{23} \\ (K_{23} + c) & (K_{33} - \omega^2 I_Y) \end{bmatrix} \begin{bmatrix} \bar{\phi} \\ \bar{\psi} \end{bmatrix} = 0 \tag{37}$$

Defining the terms:

$$\omega_R^2 = K_{22}/I_R$$

$$\omega_Y^2 = K_{33}/I_Y$$

The frequencies are given by:

$$\omega^2 = \frac{\omega_R^2 + \omega_Y^2}{2} \pm \tfrac{1}{2}\left[(\omega_Y^2 - \omega_R^2)^2 + \frac{4K_{23}(K_{23} + c)}{I_R I_Y}\right]^{\frac{1}{2}} \tag{38}$$

The frequency becomes complex when the bracketed term vanishes. Thus the condition for self-excited oscillations or flutter is:

$$-4K_{23}c \geq (\omega_Y^2 - \omega_R^2)^2 I_R I_Y + 4K_{23}^2 \tag{39}$$

Thus the non-conservative or drag-related yaw moment produced by roll must be opposite in sign to the yaw moment produced by the guidance forces in roll. This might be the case if the forward magnets carried a greater lift than the aft magnets due to shift in centre of gravity forward or a pitch-forward condition.

6.4.2 *Lateral – roll dynamics*

Another example of coupled oscillations is the lateral roll motion of a levitated model, in particular the vehicle in a 'V' or inverted 'V' guideway. A quasi-static analysis by Wilke *et al.* [33, 35] on such a configuration led to the conclusion that the vehicle would be unstable. However, a rolling moment on the magnet was neglected and, as will be shown below, inclusion of a roll moment can lead to stable lateral roll dynamics.

A sketch of the model is shown in Figure 6.11. Only the magnetic force which is

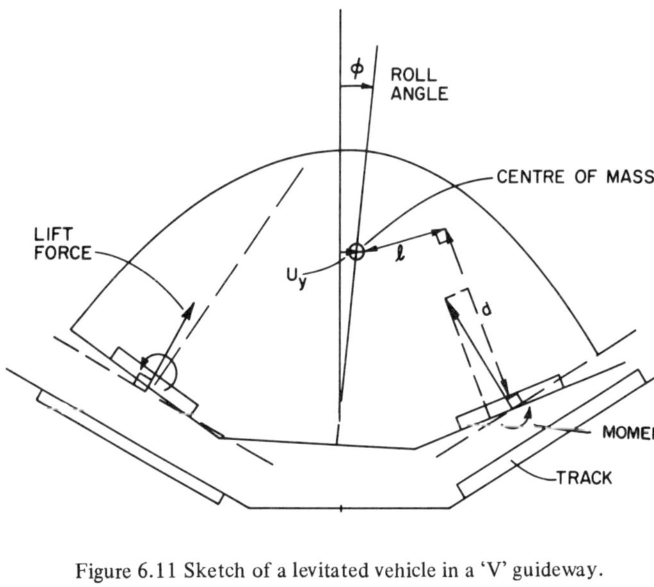

Figure 6.11 Sketch of a levitated vehicle in a 'V' guideway.

normal to the conducting sheet and a roll moment are included, both of which can be considered conservative forces. Forces in the plane of the track due to the edge effect are neglected, though in an actual model these forces might be important. Thus the force on the magnet is assumed to remain normal to the track. Coupling with heave is also neglected.

If F_0 is the total magnet force on one side of the vehicle, all lateral forces and moments will be a function of $F_0(h)$ and $\partial F_0/\partial h$, where h is the normal distance of the magnet from the track surface, α is the track angle with the horizontal.

The perturbed force and moment due to a lateral motion u_y are:

$$F'_Y = 2\frac{\partial F_0}{\partial h}\sin^2\alpha\, u_y$$
$$M'_X = -2\frac{\partial F_0}{\partial h}\sin\alpha\, lU_y \tag{40}$$

where l is the moment arm of F_0 about the centre of mass (Figure 6.11). The perturbed lateral force due to roll about the centre of mass is

$$F''_y = -2\frac{\partial F_0}{\partial h}\sin\alpha\, l\phi \tag{41}$$

which is the reciprocal effect to M'_X.

The perturbed moment due to roll is a little complicated and contains three terms shown below:

$$M''_X = \left(2\frac{\partial F_0}{\partial h}l^2 - \kappa + 2F_0 d\right)\phi \tag{42}$$

The first is the restoring moment due to the perturbed lift force in each of the magnets; the second term is the roll couple on the magnets themselves due to their finite size. The final term is due to the fact that the magnet force remains normal to the track producing an unbalanced force $F_0\phi$ parallel to the plane of the magnets and acting at a moment d (see Figure 6.11).

The characteristic equation for vibration exp $(j\omega t)$ becomes,

$$\omega^4 - \omega^2\left[\omega_0^2\left(\sin^2\alpha + \frac{l^2}{r^2}\right) + \frac{\kappa - 2F_0 d}{mr^2}\right] + \omega_0^2\sin^2\alpha\frac{(\kappa - 2F_0 d)}{mr^2} = 0 \tag{43}$$

where $\omega_0^2 = -2(\partial F_0/\partial h)/m$ and r is the radius of gyration of the vehicle in roll.

One can see that if the magnet roll couple constant κ has the value $\kappa = 2F_0 d$, one root is $\omega = 0$ and the vehicle is statically unstable. For values of $\kappa - 2F_0 d > 0$, the two normalised roots are shown in Figure 6.12.

Note that in this model no self-excited motions or flutter can occur, while divergence ($\omega^2 < 0$) is possible.

6.4.3 *Porpoising or pitch-heave dynamics*

The same concepts about conservative and non-conservative forces used in the previous sections can be applied to vertical-longitudinal motion. Consider the levitated model in Figure 6.13 in which the unsymmetrical position of the centre of gravity causes unequal lift (and hence drag) forces under forward and rear magnets. For equilibrium we must have:

$$L_1 + L_2 = mg$$
$$l_1 L_1 - l_2 L_2 = 0$$

where for simplicity the thrust force is colinear with the magnetic drag.

While one can certainly design the vehicle for zero pitch-heave coupling, in general the differential lift forces on fore and aft magnets due to heave might cause a pitching moment. The reciprocal effect is that pitch causes an unbalanced lift which produces

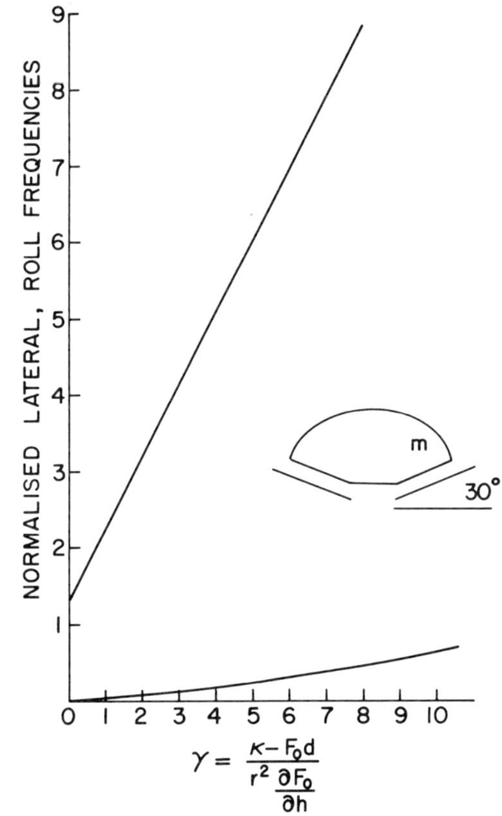

Figure 6.12 Natural frequencies of a levitated vehicle under lateral and roll motions versus magnet moment constant.

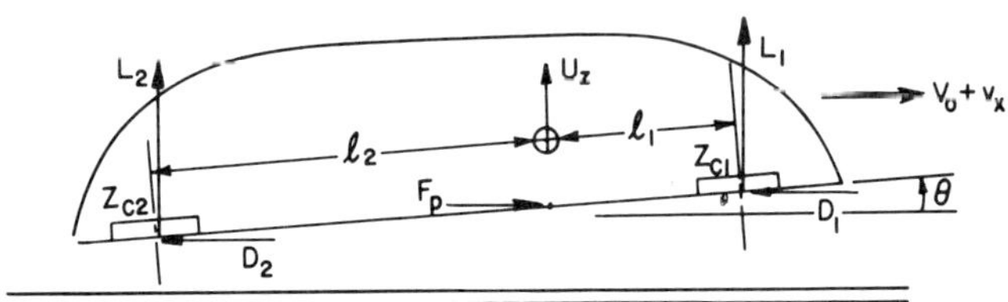

Figure 6.13 Sketch of a levitated vehicle in a guideway showing pitch and heave motions.

heave. Such effects can be represented by conservative forces resulting in the following eigenvalue problem:

$$\begin{bmatrix} (K_{11} - \omega^2 m) & K_{12} \\ K_{12} & (K_{22} - \omega^2 I) \end{bmatrix} \begin{bmatrix} \bar{U}_Z \\ \bar{\theta} \end{bmatrix} = 0$$

This analysis, however, neglects two effects; one is the contribution of the differential drag forces, due to heave and pitch, to the pitching moment, and the other is the change in longitudinal speed due to the unbalanced thrust when the drag forces change due to heave or pitch. The dynamics of the longitudinal motion are described by the equation

$$m\frac{dv_x}{dt} = -\frac{\partial D}{\partial V_0}v_x - \frac{\partial D}{\partial U_z}U_z - \frac{\partial D}{\partial \theta}\theta \tag{44}$$

where $V_0 + v_x$ is the total longitudinal velocity along the guideway and D is the sum of the drag forces D_1 and D_2 on the fore and aft magnets respectively.

The effect of the drag on the pitching moment has a number of terms. If l_1, l_2 are the distances from the vertical plane $z - x$, through the mass centre, to the fore and aft magnets respectively, then,

$$M_y' = \left(D_1 l_1 - D_2 l_2 + z_{c2} l_2 \frac{\partial D_2}{\partial z} - z_{c1}\frac{l_1 \partial D_1}{\partial z}\right) l_2 \theta \tag{45}$$

where z_{c1}, z_{c2} are the vertical distance from the x–y plane through the centre of mass to the fore and aft magnets respectively.

Additional moments arise from the heave and longitudinal motions

$$M_y'' = z_{c1}\left(\frac{\partial D_1}{\partial z}U_z + \frac{\partial D_1}{\partial v}v\right) + z_{c2}\left(\frac{\partial D_2}{\partial z}U_z + \frac{\partial D_2}{\partial v}v\right) \tag{46}$$

A complete analysis of heave-pitch and longitudinal dynamics would require a numerical procedure to determine the eigenvalues and the root locus as a function of speed and/or inertia properties. A more fruitful approach in this review might be to limit the discussion to a simplified model and examine how the root loci are affected by non-conservative forces.

Longitudinal motion is therefore neglected and consideration given to heave-pitch-coupled motion represented by the following eigenvalue problem:

$$\begin{bmatrix} (K_{11} + s^2 m) & K_{12} \\ K_{12} + \eta_1 & (K_{22} + \eta_2 + Is^2) \end{bmatrix} \begin{bmatrix} \bar{U}_z \\ \bar{\theta} \end{bmatrix} = 0 \tag{47}$$

where $s = j\omega$ and all motion is proportional to exp (st). The terms η_1 and η_2 represent the nonconservative pitching moments due to the drag force as given in the Equations (45) and (46). The characteristic equation is then given by

$$s^4 + s^2(\omega_H^2 + \omega_p^2 + \eta_2/I) + \omega_H^2\omega_p^2 + \omega_H^2\eta_2/I - K_{12}(K_{12} + \eta_1)/Im = 0 \tag{48}$$

where $\omega_H^2 = K_{11}/m$, $\omega_p^2 = K_{22}/I$. The locus of roots for $\eta_2 = 0$ and $-\infty < \eta_1 < \infty$ is given in Figure 6.14. The root locus for $\eta_1 = 0$, $\eta_2 < 0$ is also given in Figure 6.14. From this simple model, one can show that flutter or self-excited oscillation is possible if $K_{12}\eta_1 < 0$ while the effect of η_2 can only result in a divergent motion, i.e. the double root $s^2 = 0$ implies a solution (a + bt).

6.4.4 *Experimental observation*

To examine the dynamics of a magnetically-levitated model for lateral, heave, roll, yaw and pitch motions, a rotating wheel facility was constructed. The rotating wheel had a diameter of 1.2 m and has been run up to 700 r.p.m. The test track was a

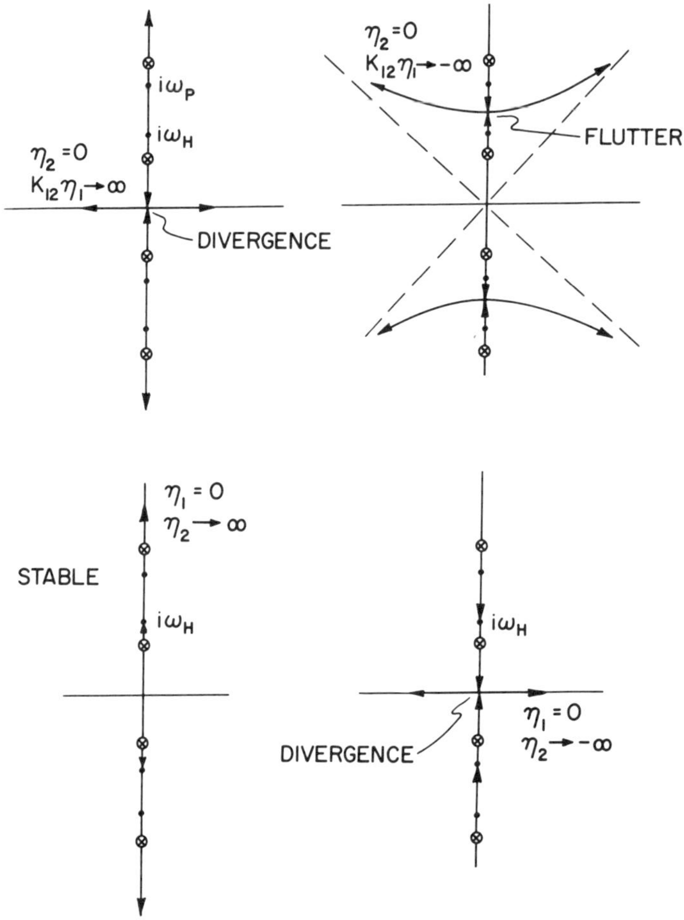

Figure 6.14 Root locus of natural frequencies versus non-conservative force parameter for heave pitch motions.

removable, thin, aluminium (2.5 mm thick) 'V'-groove-shaped rim with a 30° tilt angle with the horizontal. (Other shapes could be fitted to the wheel.) A 30 cm long model using 8 ceramic magnets was levitated at rim speeds of up to 44 m s^{-1}, at a levitation height of 5 mm (Figure 6.15). The peak magnetic field in the magnets was 800 G. The normal magnetic field distribution along a diameter is shown in Figure 6.16. This field is similar to that which would be produced by a circular pancake-type coil with circulating currents. The model mass was 0.81 kg. The drag peak occurred at 18 m s^{-1}, which checked the theoretical value based on thin track theory (Figure 6.17). The levitated model, tethered on a 3-metre, nylon line, responded to lateral perturbations with a stable, lightly damped, sinusoidal motion. Another model was used with a variable mass centre and roll inertia (Figure 6.18).

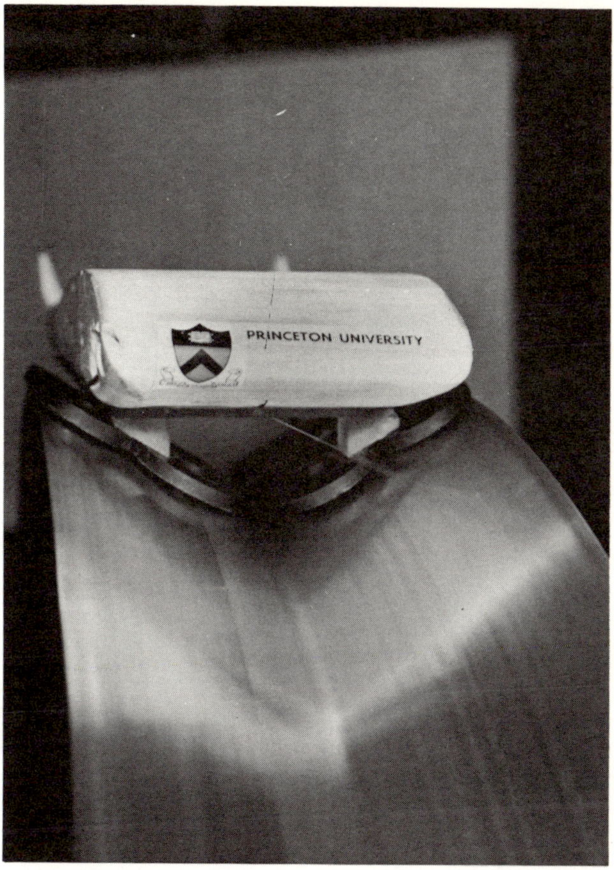

Figure 6.15 Photo of magnetically levitated model in a 'V' guideway.

Rotating wheel tests of a freely, magnetically-levitating model using permanent magnets in a U-channel were reported by the Ford research group [34]. No instabilities were reported in those tests. However, analysis of the lateral-roll dynamics of such a vehicle in a 'V' and inverted 'V' track by Wilke *et al.* [33] and the Ford group [35], have suggested the possibility of instabilities in the motion, thereby ruling out the 'V' configuration.

For experiments reported in this paper, however, stable motions of the model were observed for geometric criteria which appear to violate the limits cited by the Ford team [35]. Stable motions are shown in Figure 6.19 from a series of motion picture frames. Further, the model geometry could be changed to create a new instability in coupled lateral roll-yaw, which appears to be controlled by the magnetic drag. The stability criteria in the Ford paper [35] required the centre of mass to be below the line of intersection of the magnet planes (point A in Figure 6.18). The centre of mass in the levitated models were above this point as shown in

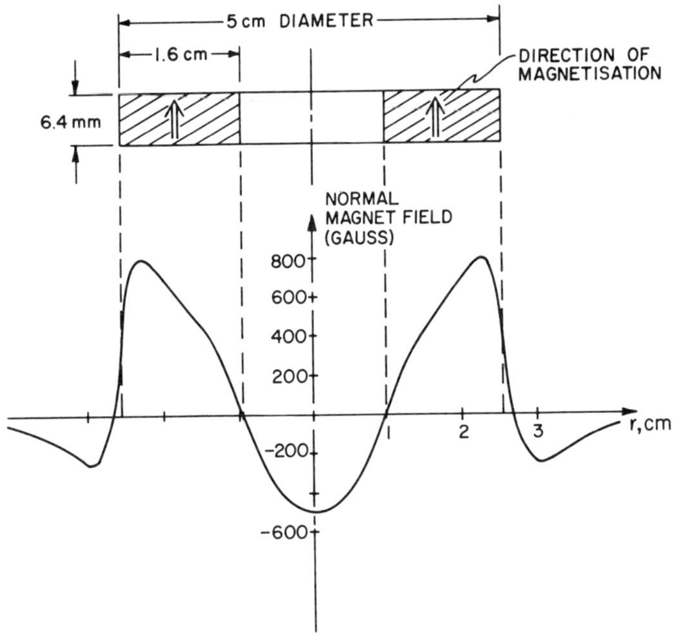

Figure 6.16 Distribution of normal magnetic field under one of the eight circular, ceramic levitation magnets.

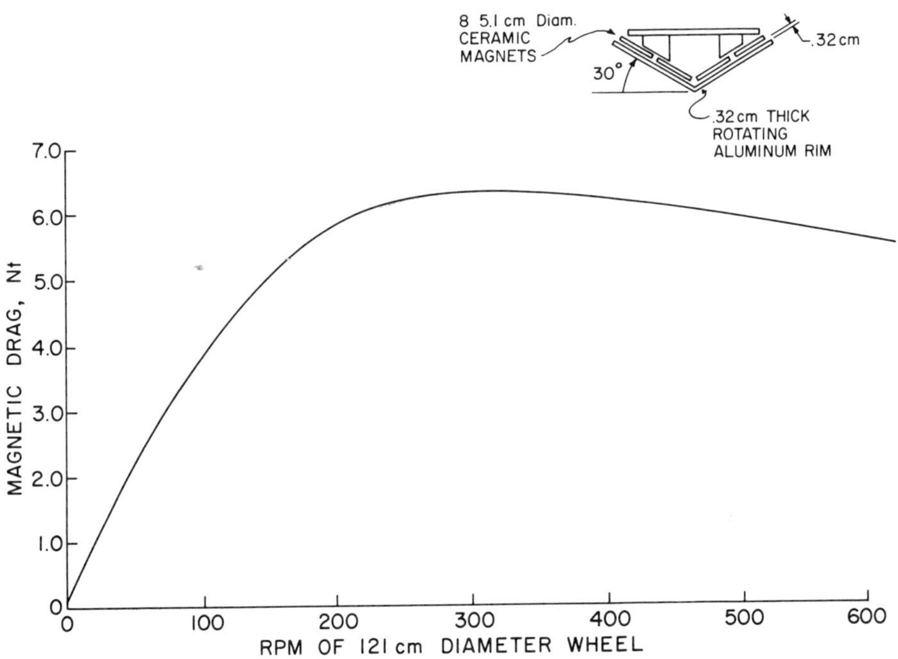

Figure 6.17 Magnetic drag versus r.p.m. for 0.8 kg levitated model.

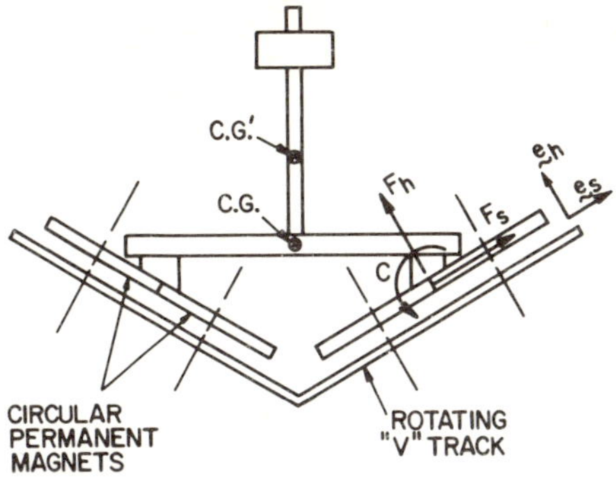

Figure 6.18 Sketch of test model for the study of yaw-roll dynamics.

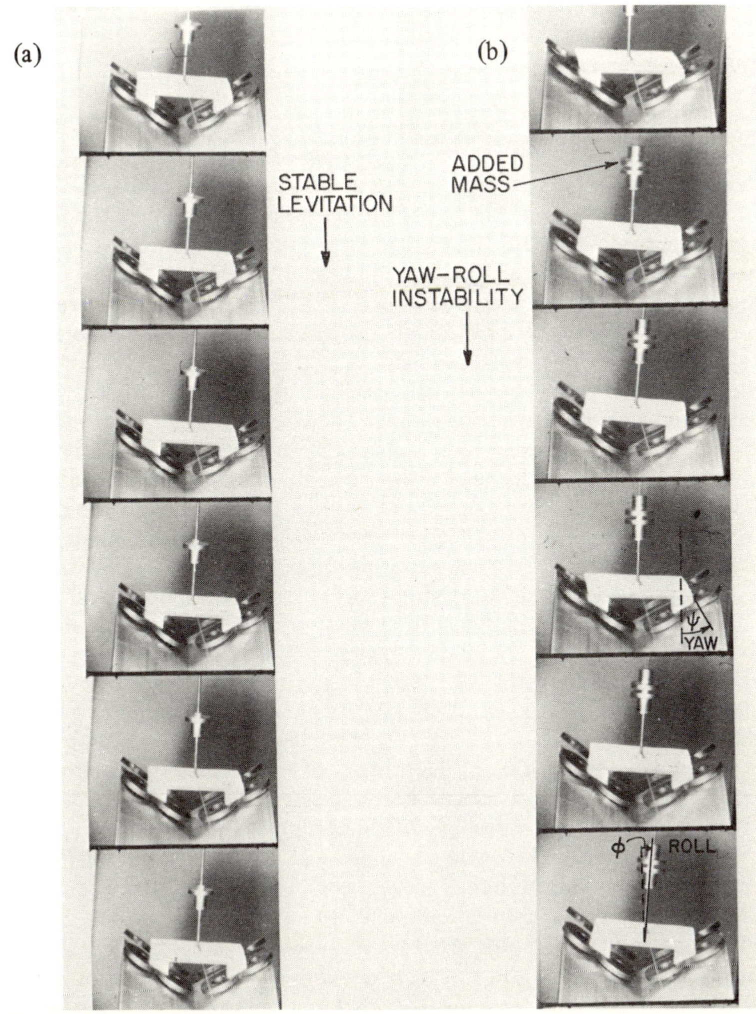

Figure 6.19 (a) Motion picture frames of stable model motion.
(b) Motion picture frames of yaw-roll instability in the magnetically levitated model.

Figure 6.18. Increases in height of the centre of mass led, however, to an apparent yaw instability. Such an instability could result from a coupling of the lateral and roll motions to yaw through the magnetic drag. Unstable motions of yaw and roll can be seen in the sequence of frames from a motion picture film.

The left side of Figure 6.19 shows stable motion of the levitated model. When an additional weight is added on the yaw-axis, thereby increasing the roll inertia, the model performed self-excited, yaw-roll oscillations as seen in the sequence of pictures on the right of Figure 6.19. The roll, as indicated by the change of position of the vertical rod on the model, was small, while the yaw motion was pronounced.

The stable motions can most likely be attributed to the roll moment created by the pairs of magnets under each of the four corners of the model. Such moments were neglected in the analysis of Wilke[17]. This explanation is plausible since the tests were run at speeds near the drag peak (Figure 6.17). These qualitative results suggest that vehicle dynamics of magnetic levitation are more complicated than the analyses given to date and that laboratory scale tests can suggest areas of research for further analysis.

6.5 Vibration problems in linear induction motors

The relative motion of masses, such as wind over a fluid surface or vehicles on a flexible bridge, is known to create large oscillations normal to the direction of motion (see for example, Fryba[36]). One might suspect that the relative motion of a linear induction motor primary and its reaction rail might exhibit similar motions. The study of this problem can be treated in two steps. First the relative motion of the primary and a rigid secondary rail can be investigated and then the problem of a moving mass on a flexible rail can be treated.

In studies on the forces in linear induction motors, the emphasis is often on the calculation of the thrust created by the magnetic forces. Except in the case of the single-sided motion, lateral forces in the double sided motor are often not of interest. The first question one can ask then about the vibration of a primary relative to the reaction rail is, 'what are the lateral magnetic restoring forces and couples on the primary for lateral motions of the primary?'. In the design of actual machines, this question is often not asked, since mechanical positioners such as follower wheels or air cushion pads are used for lateral restoring forces. Nonetheless, the lateral magnetic forces exist and it is of interest to know to what extent they can effect the lateral vibration of the primary relative to the reaction rail. This is the subject of the first part of this section.

As to the problem of moving masses on a flexible guideway, this can be a problem for the vehicle as a whole, especially for elevated guideways. However a subproblem of the moving vehicle is the motion of the primary of a linear electric machine as it moves past a flexible reaction rail, especially in the case of the double-sided induction motor. A classical analysis of the problem of a moving force on a continuous beam was given by Timoshenko in 1927, who calculated a critical speed for trains[37] beyond which the deflections of the rail would become very large. His calculations led to a value of 500 m s^{-1} for conventional rails. This problem has recently been re-examined for linear induction motor rails by Labra[38] and Keer[39]. These studies usually neglect the dynamics of the moving mass itself and concentrate instead on the mechanics of the rail. An analysis of the dynamics of a vibrating mass moving on a flexible reaction rail has been given by Moon and Dowell [40] and is reviewed in this section along with some experimental results from rotation wheel tests.

6.5.1 *Lateral dynamics of a double-sided linear motor*

In addition to the mechanical lateral guidance forces, the relative motion of primary and secondary will produce a motion-dependent force and couple on each other when the primary moves off its equilibrium position. The two-sided LIM shown in Figure 6.20 has an equilibrium position with the secondary rail midway between the two primaries. In some designs the LIM is centred, with follower wheels and springs between the primary and secondary. In other designs the centring force is provided by air cushion pads. In this section we examine the restoring forces due to magnetic fields alone.

About equilibrium, the primary can have lateral, yaw and roll motions which can change the magnetic forces. Only lateral and yaw motions, shown in Figure 6.20, are treated here. The net lateral force and couple on the rail can be written in the form:

$$\begin{bmatrix} F_Y \\ C_Z \end{bmatrix} = \begin{bmatrix} C_{11} & C_{12} \\ C_{21} & C_{22} \end{bmatrix} \begin{bmatrix} U \\ \theta \end{bmatrix} \tag{49}$$

where U is the lateral displacement and θ the yaw angle relative to the reaction rail. The total force and couple on the secondary is given by:

$$F_y = \int_0^L < f_y > \mathrm{d}x \tag{50}$$

$$C_z \mathbf{e}_z = \int_0^L \mathbf{r} \times < \mathbf{f} > \mathrm{d}x$$

where **f** is the local force and $< >$ indicates a time average over the excitation frequency of the motion. In this paper, the theory of Yamamura [41] is used, in which current sheets in the primary $K_p(x, t)$ induce current sheets in the secondary $K_s(x, t)$ where

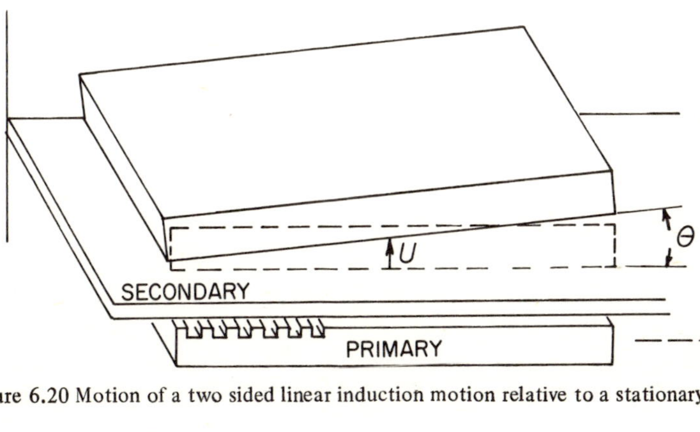

Figure 6.20 Motion of a two sided linear induction motion relative to a stationary reaction rail.

the direction of current is normal to the x–y plane in Figure 6.21 and return current paths are neglected.

In the theory of Yamamura [41], Maxwell's equation is integrated across the gap. Thus the position of the secondary in the gap does not affect the calculation of the current density K_S. Any perturbed forces due to lateral motion can only arise in the evaluation of the initial field $\mathbf{B}^0(y)$ in the gap i.e. in choosing

$$y = U + \left(x - \frac{L}{2}\right)\theta$$
$$\mathbf{r} = \left(x - \frac{L}{2}\right)\mathbf{e}_x + y\,\mathbf{e}_y \tag{51}$$

where U is the lateral displacement of the centre of mass and θ the yaw angle.

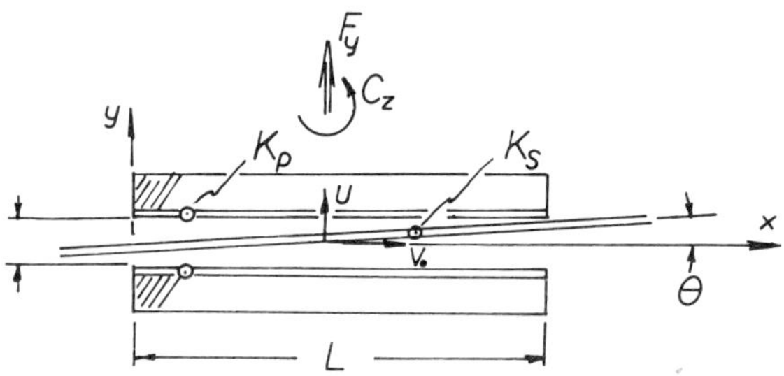

Figure. 6. 21 Sketch of two-sided linear induction motor.

For a double-sided motor with equal primary current densities on each side, the magnetic field in the gap created by the travelling current sheet

$$K_p = K_0 \exp \mathrm{j}(\omega t - kx) \tag{52}$$

is given approximately as

$$B_y^0 = B_0 \exp \mathrm{j}(\omega t - kx)$$
$$B_x^0 = -\mathrm{j}\, ky\, B_y^0 \tag{53}$$

The synchronous speed is thus $v_s = \omega/k$.

Since the force in the primary is equal and opposite to that in the secondary, the time average of the force density on the secondary is given by:

$$<\mathbf{f}> = \tfrac{1}{2}\,\mathrm{Re}[K_s(B_x^{*0}\,\mathbf{e}_y - B_y^{*0}\,\mathbf{e}_x)] \tag{54}$$

where * indicates complex conjugate.

It has been shown by Yamamura [41], that if the primary current density has the form of Equation (52), the secondary current density K_s has added terms due to end effects. When only entrance effects are included K_s is given by:

$$K_s = (K_1 + \mathrm{j}K_2)\exp \mathrm{j}(\omega t - kx) + (\kappa_1 + \mathrm{j}\kappa_2)\exp(-x/\alpha)\exp \mathrm{j}(\omega t - k_e x) \tag{55}$$

This function must satisfy the equation:

$$\frac{\partial^2 K_s}{\partial x^2} = \frac{\mu_0 \sigma \Delta}{g} \left(\frac{\partial}{\partial t} + V_0 \frac{\partial}{\partial x}\right) (K_s + 2K_p) \tag{56}$$

where g is the gap and Δ the effective secondary thickness.

It has been shown [41] for small slip that as the velocity gets large, $k_e \to k$. In this limit, the time average of the force density is given by:

$$\begin{aligned} \langle f_y \rangle &\to \frac{\mu_0 K_0}{g}\, y\, [K_1 + \kappa_1 \exp(-x/\alpha)] \\ \langle f_x \rangle &\to \frac{\mu_0 K_0}{kg} [K_2 + \kappa_2 \exp(-x/\alpha)](-1) \end{aligned} \tag{57}$$

To calculate the lateral force due to a lateral displacement U and rotation θ, the primary magnetic fields are evaluated from Equation (53) at the position:

$$\mathbf{r} = \left(x - \frac{L}{2}\right) \mathbf{e}_2 + \left[U + \left(x - \frac{L}{2}\right) \theta\right] \mathbf{e}_y$$

When this procedure is done and the integrals Equation (50) performed, the magnetic stiffness constants C_{ij} in Equation (49) take on the following values in the high speed limit:

$$\begin{aligned} C_{11} &= -\frac{\mu_0 K_0^2}{(1+\beta^2)} \left[\beta^2 + \frac{\alpha}{L} l_0\right] 2 \\ C_{12} &= \frac{\mu_0 K_0^2}{(1+\beta^2)} \frac{\alpha^2}{g} l_1\, 2, \quad C_{21} = C_{12} + T \\ C_{22} &= -\frac{\mu_0 K_0^2}{(1+\beta^2)} \left[\frac{\beta^2}{12} + \left(\frac{\alpha}{L}\right)^3 \left(l_2 + \frac{\beta}{k\alpha} l_1\right)\right] \frac{2L^3}{g} \end{aligned} \tag{58}$$

where, T is the thrust and,

$$\beta = \frac{\mu_0 \sigma \Delta}{kg} v_s \left(1 - \frac{V_0}{v_s}\right)$$

$$l_n = \frac{(-1)^n}{\alpha^{n+1}} \int_0^L \left(x - \frac{L}{2}\right)^n \exp(-x/\alpha) dx$$

Thus yaw lateral coupling leads to an unsymmetric matrix C_{ij} so that the frequencies of vibration might become complex. For harmonic motion, where the period is much greater than the excitation period, (i.e. $\Omega \ll \omega$),

$$\begin{bmatrix} U \\ \theta \end{bmatrix} = \exp(j\Omega t) \begin{bmatrix} U_0 \\ \theta_0 \end{bmatrix}$$

the characteristic equation is given by:

$$mI\Omega^4 - (IC_{11} + mC_{22})\Omega^2 + C_{11}C_{22} - C_{21}C_{12} = 0 \tag{59}$$

where m is the primary mass and I is the yaw moment of inertia about the centre of mass of the primary.

6.5.2 *Primary – flexible rail interactions*

The secondary of a two-sided linear induction motor is usually a flexible plate clamped at the base along one edge, and free at the other edge. Thus if the x, y axes lie in the plane of the plate, with x directed along the direction of travel, the transverse displacement $u_z(x, y)$ can be approximated by $u_z = u(x, t)\,\phi(y)$ where $\phi(y)$ is the displacement function for the first vibration mode of a cantilevered beam and satisfies the clamped and free-edge condition. As described in [40], this assumption along with the two-dimensional plate equation can lead to a one-dimensional model of the secondary rail in which $U(x, t)$ satisfies the following equation:

$$Db\frac{\partial^4 u}{\partial x^4} - T\frac{\partial^2 u}{\partial x^2} + Ku + \rho hb\frac{D^2 u}{Dt^2} = F(x, t)\,\phi(y_0) \qquad (60)$$

where $D = Eh^3/(1-\nu^2)\,12$, E is Young's modulus, ν is Poissons ratio, h is the plate thickness and ρ the density. Also b is the width of the infinite rail and y_0 is the point of application of the force $F(x, t)$, representing the interaction of the primary with the secondary. The operator symbol D is the convective time derivative when x is chosen as a coordinate moving with the longitudinal velocity of the motor $-V_0$ i.e.

$$\frac{D}{Dt} = \frac{\partial}{\partial t} + V_0\frac{\partial}{\partial x} \qquad (61)$$

The other variables K and T in the Equation (60), emerge from the variational method used to reduce the two-dimensional plate equation to a one-dimensional beam equation:

$$K = \frac{D}{b^3}\,12.36$$

$$T = \frac{2D}{b}\,[(1-\nu)\,4.68 - \nu\,(0.86)]$$

Equation (60) represents a beam on an elastic 'foundation'. The 'foundation' in this case is simply the bending of the cantilevered plate along the y direction. This bending also introduces an effective tension T along the x direction. This tension produces a resistance to lateral forces proportional to the curvature $\partial^2 u/\partial x^2$.

It is this type of equation that has been studied in a number of papers (e.g. [38, 39]) when $F(x, t)$ is simply a prescribed moving force. However, in order to study the complete dynamics of the rail, the motion of the LIM primary must be included. If we consider only the lateral motion of the primary, neglecting yawing motions described in the previous section, the mechanical and magnetic stiffnesses between primary and secondary are replaced by a spring of constant K_0 and the lateral force between the primary and vehicle itself is represented by a spring of constant K_1. In this analysis the lateral vehicle dynamics are neglected. When no other forces act on the primary, the equation of motion for the lateral displacement $w(t)$ of the primary is given by (Figure 6.22):

$$M_0\ddot{w} + K_1 w + K_0\,[w - u(0, t)] = 0$$

where M_0 is the mass of the primary and the third term represents the force F in Equation (60).

The combined motion of primary and secondary rail is thus governed by coupled ordinary and partial differential equations, (60) and (61). The first question one can ask about this system is whether it has real natural frequencies. To answer this question, all variables in Equations (60) and (61) are assumed to have oscillatory behaviour proportional to exp $(j\omega t)$. Finally as shown in [40], the solutions for the secondary rail displacement can be represented by a Fourier integral; i.e.

$$u(x, t) = \frac{1}{2\pi} \int_{-\infty}^{\infty} \bar{u}(k) \exp j(\omega t + kx) \mathrm{d}k$$

This assumption along with the Equations (60) and (61) leads to a characteristic equation for the natural frequencies given by:

$$0 = K_1 - \omega^2 M_0 + K_0 + K_0(K_1 - \omega^2 M_0) \frac{\phi^2(y_0)}{2\pi} \int_{-\infty}^{\infty} \frac{\mathrm{d}k}{[Dbk^4 + Tk^2 - \rho hb(w - V_0 k)^2 + K]} \quad (62)$$

The characteristic equation has the form:

$$0 = K_0 + K_0[K_1 - \omega^2 M_0) + 1G(\omega, V_0)K_0]$$

where G represents the integral in Equation (62). By manipulation of Equation (62) one can show that this function has the form;

$$G = G(\omega^2, V_0^2)$$

that is, the frequencies arc independent of the direction of travel and hence the sign of V_0. Further analysis of G reveals that the function $\omega^2(V_0^2)$ is a decreasing function of velocity for small values of V_0, i.e.

$$\omega^2 = \omega_0^2(1 - \alpha V_0^2)$$

This result appears to agree with experiments described below.

A detailed study of the roots of Equation (62) is now being undertaken by the author and Professor Dowell of Princeton University. It appears that for some values of V_0 the roots of ω^2 become complex. Whether this leads to instabilities or simply damped motions of the primary cannot be decided at the time of writing.

6.5.3 *Experiments*

The dynamics of a mass moving on a flexible cantilevered LIM reaction rail were simulated by a circular annular plate mounted on a 1.2 metre diameter rotating wheel. The flexible rail was 17.7 cm high and 0.87 mm thick and made out of

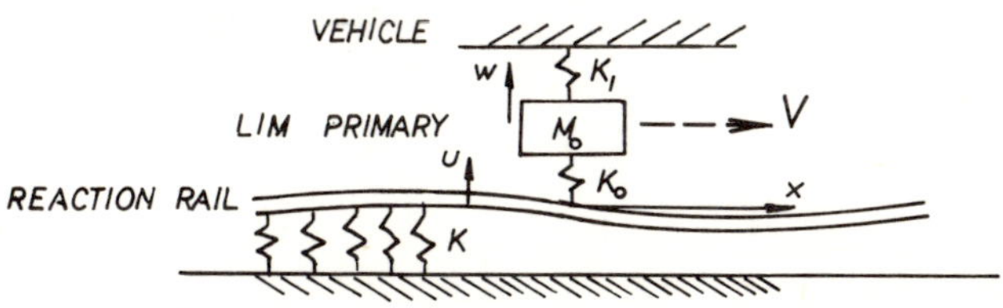

Figure 6.22 Dynamics model of a linear induction motor primary moving over a flexible reaction rail.

galvanised steel, which has the same modulus to density ratio as aluminium. Rim velocities of the plate up to 43 m s^{-1} or 710 r.p.m. were produced. No magnetic tests were performed.

The LIM primary was simulated by a mass with guide wheels to straddle the reaction rail (Figure 6.23). Between the mass and the rigid support structure, springs simulated the passive interaction of the primary with the vehicle.

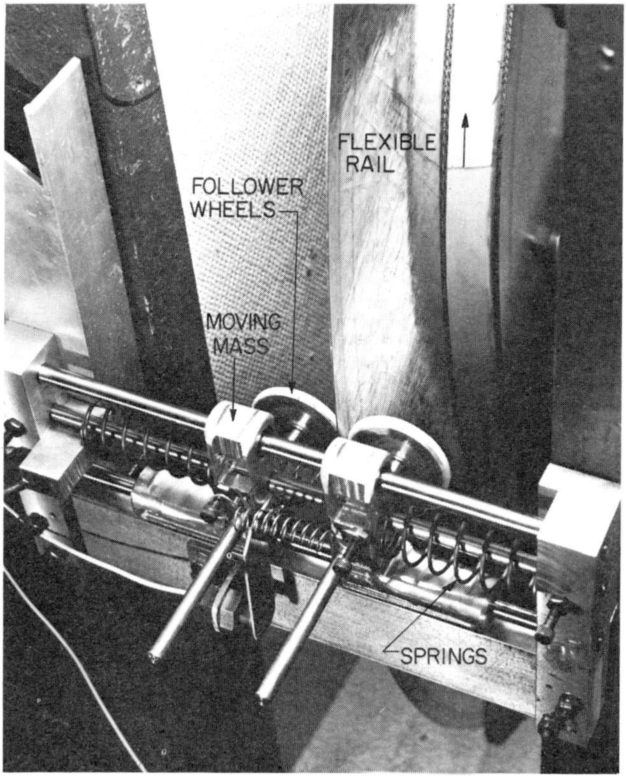

Figure 6.23 Photo of follower mass-wheel assembly on rotating, flexible reaction rail.

In the experiments the follower mass assembly was vibrated using an electromagnetic shaker. The lowest rail frequency was 30 Hz while those of the mass and springs were less than 10 Hz. The combined mass and reaction rail frequency was 18 Hz at zero velocity. As the velocity of the rail was increased however, the natural frequency dropped about 25–30% as the wheel was rotated up to 710 r.p.m., as shown in Figure 6.24. This behaviour supports the tentative conclusions of the theoretical model discussed above, that the longitudinal velocity of the LIM primary causes either a decrease in the effective lateral stiffness or an increase in the effective mass of the primary-rail system. The latter interpretation might make sense in that lateral vibration under longitudinal motion results in a greater length of rail under vibration thereby increasing the effective lateral mass of the system[42].

These questions are currently under study both experimentally and analytically for the mechanical model described above. A further contribution to the subject would be made if experiments on an actual linear induction motor primary with a flexible rail were performed. Also, the analysis could be extended to the one-sided motor of either the induction or synchronous type. In addition to the vertical and pitching dynamics of such a motor, these systems might exhibit a negative or decreasing damping with speed as in the case of the levitated wire.

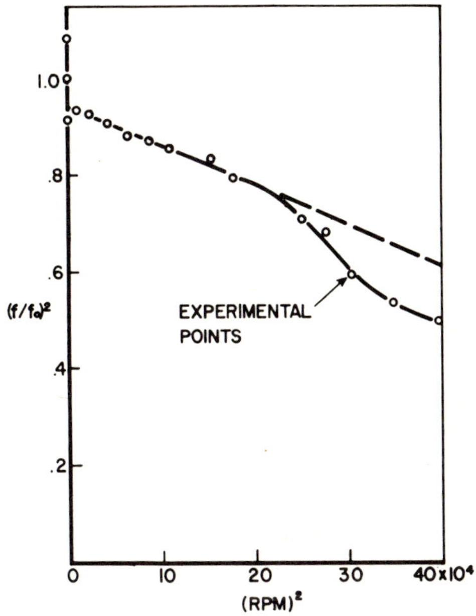

Figure 6.24 Natural frequency of a moving mass over a flexible reaction rail versus speed; experimental results from rotating wheel tests.

References

[1] Chirgwin, K. M. (1974) 'Test results from the U.S. linear induction motor research vehicle program' Proc. IEE Conf. on Linear Electric Machines, Publication No. 120 p. 236, London.

[2] Kyotani, Y. (1975) 'Development of superconducting levitated trains in Japan', (July 1975), *Cyrogenics* pp. 372–376.

[3] Borcherts, R. H. (1975) 'Repulsion magnetic suspension research – U.S. progress to date', *Cyrogenics*, (July 1975), pp. 385–393.

[4] Anon. 'Levitation Lineup – 1974', *Railway Gazette International* (October 1974), p. 377.

[5] Geary, P. J. (1964) 'Magnetic and Electric Suspension', British Scientific Instrument Research Association Report R314.

[6] Brisley, W. and Thornton, B. S. (1963) 'Electromagnetic levitation calculations for Axial symmetric systems', *Brit. J. Appl. Phys.*, **14**, pp. 682–686.

[7] Laithwaite, E. R. (1965) 'Electromagnetic levitation', *Proc. IEE*, **112**, No. 12, pp. 2361–2375.

[8] Woods, L. C., Cooper, R. K., Neil, V. K. and Taylor, C. E. (1970) 'Stability analysis of a levitated superconducting current ring stabilized by feedback and eddy current', *J. Appl. Phys.*, **41**, No. 8, pp. 3295–3305.
[9] File, J., Martin, G. D., Mills, R. G., Upham, J. L. (1968) 'Stabilised, levitated superconducting rings', *J. Appl. Phys.*, **39**, pp. 2623.
[10] Stratton, J. A. (1941) *Electromagnetic Theory*, McGraw-Hill, New York, (Earnshaw's Theorem) p. 116.
[11] Meisenholder, S. G. and Wang, T. C. (1972) 'Dynamic analysis of an electromagnetic suspension system for a suspended vehicle system', Report No. FRA-RT-73-1 (January 1972).
[12] Powell, J. R., Danby, G. R. (1967) 'High speed transport by magnetically suspended trains', ASME Paper 66-WA/RR-5, see also *Mech. Eng.* November 1967.
[13] Richards, P. L. and Tinkham, M. (1972) 'Magnetic suspension and propulsion systems for high-speed transportation', *J. Appl. Phys.*, **43**, No. 6, pp. 2680–2691.
[14] Davis, L. C. and Wilke, D. F. (1971) 'Analysis of motion of magnetic levitation systems: implications', *J. Appl. Phys.* **42**, No. 12, pp. 4779–4793.
[15] Iwamoto, M., Yamada, T. and Ohno, E. (1974) 'Magnetic damping force in electrodynamically suspended trains', 1974 Digests of the Intermag. Conf., IEEE, Toronto, Canada, Section 19, p. 14.
[16] Fink, H. J. and Hobrecht, C. H. (1971) 'Instability of vehicles levitated by eddy current repulsion – case of an infinitely long current loop' *J. Appl. Phys.* **42**, No. 9, pp. 3446–3450.
[17] Wilke, D. F. (1972) 'Dynamics, control and ride quality of a magnetically levitated high speed ground vehicle', *Transportation Research*, **6**, pp. 343–364.
[18] Coffey, H. T., Colton, J. D., Solinsky, J. C., Woodbury, J. R. (1974), 'An evaluation of the dynamics of a magnetically levitated vehicle', Report No. FRA-ORD D-74-41.
[19] Anon. (1971) 'Congress delegates see rival magnetic suspension cars' *Railway Gazette International* (November 1971).
[20] Thornton, R. D. (1974) 'The magnetoplane linear synchronous motor propulsion system', Proc. IEE. Conf. on Linear Electric Machines, Publication No. 120, p. 230, London (October 1974).
[21] Moon, F. C. (1974) 'Laboratory studies of magnetic levitation in the thin track limit', *IEEE Trans. on Magnetics*, **MAG-10**, No. 3, pp. 439–442 (September 1974).
[22] Iwasa, Y. (1973) 'Electromagnetic flight stability by model impedance simulation', *J. Appl. Phys.*, **44**, No. 2, pp. 858–862.
[23] Ohno, E., Iwamoto, M. and Yamada, T. (1973) 'Characteristic of superconductive magnetic suspension and propulsion for high-speed trains', *Proc. IEEE*, **61**, No. 5, pp. 579–586.
[24] Thornton, R. D. (1973) 'Design principles for magnetic levitation', *Proc. IEEE*, **61**, No. 5, pp. 586–598.
[25] Hammitt, A. G. (1973) *The Aerodynamics of High Speed Ground Transportation*, Western Periodicals Co., North Hollywood, California.
[26] Seckel, E. (1964) *Stability and Control of Airplanes and Helicopters*, Academic Press, New York
[27] Dickenson, B. (1968) *Aircraft Stability and Control for Pilots and Engineers*, Sir Isaac Pitman and Sons Ltd., London.
[28] Reitz, J. R. (1970) 'Forces on moving magnets due to eddy currents', *J. Appl. Phys.* **41**, No. 5, pp. 2067–2071.
[29] Moon, F. C. and Markiewicz, W. D. (1972) 'Calculations for the design of magnetically suspended vehicles', Report to Boeing-Vertol Corporation, Philadelphia, Pa. (December 1972).

[30] Crandall, S. H., Karnopp, D. C., Kurtz, E. F. Jr. and Pridmore-Brown, D. C. (1968) *Dynamics of Mechanical and Electromechanical Systems*, McGraw-Hill Book Company, New York.
[31] Guderjahn, C. A. *et al.* (1969) 'Magnetic suspension and guidance for high speed rockets by superconducting magnets', *J. Appl. Phys.*, **40**, No. 5, pp. 2133–2140.
[32] Reitz, J. R. and Davis, L. C. (1972) 'Force on a rectangular coil moving above a conducting slab', *J. Appl. Phys.*, **43**, No. 4.
[33] Wilke, D. F., Davis, L. C. and Borcherts, R. H. (1972) 'A design for a magnetically levitated high speed vehicle', Proceedings of the seventh annual meeting of IEEE industry applications society, Philadelphia, Pa, October 1972, pp. 333–342.
[34] Reitz, J. R., Borcherts, R. H., Davis, L. C. and Wilke, D. F. (1972) 'Technical feasibility of magnetic levitation as a suspension system for HSGT vehicles', Ford Motor Co. Techn. Report FRA-RT-72-50.
[35] Reitz, J. R., Borcherts, R. H., Davis, L. C., Hunt, T. K. and Wilke, D. F. (1973) 'Preliminary design studies of magnetic suspensions for high speed ground transportation', Report No. FRA-RT-73-27.
[36] Fryba, L. *Vibration of Solids and Structures under Moving Loads*, Noordhoff International Publications, Leyden, The Netherlands.
[37] Timoshenko, S. P. (1953) 'Method of analysis of statical and dynamical stresses in rail', *The Collected Papers*, McGraw-Hill Company, New York.
[38] Labra, J. J., *Response of Railroad Track to Moving Loads*, Doctoral Dissertation No. 73-19-395, New York University, University Microfilms Ann Arbor Michigan.
[39] Kerr, A. D. (1972) 'The continuously supported rail subjected to an axial force and a moving load', *Int. J. Mech. Sci.*, **14**, pp. 71–78.
[40] Moon, F. C. and Dowell, E. H. (1974) 'Dynamic interaction between a linear induction motor and elastic reaction rail', Proc. IEE Conf. on Linear Electric Machines, pp. 173–178, London (October 1974).
[41] Yamamura, S. (1972) *Theory of Linear Induction Motors*, Wiley & Sons, New York.
[42] Alexandridas, A. (1976) Doctoral Dissertation, Princeton University, Princeton N.J., USA.

7

Computation Aspects of Linear Induction Motors

J. C. SABONNADIÈRE

7.1 Introduction

The concept of propulsion without wheels involves not only the linear induction motor necessary to achieve the propulsion but also a complex system in which several tasks such as suspension, guidance and power collection must be realised. The electrical engineer who is mainly concerned with the design of the linear motor, is confronted with a variety of constraints necessary to work out the compatibility of his motor with the other requirements of the project. The performance of the linear electric motor must be compatible with the characteristics of the suspension and guidance systems. The latter are different if the levitation is obtained by an air cushion system as in the 'Aerotrain' or 'Urba' systems or by a magnetic repulsion or attraction system[1].

The analysis of the suspension and guidance systems has been well developed by Lhenry[2] and Baudon[3] under the direction of Ivanès[4, 5] and with the collaboration of the author[6, 7] in the Laboratoire d'Electrotechnique de l'Institut National Polytechnique de Grenoble. For this reason the topic of this paper is focused on the computation of the operation characteristics of the linear motor, independent of its environment. If understanding the physics of the linear motor is important for the design of all the details which make a good project, it is also of the greatest importance to have an accurate model of this motor for the prediction of its electromechanical behaviour. Indeed, as part of a complex transportation system, the linear electric motor has to achieve a given mechanical power under strict conditions of energy supply and over a fixed speed range. The electrical characteristics must also meet strict constraints on weight, volume, thermal losses and guidance facilities.

The scope of this chapter is threefold:

1. To point out the main difficulties which arise when one tries to model linear electric motors in general and the linear induction motor in particular.
2. To explain the method of solution proposed.
3. To provide a critical analysis of the results and arrive at some important conclusions on the use of the algorithm presented.

7.2 Modelling a linear motor is a hard task

The modelling of an electrical machine is never easy since the behaviour and the effects of the magnetic field are generally difficult to describe accurately in three-dimensional

Euclidian space. The art of design implies that in standard electrical machines all phenomena can be represented by an analysis in one or two dimensions; a three-dimensional representation is seldom necessary except in special cases. Among the different types of electric machines the induction motor is perhaps the most difficult to model. The difficulties arise from the strong coupling which exists between the magnetic field and the secondary currents. Such a strong coupling is necessary for efficient operation of the machine but makes flux and current density distributions difficult to compute. The linear induction motor could be said to illustrate all the possible difficulties of modelling. First it is an induction machine, but, unlike the rotating squirrel cage induction motor, its secondary is generally made of a continuous conducting sheet in which currents can flow freely in all directions. On the other hand, the magnetic circuit of the linear induction motor is an open one in which the air gap cannot be assumed very thin relative to other dimensions. Moreover, the topology of the magnetic circuit implies an asymmetric distribution of the primary winding, the three phases of which cannot be identically located on the magnetic core. This asymmetry leads to unbalanced primary currents under balanced voltage supply. All these considerations make the usual simplifying assumptions of electrical machine engineering invalid and therefore the analysis of linear induction motors must be carried out by special methods.

A truly general analysis would cover any kind of linear induction motor that might be constructed (see description by Laithwaite in a recent paper [8]), but such is beyond the scope of this chapter. The analysis here will be restricted to the double-sided linear induction motor (or single-sided with a laminated back iron yoke). The method described may be adapted (with some minor alterations), to the analysis of a single-sided linear motor with a massive back iron yoke, when the saturation of the iron may be neglected. Readers interested in special methods adapted to the single-sided linear induction motor in which the back iron yoke is unlaminated and the iron saturated may refer to the work of Foggia [9] and Sabonnadière [10].

The model of the double-sided linear induction motor consists of an unsaturated machine. Maxwell's equations for field and currents are solved in order to allow computation of the impedance and the forces for a constant voltage supply at variable frequency. Analysis of the phenomena is three-dimensional, although the magnetic circuit laminations force the magnetic flux to expand only in the plane of the laminations while the current flows only in the plane of the secondary reaction plate which is perpendicular to the magnetic plane. Solutions for field and current distributions are therefore possible in two-dimensional planes only.

7.3 Simplifying assumptions and notations

The model considered is shown in Figure 7.1 which defines the principal dimensions. The reaction rail is made of three parts with resistivities ρ, ρ_1 and ρ_2 and thicknesses c, c_1 and c_2 respectively. When the reaction rail consists of several sections with gaps between the sections, an equivalent thickness is used to represent the composite structure. Current is assumed to flow in the plane of symmetry of the rail and skin effect is neglected. Thus the current density vector has only two components J_x, J_u directed along the x and u axes respectively. The flux density inside the air gap is assumed to be located in the plane of the laminations and is assumed to be zero outside the air gap in the direction perpendicular to the axis of motion (i.e. along the

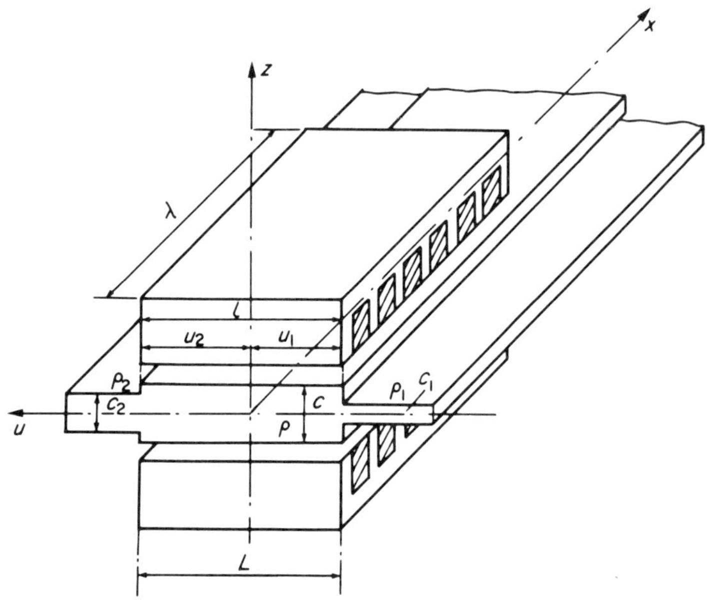

Figure 7.1 Linear motor: geometry and dimensions.

u axis). In the z-x plane the amplitude of the flux density is defined as the product of a constant B_m and a modulation function $f(x)$ which is closely related to the shape of the magnetic poles and to the primary winding distribution. The modulation function $f(x)$ is a periodic function of period L' (Figure 7.2). L' is chosen by the designer to be large compared to the length of the motor and such that $f(x)$ vanishes at both ends of the interval defined. Furthermore, $f(x)$ will be an even function (the amplitude of the field is always a positive number) with the origin of the x-axis at the middle of the motor. The analysis of the behaviour of the motor will be based on the Fourier decomposition of $f(x)$ which leads to an enumerable infinity of harmonic motors in which all quantities are periodic in time as well as in the space variable x. Under this assumption, a method of solution is first described allowing for the edge effect due to the finite size of the reaction rail, for a motor in which field and current densities are travelling electromagnetic vectors, with sinusoidal variation along the x-coordinate. The end effect is then modelled by using the Fourier decomposition of the modulation function followed by the superposition of the results for each harmonic motor to obtain the characteristics of the whole machine.

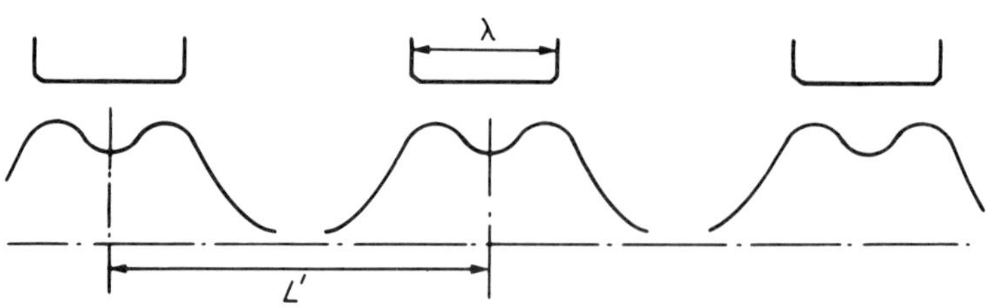

Figure 7.2 Periodic modulation function $f(x)$.

7.4 Computation of the travelling field

Using the symbols defined in Figure 7.1, $B(x, u, t)$, $J_x(x, u, t)$ and $J_u(x, u, t)$ respectively are defined as the component along z of the flux density and the components along x and u of the current densities in the reaction rail. All these quantities are sinusoidal in x with an angular frequency $\nu = 2\pi/\lambda$, the flux density being zero outside the pole in the u direction. Under these assumptions Maxwell's equations become:

(a) *Under the pole* ($u_1 < u < u_2$)

$$\frac{\partial J_u}{\partial x} - \frac{\partial J_x}{\partial u} - (V/\rho)\frac{\partial B}{\partial x} - (1/\rho)\frac{\partial B}{\partial t} = 0 \tag{1}$$

$$\frac{\partial B}{\partial x} = (\mu_0/e)(cJ_u + J_p) \tag{2}$$

$$\frac{\partial B}{\partial u} = (-\mu_0/e)\, cJ_x \tag{3}$$

Where J_p is the magnetomotive force associated with the primary magnetic field and depends on the primary current distribution, $J_p = J_{pm} \cos(\omega t - \nu x + \phi)$, ϕ arbitrary. From Equations (2) and (3) it is easy to verify the zero divergence law of the current density vector (J_x, J_u).

(b) *Outside the pole* ($u_2 < u < u_1$)

$$\frac{\partial J_{ui}}{\partial u} + \frac{\partial J_{xi}}{\partial x} = 0 \ (i = 1, 2) \tag{4}$$

$$\frac{\partial J_{ui}}{\partial x} - \frac{\partial J_{xi}}{\partial u} = 0 \ (i = 1, 2) \tag{5}$$

(c) *Boundary and interface conditions*

At the edges of the reaction rail the transverse component J_u of the current density will vanish while the longitudinal component J_x will satisfy an extreme condition. Similarly, at the interface between the different parts of the rail, the normal component of the current and the tangential component of the electric field must be continuous. These conditions are expressed by equations:

$$J_{u2}(x, L/2, t) = 0 \tag{6}$$

$$J_{u1}(x, -L/2, t) = 0 \tag{7}$$

$$\frac{\partial J_{u2}}{\partial u}(x, L/2, t) = 0 \tag{8}$$

$$\partial J_{u1}(x, L/2, t) = 0 \tag{9}$$

$$c_i J_{ui}(x, u_i, t) = cJ_u(x, u_i, t) \tag{10}$$

$$\rho_i J_{xi}(x, u_i, t) = \rho J_u(x, u_i, t) \tag{11}$$

where $i = 1, 2$.

The assumption that these quantities vary sinusoidally in time and space allows complex notation, i.e. all functions $Q(x, u, t)$ can be expressed by the product:

$$Q(x, u, t) = \bar{Q}(u) \exp [j(\omega t - \nu x)] \tag{12}$$

where j is the imaginary unit. Equations (1) to (5) may be rewritten as follows:

$$-\frac{\partial \bar{J}_x}{\partial u} = j\,\nu\,\bar{J}_u + (j/\rho)(\omega - \nu V)\bar{B}\,. \tag{13}$$

$$\bar{B} = (j\,\mu_0/\nu e)(c\bar{J}_u + \bar{J}_p) \tag{14}$$

$$\frac{\partial B}{\partial u} = (-\mu_0 c/e)\,\bar{J}_x \tag{15}$$

$$\frac{\partial \bar{J}_{ui}}{\partial x} = j\,\nu\,\bar{J}_{xi} \tag{16}$$

$$\frac{\partial \bar{J}_{xi}}{\partial u} = j\,\nu\,\bar{J}_{ui} \tag{17}$$

where $i = 1, 2$.

After some algebraic manipulation we obtain

$$\frac{\partial^2 \bar{J}_u}{\partial u^2} - [\nu^2 + j(\omega - \nu V)(\mu_0 c/\rho e)]\bar{J}_u = j(\omega - \nu V)(\mu_0/\rho e)\,\bar{J}_p \tag{18}$$

$$\frac{\partial \bar{J}_u}{\partial u} = j\,\nu\,\bar{J}_x \tag{19}$$

$$\frac{\partial^2 \bar{J}_{ui}}{\partial u^2} - \nu^2\,\bar{J}_{ui} = 0 \tag{20}$$

$$\frac{\partial \bar{J}_{ui}}{\partial u} = j\,\nu\,\bar{J}_{xi} \tag{21}$$

where $i = 1, 2$.

By introducing the slip $g = (\omega - \nu V)/\omega$ between the speed of the reaction rail and the velocity of the travelling field Equations (18) to (20) become:

$$\frac{\partial^2 \bar{J}_u}{\partial u^2} - (\nu^2 + j\,g\omega\alpha)\,\bar{J}_u = j\,g\omega\alpha/c\bar{J}_p \tag{22}$$

$$\frac{\partial^2 \bar{J}_{ui}}{\partial u^2} - \nu^2\,\bar{J}_{ui} = 0 \tag{23}$$

where $\alpha = \mu_0 c/\rho e$ is a characteristic of the electromagnetic coupling between the field and the rail. With the boundary and interface conditions:

$$\bar{J}_{u1}(x, -L/2) = 0 \tag{24}$$

$$\bar{J}_{u2}(x, L/2) = 0 \tag{25}$$

$$c_i\bar{J}_{ui}(x, u_i) = c\bar{J}_u(x, u_i) \tag{26}$$

$$\rho_i \frac{\partial \bar{J}_{ui}}{\partial u}(x, u_i) = \rho\,\frac{\partial \bar{J}_{ui}}{\partial u}(x, u_i) \tag{27}$$

Let μ^2 be the complex number defined by $\mu^2 = \nu^2 + \mathrm{j}\, g\omega\alpha$; the general solution of this classical boundary value problem is then easily obtained in terms of complex sinusoidal and exponential functions:

$$\bar{J}_u = \bar{J}_p\,(\mathrm{j}\, g\omega\alpha/\mu^2 + C\cos\mu u + D\sin\mu u) \tag{28}$$

$$\bar{J}_x = \mathrm{j}\,\mu/\nu\,\bar{J}_p\,(-C\sin\mu u + D\cos\mu u) \tag{29}$$

$$\bar{B} = \mathrm{j}\,\mu_0/\nu e\,\bar{J}_p\,[1 + \mathrm{j}\, g\omega\alpha/\mu^2 + c\,(C\cos\mu u + D\sin\mu u)] \tag{30}$$

$$\bar{J}_{ui} = \bar{J}_p\,[C_i \exp(\nu u) + D_i \exp(-\nu u)] \tag{31}$$

$$\bar{J}_{xi} = \bar{J}_p\,[-\mathrm{j}\, C_i \exp(\nu u) + \mathrm{j}\, D_i \exp(-\nu u)] \tag{32}$$

where $i = 1, 2$.

The application of interface conditions (24) to (27) to Equations (28), (31), (32) leads to the computation of the unknown constants C, D, C_1, D_1, C_2, D_2 and hence to the solution

$$\bar{J}_u = \bar{J}_p\,[(\mathrm{j}\, g\omega\alpha)/c\mu^2]\,[1 - (\delta_1\cos\mu u + \delta_2\sin\mu u)/\delta] \tag{33}$$

$$\bar{J}_x = \bar{J}_p\,[(-\mathrm{j}\, g\omega\alpha/c\mu\nu]\,[\delta_1\sin\mu u - \delta_2\cos\mu u)/\delta] \tag{34}$$

$$\bar{B} = \bar{J}_p\,(\rho\alpha/c\mu^2\nu)\,[\nu^2 - \mathrm{j}\, g\omega\alpha\,(\delta_1\cos\mu u + \delta_2\sin\mu u)/\delta] \tag{35}$$

where the coefficients δ_1, δ_2 and δ are defined by:

$$\delta = (1+\gamma_1)\{\tan\mu u_2 - \gamma_2/\tan^2\mu u_2\} - (1+\gamma_2)\{\tan\mu u_1 - \gamma_1/\tan^2\mu u_1\} \tag{36}$$

$$\delta_1 = (\tan\mu u_2 - \gamma_2/\tan^2\mu u_2)/\cos\mu u_1 - (\tan\mu u_1 - \gamma_1/\tan^2\mu u_1)/\cos\mu u_2 \tag{37}$$

$$\delta_2 = (1+\gamma_1)/\cos\mu u_2 - (1+\gamma_2)/\cos\mu u_1 \tag{38}$$

with:

$$\gamma_1 = (\rho c_1/\rho_1 c)\,\mu/\nu\,\tan\mu u_1 \cdot \tanh\,[\nu\,(u_1 + L/2)] \tag{39}$$

$$\gamma_2 = (\rho c_2/\rho_2 c)\,\mu/\nu\,\tan\mu u_2 \cdot \tanh\,[\nu\,(u_2 - L/2)] \tag{40}$$

From Equations (26) to (28) we can compute the forces exerted on the reaction rail. When the reaction rail is not symmetric with regard to the axis of the magnetic poles, two forces are exerted on the rail: a thrust force oriented in the direction of motion and a lift force which tends to eject the rail from the air gap. The lift force is not very important under steady state operation but can cause disturbance on the dynamics of the suspension system in practice. The computation of these forces is made by the application of the Laplace force evaluation method which gives:

$$F_x = c\int_{u_1}^{u_2}\int_{-\lambda/2}^{\lambda/2} B(x,u,t)\,[J_u(x,u,t)\,\mathrm{d}x\,\mathrm{d}u] \tag{41}$$

$$F_u = c\int_{u_1}^{u_2}\int_{-\lambda/2}^{\lambda/2} B(x,u,t)\,[J_x(x,u,t)\,\mathrm{d}x\,\mathrm{d}u] \tag{42}$$

The integration of the complex functions $\bar{J}_x, J_u$, and B and the use of the well-known property of the product of sinusoidal functions yields:

$$F_X = c\lambda/4 \int_{u_1}^{u_2} (\bar{B}\bar{J}_u{}^* + \bar{B}\bar{J}_u{}^*)\, du \tag{43}$$

$$F_u = c\lambda/4 \int_{u_1}^{u_2} (\bar{B}\bar{J}_X^* + \bar{B}^*\bar{J}_X)\, du \tag{44}$$

These integrals give complicated expressions for F_X and F_u which are best evaluated by a computer. For instance F_X is given by:

$$F_X = (\rho\alpha\lambda/\nu)\left[\frac{g\omega\alpha/c}{\nu^4 + (g\omega\alpha)^2}\right] \left\langle \nu^2 \left(\frac{u_1 - u_2}{2}\right) + \mathrm{Re}\left\{\mu/\delta \sin\mu \left(\frac{u_1 - u_2}{2}\right) \left[\delta_1 \cos\mu\left(\frac{u_1+u_2}{2}\right) + \delta_2 \sin\mu\left(\frac{u_1+u_2}{2}\right)\right]\right\}\right\rangle \tag{45}$$

In this formula, the first term in the bracket, represents the value of the thrust in the absence of edge effect[11], while the second term is the effect of the finite width of the reaction rail. Historically, early theories on the linear motor, assuming an infinite extent of the reaction rail, gave only the first term of Equation (45) which led, in certain cases, to important errors in the thrust and gave no indication of lift force. Numerical calculations using algorithms derived from Equations (43) and (44) have shown that at rest, the lift force may rise to about 20% of the thrust force and cannot be neglected in the simulation of dynamic behaviour of the vehicle.

The knowledge of the total flux density in the air gap allows the computation of the flux through the coils of the primary winding and thus the calculation of the impedance at every slip value. The voltage at the output of the motor is related to the line current by the equation:

$$u(t) = R_i + L_f \frac{di}{dt} + \frac{d\phi}{dt} \tag{46}$$

where R is the resistance of the primary winding, L_f is the leakage inductance which is equal to the sum of the end winding and slot leakage inductances, and ϕ represents the flux through the coil of each phase. Using complex notation Equation (46) becomes:

$$\bar{U}_1 = R\bar{I}_1 + j\,\omega\, L_f + j\,\omega\,\bar{\phi} \tag{47}$$

where the flux ϕ is given by:

$$\phi = \int_{-\lambda/2}^{\lambda/6} \int_{u_1}^{u_2} B(x, u, t)\, dx\, du \tag{48}$$

After evaluation of the definite integral Equation (48), it will be evident that, as in the case of forces, there is a term corresponding to the assumption of an infinite rail and there is a second term which takes into account the edge effect due to the finite width of the reaction rail.

After solving the equations of the motor under the assumption of sinusoidal behaviour of field and current with time and space variables, the more difficult problem of end effect can be tackled.

7.5 The end effect

The previous paragraph dealt with a machine in which the number of poles and the nominal speed, make the end effect on the output characteristics negligible. Thus the assumption that a real motor is a slice of an infinite one with periodic excitation remains valid. Unfortunately, reality is a little more complicated still and the linear motors encountered in industrial high speed applications are always subject to end effect. The end effect is created by the motion of the conducting sheet in a region of space in which the flux density has a sharp variation. The electromotive force due to this field variation generates eddy currents, which in turn produce an opposing magnetic field. This effect leads to the reduction of the total flux under the pole and thus to a lowering of the thrust. When the extent of the poles is infinitely long, as it appears in rotating machines, the flux density is a precisely periodic wave obtainable from the product of a constant amplitude B_m and the periodic function exp $[j\,(\omega t - \nu x)]$. Experimental data on the field amplitude and phase values reveals the existence of a travelling field at the centre of the stator and a stationary field at both ends of the magnetic circuit. The non-linear variation of the phase angle at the extremities of the stator will be neglected. In addition, it is assumed that the z-component of the flux density in the air gap is the product of a travelling wave and a modulation function, thus:

$$B(x, u, t) = B_m(u)\ [f(x)]\ \cos(\omega t - \nu x) \qquad (49)$$

It should be noted that the amplitude of B varies along the air gap. The modulation function also used by Jufer[12] and Oberretl[13] represents the exact distribution of the modulus of the flux density in the air gap. The function $f(x)$ is obtained either from experimental data used in reduced coordinates or by conformal mapping techniques which model both the exact shape of the magnetic surfaces and the number of poles of the winding arrangements. Figure 7.3 shows the function $f(x)$ for a four-pole motor. The method of computation of $f(x)$ is described in more detail in the appendix at the end of this chapter. As previously mentioned, $f(x)$ is an even periodic function, the period L' being chosen to assure it a zero value at each end of the interval. The decomposition of $f(x)$ into a Fourier series involves the modulation of the flux density amplitude B_m into an infinite number of harmonics, each with a pulsation which is an integer multiple of the basic pulsation $\nu' = 2\pi/L'$.

$$B = B_m \cos(\omega t - \upsilon x) \sum_{n=0}^{\infty} a_n \cos \nu' x \qquad (50)$$

The right hand side of Equation (50) can be transformed into a sum of travelling fields which move at different velocities:

$$B = B_m \sum_{n=0}^{N} \{a_n/2\}\{\cos[\omega t - (\nu - n\nu')x] + \cos[\omega t - (\nu + n\nu')x]\} \qquad (51)$$

where the integer N is the maximum number of harmonics retained in the calculation depending on the accuracy required.

The principle of the method from here on is as follows. Assume that quantities, J_x, J_u are of the same form as B and compute the field distribution and current

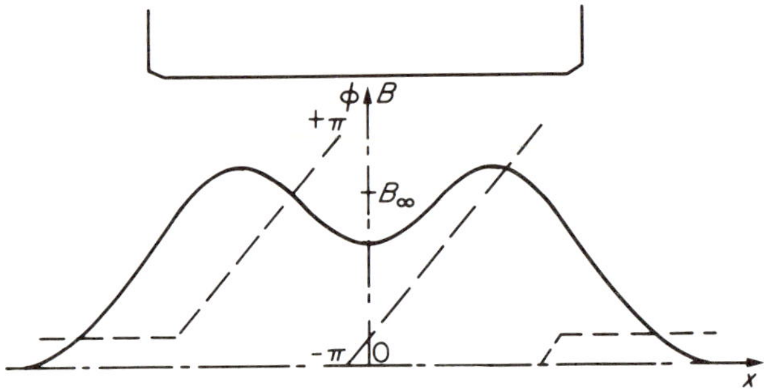

Figure 7.3 Phase and amplitude of the flux density in a 4-pole motor.

densities along the u axis. Then from the superposition of the harmonic solution the final solution is obtained. This method leads to the following idealised model:

1. The model motor has an infinitely long stator of pole pitch L' and the same leakage reactance as the real motor.
2. This motor consists of N fictitious motors, each generating a travelling field at an angular frequency $(\nu - n\nu')$.
3. These fictitious motors are connected in series and have the same primary field excitation as the real motor, but each is multiplied, in the case of the nth motor, by the coefficient a_n of the Fourier decomposition of the modulation function $f(x)$.
4. For each virtual motor the method of computation described in Section 7.4 is applied. For the n-th harmonic the field $B(n)$ and the current densities $J_u(n)$ and $J_x(n)$ are computed as periodic functions of angular frequency $\nu_n = \nu \pm n\nu'$. A harmonic slip g_n is defined in term of the slip g of the real motor and the rank of the harmonic motor n: $g_n = [1 - (1 - g)\,\nu n/\nu]$.

By principle of superposition, the field and current densities in the real motor are the sum of harmonic values:

$$B = \sum_{n=0}^{N} B(n) + B(-n) \tag{52}$$

$$J_u = \sum_{n=0}^{N} J_u(n) + J_u(-n) \tag{53}$$

$$J_x = \sum_{n=0}^{N} J_x(n) + J_x(-n) \tag{54}$$

From this point the forces and impedances can be found using Equations (43), (44) and (45), (46). Again, if the harmonic motors are connected in series, the force is the sum of elementary forces

$$F_x = \sum_{n=0}^{N} [F_x(n) + F_x(-n)] \tag{55}$$

$$F_y = \sum_{n=0}^{N} [F_u(n) + F_u(-n)] \tag{56}$$

The elementary forces $F_u(n), F_x(n), F_u(-n), F_x(-n)$ are calculated from the formulae derived in Section 7.4. Similarly, the impedance $Z(\omega, g)$ of the real motor is the sum of the primary impedance and all the elementary impedances of the harmonic motors:

$$Z(g, \omega) = R + j\omega L_f + \sum_{n=0}^{N} [Z(n) + Z(-n)] \tag{57}$$

The results of this section allow the computation of the motor output under constant current operation. The analysis at constant voltage requires a knowledge of the leakage inductance. This is dealt with as follows.

7.6 Leakage inductance

The meaning of leakage inductance in the context of this chapter is slightly different from the classical concept of leakage inductance used in rotating induction machines. Here, by leakage inductance is meant the inductance associated with all flux not crossing the air gap. All the flux which does cross the air gap is included in the calculation of the flux ϕ defined in Section 7.3. The leakage inductance thus consists of the slot leakage inductance and the end winding inductance. The slot inductance for a linear motor is identical with that of rotating machines except for some differences due to a modification of the geometry and the method of calculation is described in classical textbooks on rotating machines. The leakage flux distribution of the end winding may be, on the other hand, perceptibly different in rotating and linear machines and therefore must be analysed specially. Two kinds of winding are generally used in linear motor design: (a) gramme ring winding where the width of the stator is a little larger than the pole pitch and (b) multiple layer winding in all cases where the latter condition is not fulfilled. In both cases the self and mutual inductances will be obtained from the vector potential by energy considerations. The coupling energy of two coils is given by:

$$W_{ij} = \iiint_{D_j} A_i I_j dv = 1/2\, M_{ij} I_i I_j \tag{58}$$

where D_j is the domain in which the current I_j is flowing while A_i is the vector potential due to the current I_i. The problem now is to find the vector potential that differs in the two types of winding.

7.6.1 *Gramme winding*

The geometry of the end winding is defined in Figure 7.4 for one half of the motor with one slot per pole and phase. Coils numbered 3 and 4 may be considered as being located near an infinite-area magnetic surface because the effect of the corner is assumed to be negligible. This assumption is not valid for coils 1, 2, 5 and 6 since the

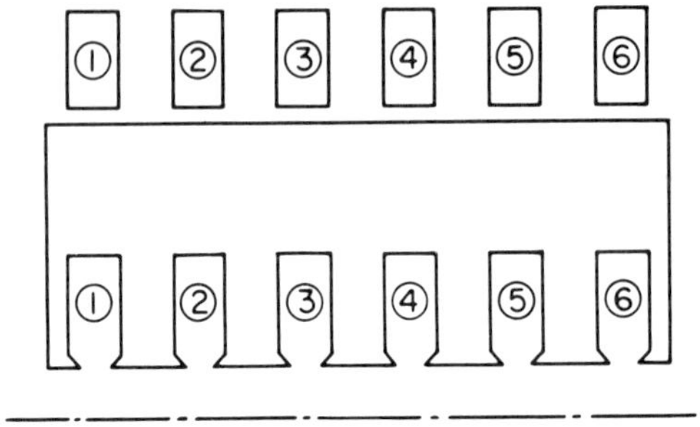

Figure 7.4 Gramme ring wound linear motor.

vector potential is affected by the closest corner. The vector potential of an infinitely thin conductor carrying a current density I is obtained from image theory by the application of conformal mapping [10]. The mapping $z = st^{3/2}$ of the t-plane to the z-plane is the suitable transformation to obtain the vector potential of the configuration of Figure 7.5. By integrating the previous infinitesimal vector potential over the range of the coils, the map of Figure 7.6 is obtained, representing the end winding flux of a

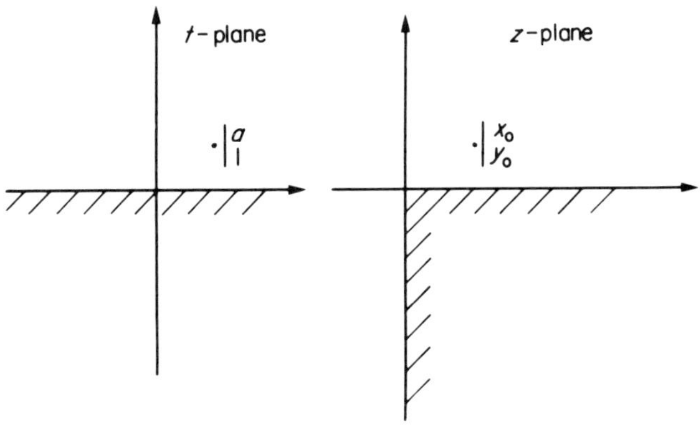

Figure 7.5 The $z - st^{3/2}$ conformal transform.

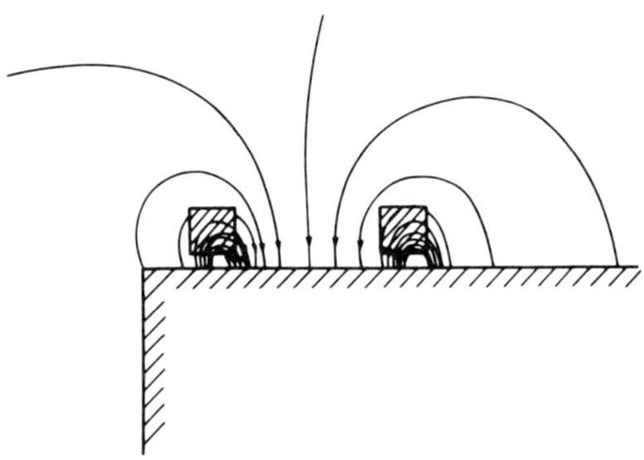

Figure 7.6 Flux plot for 2 conductors/phase.

phase [11]. The inductance of a phase consisting of $2m$ conductors in which m conductors C_j are carrying current that flows into the paper and m conductors D_k in which the current flows out of the paper is given by:

$$L = \frac{N_c^2 \times l^2 \mu_0}{16\pi^2 S^2} \sum_{j=1}^{m} \sum_{k=1}^{m} \{K(C_j, D_k) + K(D_j, C_k) - K(C_j, C_k) - K(D_j, D_k)\} \quad (59)$$

where N_c is the number of turns, l is the length of the conductor outside the slot, S is the section of the conductor, and K represents a complex-valued function of eight complex variables, built from the plane coordinate of the vertices of the conductors. The complex function K is the result of successive surface integrations made to obtain the vector potential and from it the inductance by Equation (58). The function K is deduced from the elementary function

$$H(u, v) = -(1/24)\{(v^3 + u^3)^4 \log(v + u) + (v^3 - u^3)^4 \log(v - u)\} \quad (60)$$

by the algorithm described below.

From **H** the function K_0 is:

$$\begin{aligned} K_0 = \mathrm{Re}\,\{&\mathbf{H}(v_1, u_1) - \mathbf{H}(v_2, u_2) - \mathbf{H}(v_1, u_3) + \mathbf{H}(v_1, u_4) \\ &-\mathbf{H}(v_2, u_1) + \mathbf{H}(v_2, u_2) + \mathbf{H}(v_2, u_3) - \mathbf{H}(v_2, u_4) \\ &-\mathbf{H}(v_3, u_1) + \mathbf{H}(v_3, u_2) + \mathbf{H}(v_3, u_3) - \mathbf{H}(v_3, u_4) \\ &+\mathbf{H}(v_4, u_1) - \mathbf{H}(v_4, u_2) - \mathbf{H}(v_4, u_3) + \mathbf{H}(v_4, u_4)\} \end{aligned} \quad (61)$$

The complex variable v_i (respectively u_i) is defined from coordinates x_i, y_i (respectively x_{oi}, y_{oi}) of the first conductors C_j (respectively the second conductor D_k) which occur in the definition of function K by $v_i = (x_i + jy_i)^{1/3}$ (respectively $u_i = (x_{oi} + jy_{oi})^{1/3}$).

Figure 7.7 defines to coordinates x_i, y_i, x_{oi}, y_{oi}, etc.

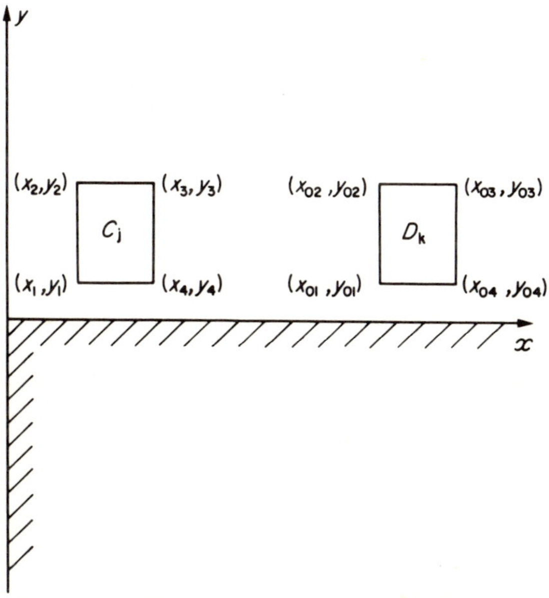

Figure 7.7 Definition of coordinates.

The function K is obtained from the difference:

$$K = K_0\,(v_1, v_2, v_3, v_4, u_1, u_2, u_3, u_4) - K_0\,(v_1^*, v_2^*, v_3^*, v_4^*, u_1, u_2, u_3, u_4) \quad (62)$$

where v_1^* is the complex conjugate of v_i.

The expression for the mutual inductance between two phases is more complicated. It is found by the principle of integration of the vector potential created by the conductors of a phase on the domain defined by the conductors of the other phases. This integration leads to formulae which are similar to Equations (59) to (62) but with a larger number of terms.

7.6.2 *Multiple layer windings*

The geometric representation of this type of winding is given in Figure 7.8; for the

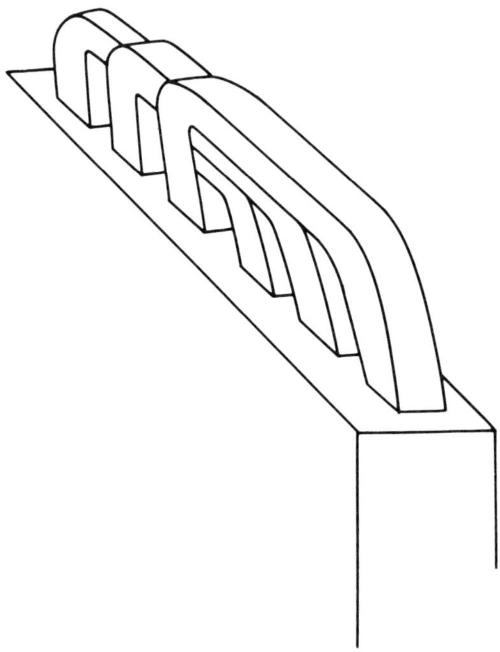

Figure 7.8 Three-layer wound stator.

single-layer winding the schematic is given in Figure 7.9. The application of magnetic image theory developed by Carpenter[15] leads to the mutual inductance of two coils by Equation (63) developed by Nicolas[16].

$$\mathbf{M} = \frac{\mu_0}{2}\left\{f\left[X\left(\frac{b+d}{f}\right) + X\left(\frac{b-d}{f}\right) - 2X\left(\frac{d}{f}\right)\right] - a\left[\log\frac{1}{(d^2+f^2)}\right.\right.$$

$$\left. + \tfrac{1}{2}\log[(b+d)^2+f^2] + \tfrac{1}{2}\log[(b-d)^2+f^2]\right] + 2\,(4a^2+f^2)^{\frac{1}{2}}\left[X\left(\frac{b+d}{(4a^2+f^2)^{\frac{1}{2}}}\right)\right.$$

$$+X\left(\frac{b-d}{(4a^2+f^2)^{\frac{1}{2}}}\right)-2X\left(\frac{d}{(4a^2+f^2)^{\frac{1}{2}}}\right)\Bigg]-[(b+d)^2+f^2]^{\frac{1}{2}}\left[1-X\left(\frac{2a}{[(b+d)^2+f^2]^{\frac{1}{2}}}\right)\right.$$

$$\left.+2X\left(\frac{a}{[(b+d)^2+f^2]^{\frac{1}{2}}}\right)\right]-[(b-d)^2+f^2]^{\frac{1}{2}}\left[1-X\left(\frac{2a}{[(b-d)^2+f^2]^{\frac{1}{2}}}\right)\right.$$

$$\left.+2X\left(\frac{d}{[(b-d)^2+f^2]^{\frac{1}{2}}}\right)\right]-(d^2+f^2)^{\frac{1}{2}}\left[1+3X\left(\frac{2a}{(d^2+f^2)^{\frac{1}{2}}}\right)-2X\left(\frac{a}{(d^2+f^2)^{\frac{1}{2}}}\right)\right]\Bigg\} \quad (63)$$

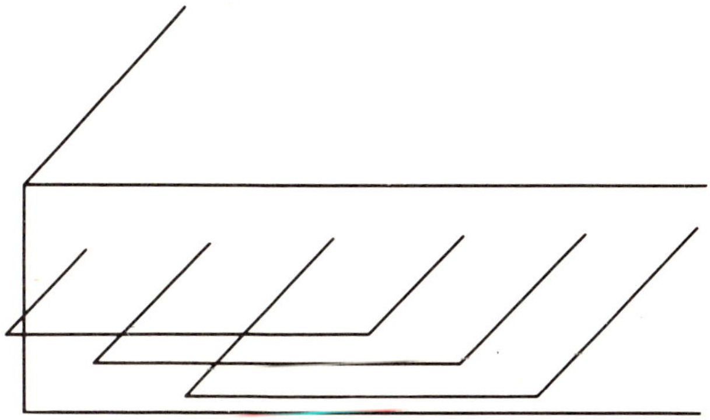

Figure 7.9 A model of the winding shown in Figure 7.8.

where $X(x,y) = x/y \sin h^{-1} (x/y) - [1 + (x/y)^2]^{-1/2}$ for which the notation is defined in Figure 7.10.

In both cases the formulae given in this section have been programmed on a computer and verified against experimental data. For Gramme ring windings one

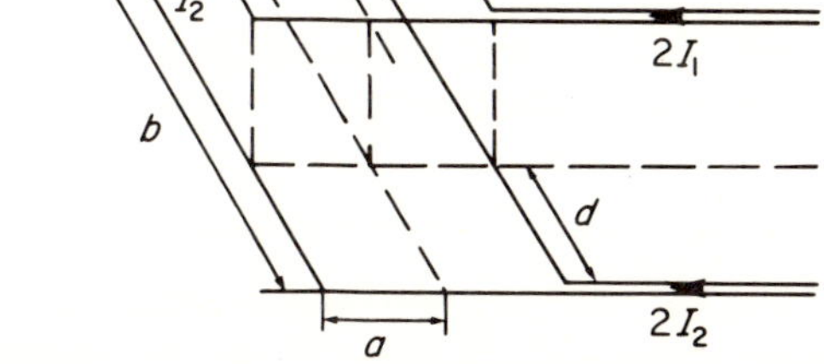

Figure 7.10 Notation for the evaluation of the mutual inductance between coils in 2 phases.

coil has been located near the corner of an open magnetic circuit (Figure 7.11). The computed and measured self-inductance are shown in Figure 7.12. For the multiple layer winding Nicolas[16, 17] has compared calculated and measured values of the

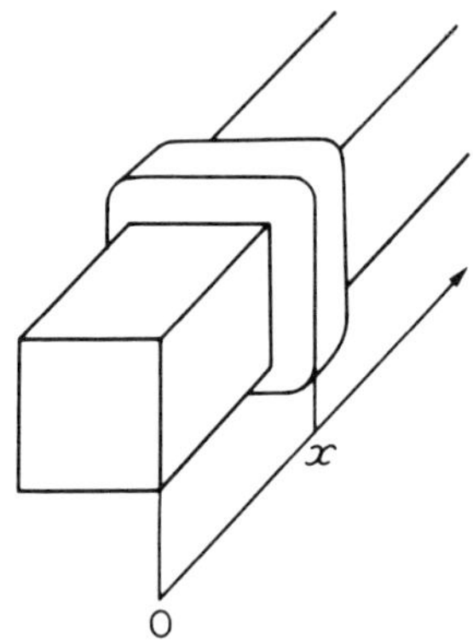

Figure 7.11 Experimental device for the measurement of self-inductance.

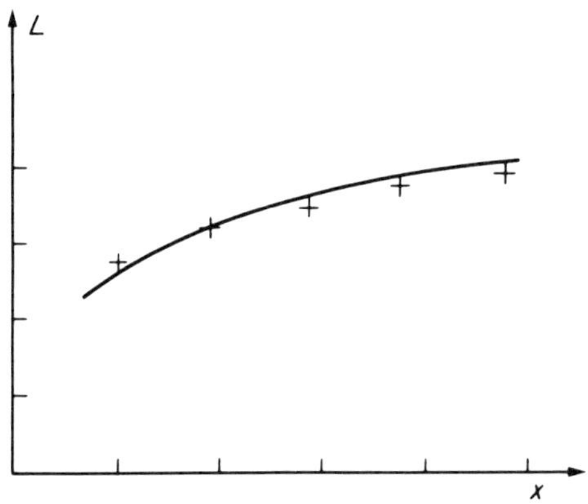

Figure 7.12 The self-inductance of the device shown in Figure 7.11:—— computed, ++++ measured.

three-phase inductance matrix of a small linear motor and obtained the following results for the computed matrix (a) and the measured one (b):

$$(a) = \begin{vmatrix} 320 & +72 & -195 \\ +72 & 320 & -195 \\ -195 & -195 & 320 \end{vmatrix}, (b) = \begin{vmatrix} 320 & +54 & -170 \\ +54 & 320 & -170 \\ -170 & -170 & 320 \end{vmatrix}$$

All the parts of the analysis have now been described. A discussion of its feasibility in practice and a comparison with experimental evidence now follows.

7.7 The theory is a good one on paper but how does it work?

The method described here has been programmed and implemented on a computer in order to test its feasibility, its accuracy and generally its value in terms of computer time consumption. Before comparison with experimental results, it should be mentioned that if the algorithm and the computer programme are very expensive in computer time, it raises some difficulties in programming. All the characteristics are given in terms of rapidly converging formulae and thus their computation is rapid, but the formulae are very complicated, having a lot of functions of several complex variables that are always a source of trouble in programming. The economy of the method in terms of computer time has proved to be most encouraging. In specific tests, its accuracy on several motors of various types and sizes has been adequate, especially as the comparisons were carried out on motors covering a wide range of power. Experimental data were provided either by manufacturers, by laboratory tests or from results published in the literature. Most of the comparisons presented below are extracted from the thesis of Nicolas[16] presented in April 1975. The first comparison concerns a small motor on which the experimental tests were made by Foggia[9] and Nicolas[16]. This 220-volt motor has a mechanical output of about two kilowatts. Figure 7.13 shows the variation of thrust with speed at constant current and is in good

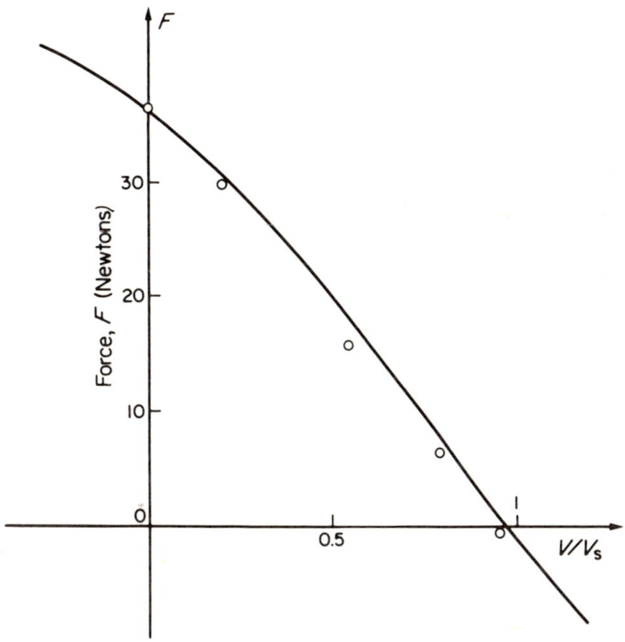

Figure 7.13 Thrust/speed graph: —— computed, ooooo measured

agreement with the theory, especially near the point of zero thrust. Figure 7.14 shows the variation in flux density inside the air gap at standstill. The errors can be attributed partly to the assumptions made and partly to the uncertainty in the exact location of the test point in the air gap.

The next comparisons, Figure 7.15, were made on a motor studied by Elliot[18]. Both construction details and experimental performance data were available for this motor. It is interesting to note that for small values of slip the theoretical method

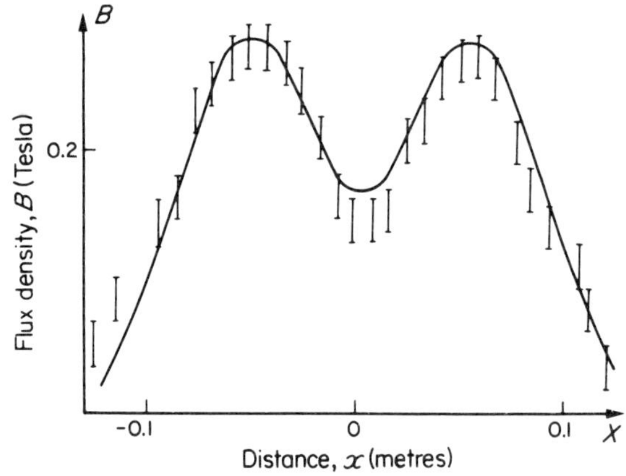

Figure 7.14 Flux density $B(x)$: —— computed, I measured.

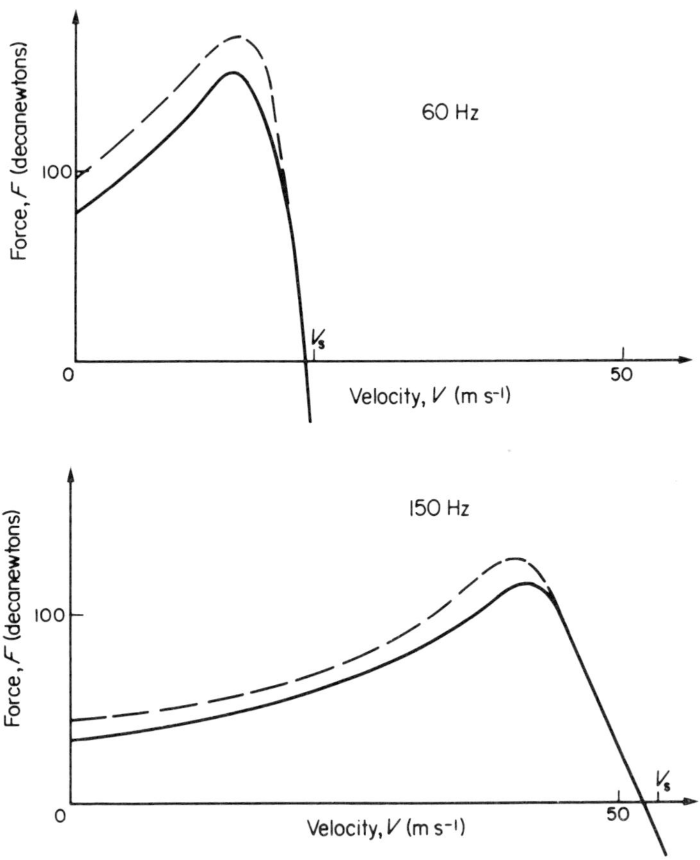

Figure 7.15 Two thrust/speed characteristics (Elliot) – – – – – experimental, ——— computated.

gives good results; there is, however, some loss of accuracy at points where the reaction field is important. This is a consequence of the assumptions which lead to a biased estimation of the reaction field with a negative deviation. Nonetheless, in the range of velocities around the nominal point of operation the method gives reliable results.

Finally Figures 7.16, 7.17, 7.18 and 7.19 are devoted to the complete analysis of

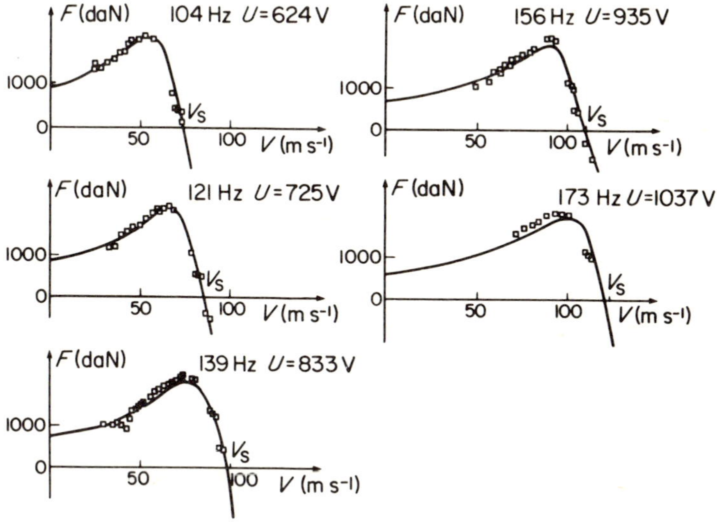

Figure 7.16 Garrett motor: constant voltage thrust/speed characteristic □□□□□ for experiment, ——— by computation.

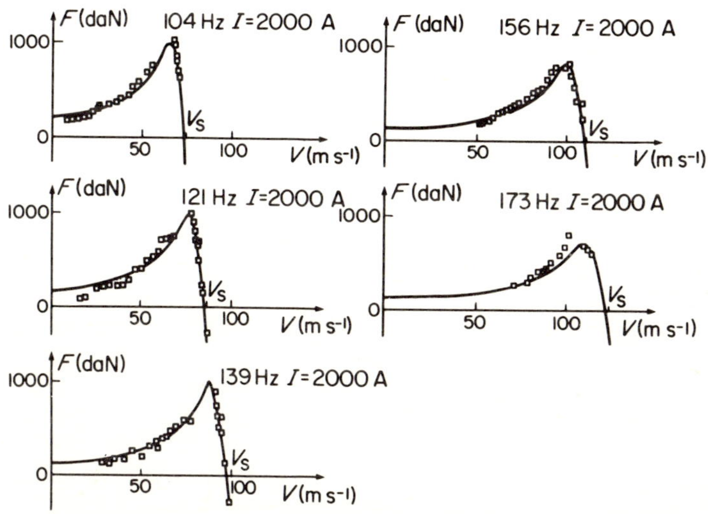

Figure 7.17 Garrett motor: constant current operation, thrust/speed curve □□□□□ experiment, ——— computation.

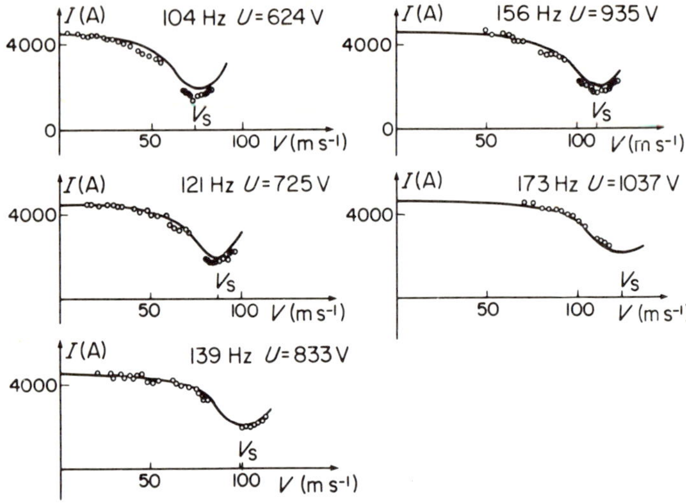

Figure 7.18 Garrett motor: input current at constant voltage operation ooooo experiment, ——— computation.

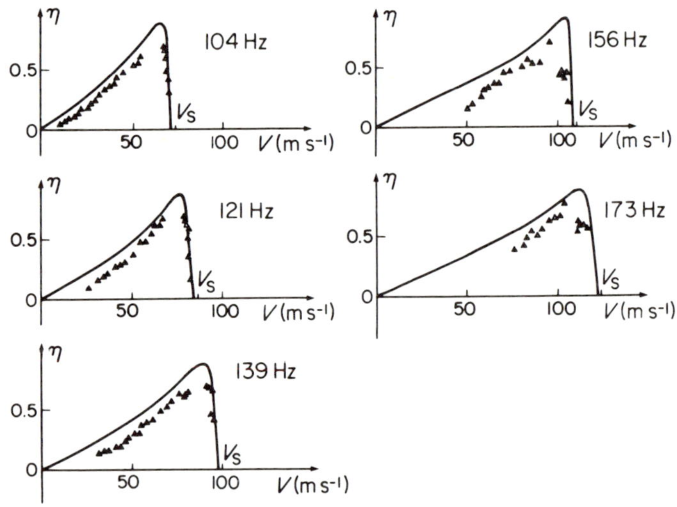

Figure 7.19 Garrett motor - efficiency/speed characteristic. △△△△△ experiment, ——— computation.

the LIMRV motor built and tested by the Garrett Company and the Federal Department of transportation in the USA. The experimental results were presented by Chirgwin[19] at the IEE Conference on linear motors in London in November 1974.

The computations were performed by Nicolas on a small computer. Figure 7.16 shows the thrust/speed curves for five different values of the supply frequency. For these results taken at constant voltage there is an almost unexpected degree of agreement over the range of speeds for which experimental data are available. However, at large values of slip, for which the theory appears to show some error in the previous example, no experimental data on this motor was available. Figure 7.17 is similar to Figure 7.16, but the results are taken at constant current. Figure 7.18 is a comparison between input current at constant voltage as a function of speed and confirms the previous results. Inspection of the latter results shows that the theory is sufficiently accurate for practical prediction of thrust at fixed input current and provides a good estimation of the impedance. Examination of Figure 7.19 on which the efficiency η is plotted as a function of speed, reveals some disagreement between computed and experimental data. The differences can be explained in the same way as that described in the estimation of parameters such as the primary resistance or exact shape of the end winding, or the resistivity of the reaction rail. Although a rough estimate of these parameters does not greatly affect the modulus of the impedance, it is of the greatest importance in respect of its argument. An error in estimating the argument of the impedance affects the power factor, which in turn alters the value of efficiency.

The last evaluation was made by Pacaut [20] on a single sided linear motor with a laminated back iron yoke and a variable number of poles on the primary. These results are a confirmation of the test results shown in Figure 7.15, that is to say, a good agreement at high speeds with some divergence at low speeds (Figure 7.20).

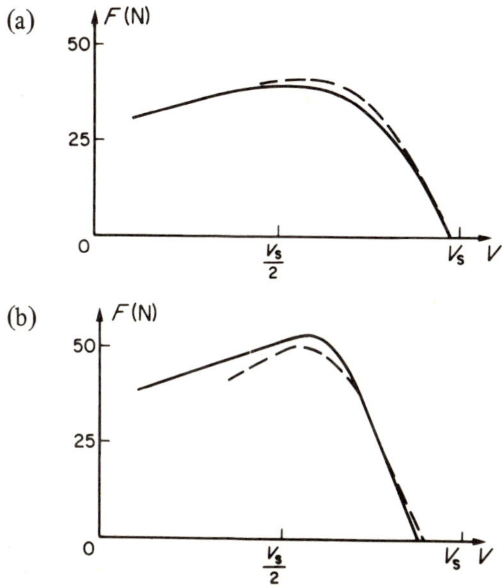

Figure 7.20 Pacaut motor: (a) two poles, (b) four poles ----- experiment, ——— computation.

7.8 Conclusions

The importance of a reliable model for the prediction of the electrical and mechanical output characteristics of a linear induction motor is now widely accepted by the engineers involved in advanced transportation systems.

A method of computation has been explained that relates the modulation function technique for end-effect analysis and the three-dimensional solution of the edge effect by means of specific formula. The additional methods of evaluation of the leakage inductance is necessary to provide a complete programme for the analysis of the motor under constant voltage operation. The comparison between the performance predicted by the theory and the experimental data has been made on several motors of different type, size, and nominal speed. The most significant results only were presented here to bring out the main features of the method. This survey points out that the theoretical method is economical in terms of computer time, a little difficult with regard to programming and is reliable in the higher speed range, but it is somewhat inaccurate at standstill. A more accurate method, valid in the whole speed range must consist of a truly three-dimensional analysis which takes into account the flux variations in the three dimensions. Such a method is likely to be very expensive and would be of interest solely for extending the accuracy of the cheaper and quite reliable method just described. Such three-dimensional analysis made by the author and analysed by Nicolas [16] will, if the advance in high speed transportation systems continues, be published with more details after some refinements have been made on the computer programme.

Appendix: Evaluation of the modulation function

The problem of evaluation of the modulation function involves the calculation of the flux density created by a current-carrying conductor located in an air gap of finite length. The schematic of the problem is shown in Figure 7.21 for the upper half of the air gap. The evaluation of the field created by the conductor carrying the current *I* is made by division of the region into 5 zones as defined in Figure 7.21. The analysis of the field is dependant on the zone of location of the conductor.

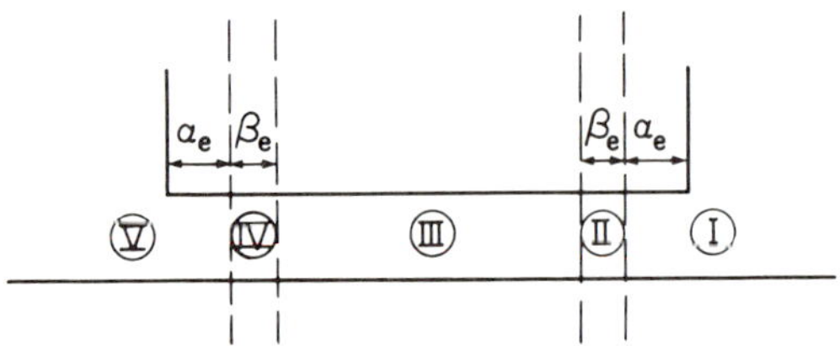

Figure 7.21 The five zones model.

(i) *Conductor in zone* I

In zone I and II the field is calculated for a conductor located in an air gap which is assumed to be infinite on the left.

In zone III the field is constant and has the value of the field at the boundary of zones II and III.

In zones IV and V the field is calculated assuming the air gap to be infinite on the right.

(ii) *Conductor in zone* II

In zones I and II the field is again computed by assuming the air gap to be infinite on the left.

In zones II and III the field of a conductor in an infinite air gap is evaluated. An additional condition ensures the continuity of the field and of its first derivative on both sides of the conductor.

In zones IV and V the air gap is assumed to be infinite on the right and the field is assumed to be continuous at the boundaries of regions III and IV.

(iii) *Conductor in zone* III

Zone I: the air gap is assumed to be infinite on the left, and the field is to be continuous at the boundaries of zones I and II.

Zones II, III and IV: the field appropriate to a conductor in an infinite air gap is calculated.

Zone V: the air gap is assumed to be infinite on the right and the field is continuous at the boundaries of regions I and II.

(iv) *Conductor in zones* IV *and* V

It is obvious that the solution for a conductor located in zones IV and V will be similar to the solution obtained for a conductor respectively in regions II and I. The solution of the field in the geometries described above will be obtained by superposition of the three basic solutions.

(a) *Infinite air gap* The solution of the field created by a conducting sheet containing a uniform current I is given by the classical formula (see Figure 7.22).

$$B_z = \frac{\mu_0 I}{4\pi b} \log \left\{ \frac{\cosh\,[\pi\,(x - x_0 - b)/e] + \cos \pi\, z/e}{\cosh\,[\pi\,(x - x_0 + b)/e] + \cos \pi\, z/e} \right\} \tag{64}$$

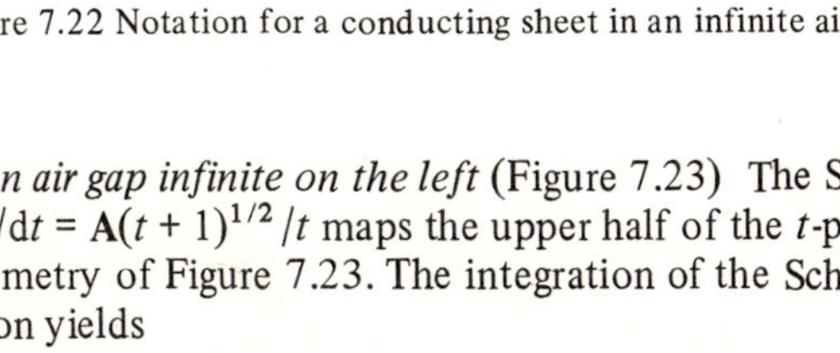

Figure 7.22 Notation for a conducting sheet in an infinite air gap.

(b) *Conductor in an air gap infinite on the left* (Figure 7.23) The Schwartz-Christoffel transformation $\mathrm{d}z/\mathrm{d}t = A(t+1)^{1/2}/t$ maps the upper half of the t-plane on the z-plane defined by the geometry of Figure 7.23. The integration of the Schwartz-Christoffel differential equation yields

$$z = \frac{2e}{\pi}\left\{(t+1)^{1/2} - \tfrac{1}{2}\log\left[\frac{1+(t+1)^{1/2}}{-1+(t+1)^{1/2}}\right]\right\} \tag{65}$$

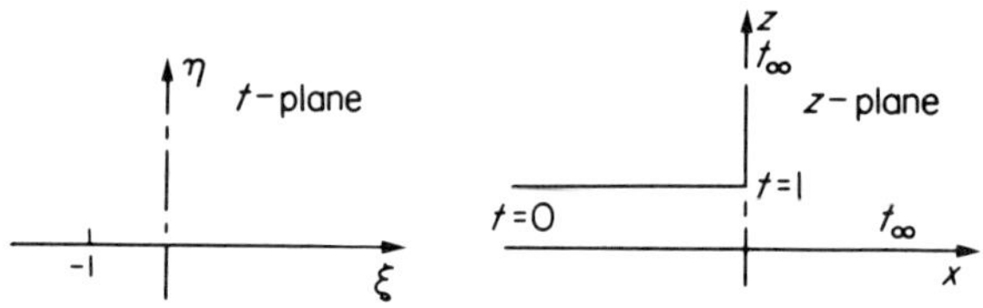

Figure 7.23 Conformal mapping with infinite air gap on the left.

The complex potential created in the t-plane by a conductor at the point t_0 is:

$$W = \frac{I}{2\pi} \log (t - t_0)(t - t_0) \tag{66}$$

and provides the field [14]

$$H_z = -\operatorname{Re}\left(\frac{\mathrm{d}W}{\mathrm{d}z}\right) \tag{67}$$

Numerical inversion of Equation (65) gives, for each point in the air gap, the corresponding value of t and thus the value of H_z derived from this value of t by Equation (67).

(c) *Conductor in an air gap infinite on the right* (Figure 7.24) The Schwartz-Christoffel transformation is now $\mathrm{d}z/\mathrm{d}t = A(1-t)^{1/2}/t$ which leads to the mapping

$$z = -\frac{2e}{\pi}\left\{(1-t)^{1/2} - 1/2 \log\left[\frac{1+(1-t)^{1/2}}{1-(1-t)^{1/2}}\right]\right\} \tag{68}$$

Calculation of the field is effected as for (b).

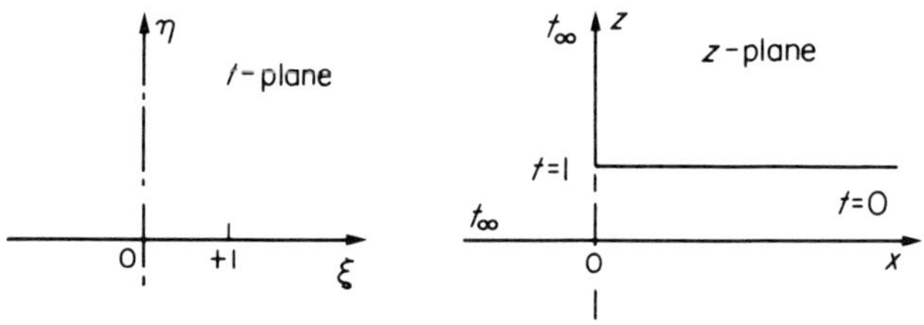

Figure 7.24 Conformal mapping with infinite air gap on the right.

The coefficients α_e and β_e of Figure 7.21 will be chosen to ensure a given accuracy, but generally it appears that, beyond a distance equal to twice the air gap length, the field is virtually equal to the value NI/e given by Ampere's theorem (Figure 7.25).

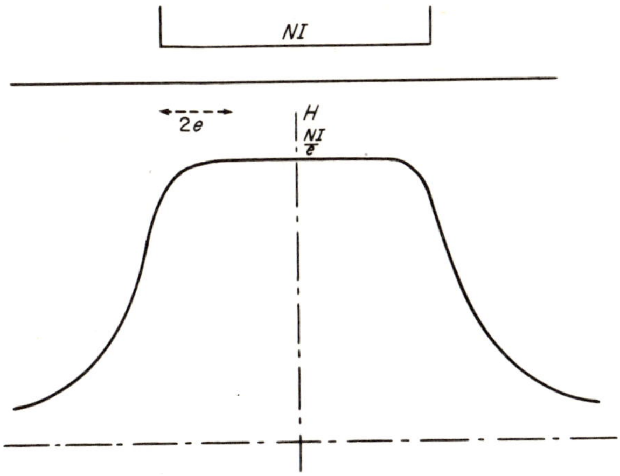

Figure 7.25 Field plot for a two-pole motor.

References

[1] Ivanes, M., Baudon, Y. and Lhenry, M. (1972) 'Suspension non aérodynamiques', Rapport Contrat D.G.R.S.T. No. 70-7-2520.

[2] Lhenry, M. (1973) 'Contribution à l'étude des suspensions magnétiques, lévitation et guidage', Thèse de Docteur-Ingénieur, Grenoble 1973.

[3] Baudon, Y. (1974) 'Contribution à l'étude des forces dans les systèmes de sustentation à coussin magnétique', Thèse de Docteur-Ingénieur, Grenoble 1974.

[4] Baudon, Y. and Ivanes, M. (1974) 'Méthode de partition pour le calcul des courants induits dans les sustentations par attraction', Symposium International sur les moteurs électriques linéaires, Lyon, May 1974.

[5] Lhenry, M., Gilles, G. and Ivanes, M. (1974) 'Magnetic suspension of transport vehicles. Control aspect of guided land transport', IEE Conference Brighton, August 1974.

[6] Baudon, Y., Ivanes, M. and Sabonnadière, J. C. (1973) 'Courants de Foucault produits par un electroaimant en mouvement', *Comptes rendus à l'Académie des Sciences*, Paris, **276**, Série B, pp. 817–820.

[7] Baudon, Y.. Ivanes, M. and Sabonnadière, J. C. (1973) 'Evaluation des forces de lévitation et de trainée dans les suspensoins magnétiques par électroaimant', Symposio sui motori lineare, Capri, June 1973.

[8] Laithwaite, E. R. (1975) 'Linear electric machines, A personal view', *Proceedings IEEE*, February 1975, pp. 250–290.

[9] Foggia, A. (1974) 'Contribution à l'étude de l'induction magnétique dans les convertisseurs électromécaniques linéaires', Thèse de Doctorat es Sciences, Lyon.

[10] Foggia, A., Sabonnadière, J. C. and Silvester, P. (1975) 'Finite element solution of saturated travelling field problems', IEEE Winter Power Meeting, January 1975, paper No. 7 75 030-2.

[11] Sabonnadière, J. C. (1969) 'Contribution à l'étude des moteurs asynchrones linéaires', Thèse de Doctorat es Sciences, Grenoble 1969.

[12] Jufer, M. and Wavre, N. (1972) 'Le moteur linéaire. Développement théorique et applications', *Bulletin de l'Association Suisse des Electriciens*, T. 63, No. 15, pp. 844–856.

[13] Oberretl, K. (1973) 'Dreidimensionale Berechnung des Lineamotors mit Berücksichtigung der Endeffekte und der Wicklungsverteching', Symposio sui motori lineare, Capri, June 1973.

[14] Binns and Lawrenson (1965) *Analysis and Computation of Electric and Magnetic Field Problems,* Pergamon Press, Oxford.

[15] Carpenter, J. (1968) 'The application of the method of images to machine end winding fields', *Proc. of the Inst. Elec. Eng.*, **107.c**, October 1960.

[16] Nicolas, A. (1975) 'Etude des phénomènes électromagnétiques dans les moteurs asynchrones linéaires', Thèse de Docteur-Ingénieur, Lyon 1975.

[17] Nicolas, A. and Sabonnadière, J. C. (1974) 'Computation of constant voltage operation of linear induction motors', IEE Conference on linear electric machines, London 1974.

[18] Elliot, D. (1972) 'Numerical analysis method for linear induction machines', 12th Symposium on the engineering aspects of magnetohydrodynamics, Argonne, Illinois, March 1972.

[19] Chirgwin, K. M. (1974) 'Tests results from the U.S. linear induction motor research vehicle program', IEE Conference on linear machines, London 1974.

[20] Pacaut, R. (1975) 'Etude des forces dans une machine linéaire à induction', Thèse de 3ème Cycle, Grenoble 1975.

[21] Kant, M. (1970) 'Application de la théorie du champ électromagnétique aux machines linéaires à induction', *Revue de Physique Appliquée*, Octobre 1970.

8

Analysis of Linear Induction Motors Using a Space Harmonic Technique

S. NONAKA K. YOSHIDA

8.1 Introduction

Linear induction motors (LIMs) have recently become of major interest as one of the most suitable propulsion systems for high-speed ground transportation [1].

At this stage, it is proposed to establish an analysis method and performance equations suitable for digital computer calculations, in order to evaluate the characteristics and to be able to design an optimum LIM. In the analysis of LIMs in which end effect is taken into account, Yamamura has used a Fourier transformation technique [2], Nasar a relaxation method [3] and Kliman and Elliott a mesh/matrix method [4]. Their analyses, however, seem to be insufficient for practical designs.

The authors have proposed a space harmonic analysis of LIMs by which, not only the end effect, but also the actual design parameters can be taken into account [5, 6]. The method is applied to the short-primary double-sided LIM where the secondary conducting plate is running symmetrically, or displaced from the centre of the air gap, between the two primaries, and the performance equations suitable for numerical calculations are derived directly from Maxwell's equations [7–9]. Independently of these studies, similar analytical approaches have been attempted by North [10] and Oberretl [11].

In this chapter, a space harmonic analysis of LIMs leads to a clear theoretical indication that it is possible to improve extensively the performance of high-speed LIMs by making a proper selection of the design parameters [12]. Experimental verification of the theory is included.

8.2 Space harmonic analysis of LIMs

8.2.1 *Analytical model of short-primary double-sided LIM*

(a) *Even number of poles on each side*

(i) *Analytical model* Figure 8.1 shows a model for the analysis of short-primary, double-sided LIMs. The actual length of the primary laminated cores is taken to be approximately L, which is the length of the primary windings. In the model, however, the primary cores are extended to infinity in both directions of the x coordinate. The primary winding blocks are separated by a sufficiently long distance L_c that the magnetic interaction between two adjacent windings can be neglected. Thus the repeatable section is $2L_1 = L + L_c$ as shown. The m.m.f. wave produced by the

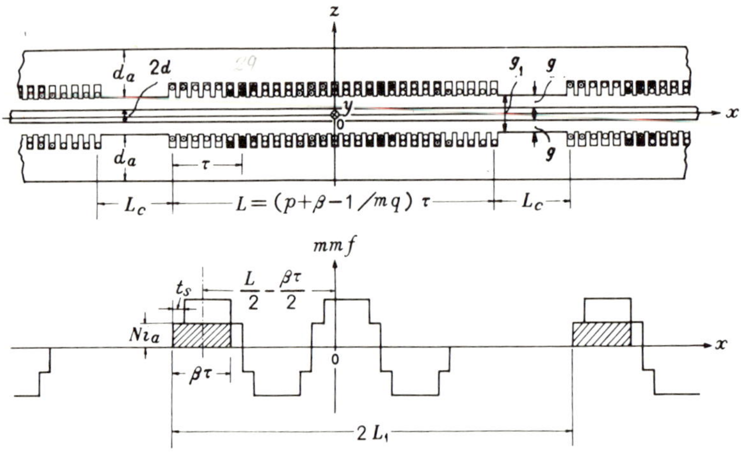

Figure 8.1 Model of short-primary LIM and distribution of the phase a m.m.f. ($m = 3$, $p = 4, q = 2$, $\beta = 5/6$).

primary currents consists of the fundamental space component of equivalent pole pitch L_1 and an infinite number of space harmonic components.

In Figure 8.1, m is the number of phases, p the number of poles on each side, q the number of slots per pole and phase, τ the pole pitch, $\beta\tau$ the coil pitch, t_s the slot pitch, $2d$ the thickness of the secondary conducting plate, g the mechanical clearance of each air gap, g_1 the length of primary-to-primary air gap, N the number of turns in each coil and i_a the instantaneous value of the phase (the first phase a) current. A normal double-layer winding with half-filled end-slots is considered.

(ii) *The m.m.f. distribution* The m.m.f. waveform of the first coil of the first pole coil group of the phase a is indicated by hatching in Figure 8.1. If the origin of the x coordinate is at the centre of the length L, the Fourier series expansion of this m.m.f. is:

$$f_{a,\,p1-q1}(x) = Ni_a \left\{ \frac{\beta\tau}{2L_1} + \frac{2}{\pi} \sum_{n=1}^{\infty} \frac{1}{n} k_{pn} \cos \frac{\pi}{\tau_{1n}} \left[x + \left(q - \frac{1}{mq} \right) \frac{\tau}{2} \right] \right\} \tag{1}$$

where $n = 1, 2, 3, \ldots$ indicates the order of the space harmonic component and $\tau_{1n} = L_1/n$ the equivalent pole pitch of the nth space harmonic m.m.f., and

$$k_{pn} = \sin \frac{\pi}{2} \frac{\beta\tau}{\tau_{1n}} \tag{2}$$

By summation from the first coil to the qth coil, the m.m.f. of the first pole of phase a is obtained as follows:

$$f_{a,\,p1}(x) = qNi_a \left\{ \frac{\beta\tau}{2L_1} + \frac{2}{\pi} \sum_{n=1}^{\infty} \frac{1}{n} k_{pn} k_{dn} \cos \frac{\pi}{\tau_{1n}} \left[x + \left(p - \frac{1}{m} \right) \frac{\tau}{2} \right] \right\} \tag{3}$$

where,

$$k_{\mathrm{dn}} = \frac{\sin q \dfrac{\pi}{2} \dfrac{t_{\mathrm{s}}}{\tau_{1\mathrm{n}}}}{q \sin \dfrac{\pi}{2} \dfrac{t_{\mathrm{s}}}{\tau_{1\mathrm{n}}}} \tag{4}$$

Further, by summation from the first pole to the pth pole, the m.m.f. of phase a can be expressed as:

$$f_{\mathrm{a}}(x) = -pqNi_{\mathrm{a}}\frac{2}{\pi}\sum_{n=1}^{\infty}\frac{1}{n}k_{\mathrm{wn}}k_{\mathrm{Lpn}}\sin\frac{\pi}{\tau_{1\mathrm{n}}}\left(x + \frac{m-1}{m}\frac{\tau}{2}\right) \tag{5}$$

where,

$$k_{\mathrm{wn}} = k_{\mathrm{pn}}k_{\mathrm{dn}} \tag{6}$$

$$k_{\mathrm{Lpn}} = \frac{\sin p \dfrac{\pi}{2} \dfrac{\tau}{\tau_{1\mathrm{n}}}}{p\cos \dfrac{\pi}{2} \dfrac{\tau}{\tau_{1\mathrm{n}}}} \tag{7}$$

Here, it is assumed that the m-phase current is balanced. Then the vth phase current is expressed by:

$$i_v = 2^{\frac{1}{2}}I_1 \sin\{\omega_1 t - (v-1)\,\pi/m\} \tag{8}$$

where I_1 denotes the effective value of each phase current, $\omega_1 = 2\pi f_1$, the angular supply frequency, and $v = 1, 2, \ldots m$. Therefore, the resultant m.m.f. $f(x, t)$, which is set up by the m-phase current flowing in the m-phase winding on one side of the primary, is given by summation over v as follows:

$$f(x,t) = -\sum_{v=1}^{m} pqNi_v(t)\frac{2}{\pi}\sum_{n=1}^{\infty}\frac{1}{n}k_{\mathrm{wn}}k_{\mathrm{Lpn}}\sin\frac{\pi}{\tau_{1\mathrm{n}}}\left\{x + \frac{m-1}{m}\frac{\tau}{2} - (v-1)\frac{\tau}{m}\right\} \tag{9}$$

Substituting Equation (8) into Equation (9) and replacing $\{\omega_1 t + (m+1)\,\pi/2m\}$ by $\omega_1 t$:

$$f(x,t) = \frac{2^{\frac{1}{2}}}{\pi}\, m\frac{N_{\mathrm{ph}}}{2}I_1 \sum_{n=1}^{\infty}\frac{1}{n}k_{\mathrm{wn}}k_{\mathrm{Lpn}}\left\{k_{\mathrm{mfn}}\cos\left(\omega_1 t - \frac{\pi}{\tau_{1\mathrm{n}}}x\right) - k_{\mathrm{mbn}}\cos\left(\omega_1 t + \frac{\pi}{\tau_{1\mathrm{n}}}x\right)\right\} \tag{10}$$

where N_{ph} is the number of series turns per phase, assuming that the two sides of the double-sided primary windings are in series and star-connected; then $N_{\mathrm{ph}} = 2pqN$, and k_{mfn} and k_{mbn} are, respectively:

$$k_{\mathrm{mfn}} = \frac{\sin\dfrac{\pi}{2}\left(1 - \dfrac{\tau}{\tau_{1\mathrm{n}}}\right)}{m\sin\dfrac{\pi}{2m}\left(1 - \dfrac{\tau}{\tau_{1\mathrm{n}}}\right)} \quad \text{and } k_{\mathrm{mbn}} = \frac{\sin\dfrac{\pi}{2}\left(1 + \dfrac{\tau}{\tau_{1\mathrm{n}}}\right)}{m\sin\dfrac{\pi}{2m}\left(1 + \dfrac{\tau}{\tau_{1\mathrm{n}}}\right)} \tag{11}$$

(b) *Odd number of poles on each side*

(i) *Analytical model* LIMs with an odd number of poles are also feasible. In fact, two 5-pole double-sided LIMs are mounted in tandem under the tracked levitated research vehicle (TLRV) of the US Department of Transportation, which was formerly called the TACRV.

Figure 8.2 shows an analytical model of the short-primary LIM with an odd

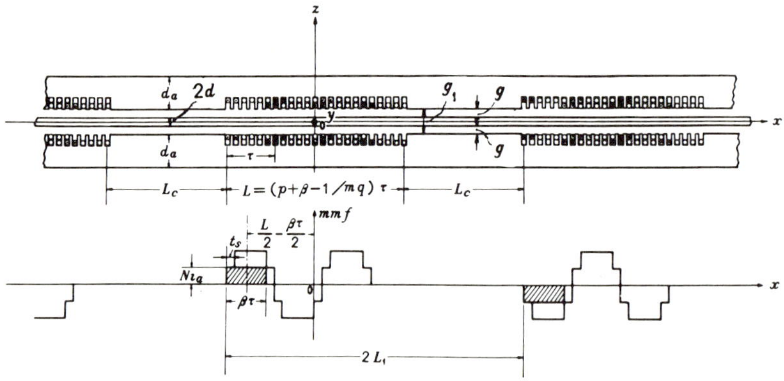

Figure 8.2 Model of short-primary LIM and distribution of the phase a m.m.f. ($m = 3$, $p = 3$, $q = 2$, $\beta = 5/6$).

number of poles[13]. Adjacent primary windings are arranged for the m.m.f. distributions to take a reverse direction to each other. Consequently, the space period of the m.m.f. distribution is equal to $2(L + L_c)$.

(ii) *The m.m.f. distribution* The Fourier series for the m.m.f. of the first coil of the first pole coil group of the phase a, as shown by hatching in Figure 8.2 can be expressed:

$$f_{a,\,p1-q1}(x) = Ni_a \frac{4}{\pi} \sum_{n=1,\,3}^{\infty} \frac{1}{n} k_{pn} \cos \frac{\pi}{\tau_{1n}} \left\{ x + \left(p - \frac{1}{mq} \right) \frac{\tau}{2} \right\} \tag{12}$$

where $n = 1, 3, 5, \ldots$, indicates the order of the space harmonic component and $\tau_{1n} = 2L_1/n$ the equivalent pole pitch of the nth space harmonic m.m.f. which is seen to be twice that of the model shown in Figure 8.1, and k_{pn} is given by the same expression as Equation (2).

Then the m.m.f. of the first pole of the phase a is:

$$f_{a,\,p1}(x) = qNi_a \frac{4}{\pi} \sum_{n=1,\,3}^{\infty} \frac{1}{n} k_{pn} k_{dn} \cos \frac{\pi}{\tau_{1n}} \left\{ x + \left(p - \frac{1}{m} \right) \frac{\tau}{2} \right\} \tag{13}$$

where k_{dn} is also expressed as in Equation (4).

When p is an odd number, carrying out the summation from the first pole to the pth pole:

$$f_a(x) = pqNi_a \frac{4}{\pi} \sum_{n=1,\,3}^{\infty} \frac{1}{n} k_{wn} k_{Lpn} \cos \frac{\pi}{\tau_{1n}} \left(x + \frac{m-1}{m} \frac{\tau}{2} \right) \tag{14}$$

where,

$$k_{\mathrm{Lpn}} = \frac{\cos p \dfrac{\pi}{2}\dfrac{\tau}{\tau_{1\mathrm{n}}}}{p \cos \dfrac{\pi}{2}\dfrac{\tau}{\tau_{1\mathrm{n}}}} \tag{15}$$

The analytical model shown in Figure 8.2 is also generally applicable to an even number of poles on each side.

When p is an even number, similarly:

$$f_{\mathrm{a}}(x) = -pqNi_{\mathrm{a}}\frac{4}{\pi}\sum_{n=1,3}^{\infty}\frac{1}{n}k_{\mathrm{wn}}K_{\mathrm{Lpn}}\sin\frac{\pi}{\tau_{1\mathrm{n}}}\left(x + \frac{m-1}{m}\frac{\tau}{2}\right) \tag{16}$$

where k_{Lpn} is given as in Equation (7).

Consequently, the resultant m.m.f. $f(x, t)$ can be expressed as follows:

$$f(x, t) = 2\frac{2^{\frac{1}{2}}}{\pi} m\frac{N_{\mathrm{ph}}}{2} I_1 \sum_{n=1,3}^{\infty}\frac{1}{n}k_{\mathrm{wn}}k_{\mathrm{Lpn}}\left\{k_{\mathrm{mfn}}\cos\left(\omega_1 t - \frac{\pi}{\tau_{1\mathrm{n}}}x\right) \pm k_{\mathrm{mbn}}\cos\left(\omega_1 t + \frac{\pi}{\tau_{1\mathrm{n}}}x\right)\right\} \tag{17}$$

where $N_{\mathrm{ph}} = 2pqN$, k_{mfn} and k_{mbn} are given by the expressions as in Equation (11), the plus sign being applicable for odd values of p and the minus sign for even values.

While for an even value of p either Equation (10) or (17) can be used, for an odd number a solution can only be obtained using Equation (17).

8.2.2 *Primary surface current*

The surface current loading j_1, which produces the same m.m.f. as does $f(x, t)$, given by Equation (10) can be introduced:

$$j_1 = \sum_{n=1}^{\infty} j_{1,\mathrm{n}} = \sum_{n=1}^{\infty}(j_{1,\mathrm{fn}} + j_{1,\mathrm{bn}}), \tag{18}$$

where,

$$j_{1,\mathrm{fn}} = J_{1,\mathrm{fn}} \exp \mathrm{j}\,(\omega_1 t - \pi x/\tau_{1\mathrm{n}}) \tag{19}$$

$$j_{1,\mathrm{bn}} = J_{1,\mathrm{bn}} \exp \mathrm{j}\,(\omega_1 t + \pi x/\tau_{1\mathrm{n}}) \tag{20}$$

that is, $j_{1,\mathrm{n}}$ indicates the nth space harmonic surface current, $j_{1,\mathrm{fn}}$ the nth forward-travelling surface current, $j_{1,\mathrm{bn}}$ the nth backward-travelling surface current and their travelling speed $v_{1\mathrm{n}} = 2\tau_{1\mathrm{n}}f_1$.

The maximum surface current densities $J_{1,\mathrm{fn}}$ and $J_{1,\mathrm{bn}}$ are related to the actual primary winding current I_1 [14], by the following equations:

$$J_{1,\mathrm{fn}} = 2^{\frac{1}{2}} m\frac{k_{\mathrm{wn}}k_{\mathrm{Lpn}}N_{\mathrm{ph}}}{2L_1}k_{\mathrm{mfn}}I_1 \tag{21}$$

$$J_{1,\mathrm{bn}} = 2^{\frac{1}{2}} m\frac{k_{\mathrm{wn}}k_{\mathrm{Lpn}}N_{\mathrm{ph}}}{2L_1}k_{\mathrm{mbn}}I_1 \tag{22}$$

Similarly, the surface current j_1 corresponding with $f(x, t)$ given by Equation (17) becomes:

$$j_1 = \sum_{n=1,3}^{\infty} j_{1,n} = \sum_{n=1,3}^{\infty} (j_{1,fn} \mp j_{1,bn}) \tag{23}$$

where the minus and plus signs are applied for odd and even values of p, respectively. $j_{1,fn}, j_{1,bn}, J_{1,fn}$ and $J_{1,bn}$ are given by the same expressions as those in Equations (19) to (22), respectively.

8.2.3 *Two-dimensional model excited by each space harmonic of surface current*

Since the following analysis is similar both for the surface currents given by Equation (18) and for Equation (23), the theoretical procedure shown here is worked only for the former.

(a) *Single secondary conducting plate* There is no restriction relating to the short primary for each surface current component. Thus a two-dimensional model as shown in Figure 8.3 can be used.

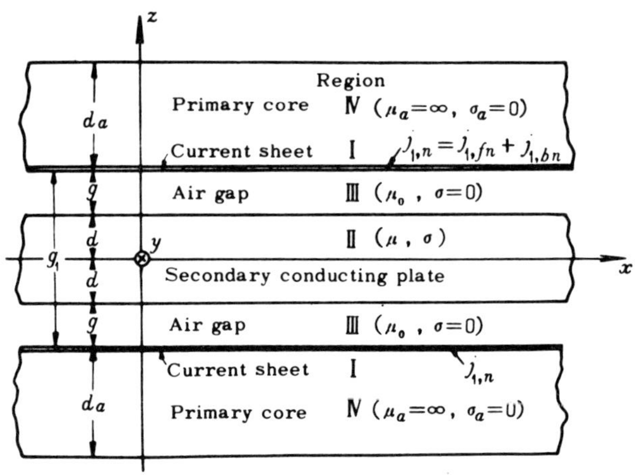

Figure 8.3 Two-dimensional model for analysis of the motor excited by the nth space harmonic of surface current $j_{1,n}$.

In such a model, it is assumed that the nth harmonic surface current $j_{1,n}$ flows uniformly in the y-direction and that the secondary conducting plate with conductivity σ and permeability μ travels at a constant speed v_2 in the x-direction, independently of the harmonic order.

In the two-dimensional model, the governing equation of the magnetic field due to $j_{1,n}(=j_{1,fn}+j_{1,bn})$ is expressed by the y-component of the vector potential as follows:

$$\frac{\partial^2 A_{y,n}}{\partial x^2} + \frac{\partial^2 A_{y,n}}{\partial z^2} = \sigma\mu \left(\frac{\partial A_{y,n}}{\partial t} + v_2 \frac{\partial A_{y,n}}{\partial x} \right) \tag{24}$$

On the basis of the boundary condition at $z = d + g$, the vector potential for both forward- and backward-travelling components can be expressed in the following form of separated variables.

$$A_{y,\,n} = A_{y,\,fn} + A_{y,\,bn} = \{A_{y,\,fn}(z)\exp(-j\pi x/\tau_{1n}) + A_{y,\,bn}(z)\exp(j\pi x/\tau_{1n})\}\exp(j\omega_1 t) \tag{25}$$

The solution in each region is obtained in the same way. This is given in detail in reference[15]. The resultant magnetic fields due to j_1 can be obtained by summation of those solutions. Hence the x- and z-components of the flux density in the air gap (region III) at the surface of the primary are obtained:

$$B_{x1}^{III} = \sum_{n=1}^{\infty} B_{x1,\,n}^{III} = \sum_{n=1}^{\infty}\left[-\frac{\partial A_{y,\,n}^{III}}{\partial z}\right]_{z=d+g}$$

$$= -\mu_0 \sum_{n=1}^{\infty} (j_{1,\,fn} + j_{1,\,bn})$$

$$= -\mu_0 \sum_{n=1}^{\infty} 2^{\frac{1}{2}} m \frac{k_{wn}k_{Lpn}N_{ph}}{2L_1} I_1\,[k_{mfn}\exp(-j\pi x/\tau_{1n})$$

$$+ k_{mbn}\exp(j\pi x/\tau_{1n})]\exp(j\omega_1 t) \tag{26}$$

$$B_{z1}^{III} = \sum_{n=1}^{\infty} B_{z1,\,n}^{III} = \sum_{n=1}^{\infty}\left[\frac{\partial A_{y,\,n}^{III}}{\partial x}\right]_{z=d+g}$$

$$= -j\mu_0 \sum_{n=1}^{\infty} (K_{Hfn}\,j_{1,\,fn} - K_{Hbn}\,j_{1,\,bn})$$

$$= -j\mu_0 \sum_{n=1}^{\infty} 2^{\frac{1}{2}} m \frac{k_{wn}k_{Lpn}N_{ph}}{2L_1} I_1\,[k_{mfn}K_{Hfn}\exp(-j\pi x/\tau_{1n})$$

$$- k_{mbn}K_{Hbn}\exp(j\pi x/\tau_{1n})]\exp(j\omega_1 t) \tag{27}$$

where,

$$K_{Hfn} = \coth\frac{\pi}{\tau_{1n}}(g + d\lambda'_{fn}) \tag{28}$$

λ'_{fn} is determined by the following equations:

$$\frac{\pi}{\tau_{1n}} d\lambda'_{fn} = \tanh^{-1}\left(\frac{\mu_0}{\mu}\lambda_{fn}\tanh\frac{\pi}{\tau_{1n}} d\lambda_{fn}\right) \tag{29}$$

$$\lambda_{fn} = (1 + j\sigma\mu s_{fn} v_{1n}\tau_{1n}/\pi)^{\frac{1}{2}} \tag{30}$$

$$s_{fn} = (v_{1n} - v_2)/v_{1n} \tag{31}$$

and similarly,

$$K_{\mathrm{Hbn}} = \coth \frac{\pi}{\tau_{1\mathrm{n}}} (g + d\lambda'_{\mathrm{bn}}) \tag{32}$$

$$\frac{\pi}{\tau_{1\mathrm{n}}} d\lambda'_{\mathrm{bn}} = \tanh^{-1} \left(\frac{\mu_0}{\mu} \lambda_{\mathrm{bn}} \tanh \frac{\pi}{\tau_{1\mathrm{n}}} d\lambda_{\mathrm{bn}} \right) \tag{33}$$

$$\lambda_{\mathrm{bn}} = (1 + j\, \sigma\mu s_{\mathrm{bn}} v_{1\mathrm{n}} \tau_{1\mathrm{n}}/\pi)^{\frac{1}{2}} \tag{34}$$

$$s_{\mathrm{bn}} = (v_{1\mathrm{n}} + v_2)/v_{1\mathrm{n}} \tag{35}$$

(b) *Sandwich type composite secondary* In the design of LIMs, it is important to examine the effect of the configurations, materials and physical dimensions of secondary conducting plates on the motor characteristics.

Here the secondary of a sandwich structure is shown to consist of compound conducting plates, as shown in Figure 8.4. In the model, the region II-1 represents, in general, a ferromagnetic plate with conductivity σ_1, permeability μ and thickness $2d_1$ and each of two regions II-2 represents a non-magnetic conducting plate with constants σ_2, μ_0 and d_2 respectively. As a particular case, by substituting $\sigma_1 = 0$ and $\mu = \mu_0$, region II-1 holds for a non-magnetic and non-conducting material such as solid concrete or a hollow duct, as shown later.

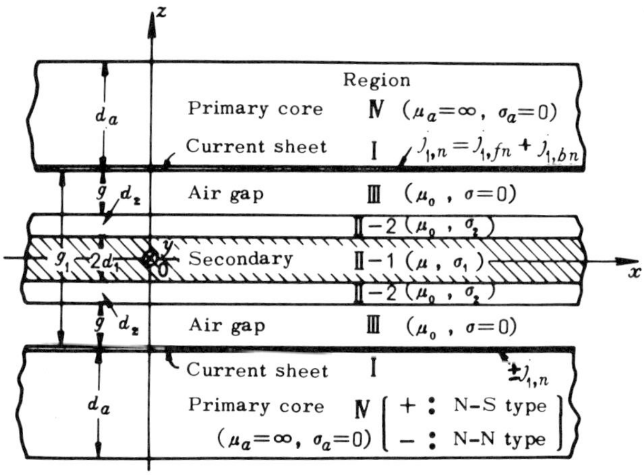

Figure 8.4 Two-dimensional model of the motor with a composite sandwich secondary.

Further, two cases are considered according to the polarity of the excitation of the double-sided primary. That is to say, one is a usual N-S type in which field poles on both primaries are faced N-S opposite each other, and another is a N-N type in which poles set up N against N. In the N-N type, the surface currents of both primaries are not in phase, but differ by π in electrical angle; hence if the current at one primary is taken as $j_{1,\mathrm{n}}$, the current at the other primary becomes $-j_{1,\mathrm{n}}$ as shown in Figure 8.4.

The solutions of the field equations due to each surface current component are obtained as described fully in an earlier paper [16]. Consequently, the x-component of the air gap flux density B_{x1}^{III} is given by the same equation as Equation (26). The z-component of the air gap flux density B_{z1}^{III} can be written by the same expression as Equation (27) with the following transformation in terms of K_{Hfn} and K_{Hbn}.

$$K_{Hfn} = \coth \frac{\pi}{\tau_{1n}} (g + d_2 \lambda'_{2\,fn}) \tag{36}$$

$$K_{Hbn} = \coth \frac{\pi}{\tau_{1n}} (g + d_2 \lambda'_{2bn}) \tag{37}$$

where λ'_{2fn} is determined from the following equations:

$$\frac{\pi}{\tau_{1n}} d_2 \lambda'_{2fn} = \tanh^{-1} \left\{ \lambda_{2fn} \tanh \frac{\pi}{\tau_{1n}} (d_2 \lambda_{2fn} + d_1 \lambda'_{1fn}) \right\} \tag{38}$$

$$\frac{\pi}{\tau_{1n}} d_1 \lambda'_{1fn} = \begin{cases} \tanh^{-1} \left(\dfrac{\mu_o}{\mu} \dfrac{\lambda_{1fn}}{\lambda_{2fn}} \tanh \dfrac{\pi}{\tau_{1n}} d_1 \lambda_{1fn} \right) & (39) \\ & \text{for N-S type} \\ \tanh^{-1} \left(\dfrac{\mu_o \lambda_{1fn}}{\mu\ \lambda_{2fn}} \coth \dfrac{\pi}{\tau_{1n}} d_1 \lambda_{1fn} \right) & (40) \\ & \text{for N-N type} \end{cases}$$

$$\lambda_{1fn} = (1 + j\sigma_1 \mu s_{fn} v_{1n} \tau_{1n}/\pi)^{\frac{1}{2}} \tag{41}$$

$$\lambda_{2fn} = (1 + j\sigma_2 \mu_o s_{fn} v_{1n} \tau_{1n}/\pi)^{\frac{1}{2}} \tag{42}$$

Similarly, λ'_{2bn} can be determined by substituting subscript fn by subscript bn in Equations (38) to (42).

8.2.4 *Derivation of performance equations*

(a) *Primary induced voltage* At the primary current sheet, the y-component e.m.f.s induced by the nth forward- and backward-travelling air gap flux are given by, respectively,

$$E_{y1,\,fn} = v_{1n} B_{z1,\,fn}^{III} = -jv_{1n} \mu_o K_{Hfn} J_{1,\,fn} \exp\,[j(\omega_1 t - \pi x/\tau_{1n})] \tag{43}$$

$$E_{y1,\,bn} = -v_{1n} B_{z1,\,bn}^{III} = -jv_{1n} \mu_o K_{Hbn} J_{1,\,bn} \exp\,[j(\omega_1 t + \pi x/\tau_{1n})] \tag{44}$$

Equations (43) and (44) have a significant meaning that the frequency of the e.m.f. at any point x is equal to the supply frequency f_1, in spite of the space harmonic order. The effective value of the induced e.m.f. in each phase of the actual primary winding is now considered. The effective number of turns per phase for the nth harmonic component is equal to $k_{wn} k_{Lpn} N_{ph}$ and the effective length of a half coil is equal to the stack height of the primary core h. Then:

$$\dot{E}_{1,\,fn} = \frac{1}{(2)^{\frac{1}{2}}} (-jv_{1n} \mu_o K_{Hfn} J_{1,\,fn}) \times 2hk_{wn} k_{Lpn} N_{ph}. \tag{45}$$

By substitution of Equation (21) in Equation (45), the induced voltage can be related to the actual winding current I_1, thus:

$$\dot{E}_{1,\,\mathrm{fn}} = -2mhf_1\mu_0(1/n)\,(k_{\mathrm{wn}}k_{\mathrm{Lpn}}N_{\mathrm{ph}})^2\ k_{\mathrm{mfn}}jK_{\mathrm{Hfn}}\dot{I}_1\,. \tag{46}$$

Similarly,

$$\dot{E}_{1,\,\mathrm{bn}} = -2mhf_1\mu_0(1/n)\,(k_{\mathrm{wn}}k_{\mathrm{Lpn}}N_{\mathrm{ph}})^2\ k_{\mathrm{mbn}}jK_{\mathrm{Hbn}}\dot{I}_1\,. \tag{47}$$

Therefore, the resultant e.m.f. $\dot{E}_1$ becomes

$$\dot{E}_1 = \sum_{n=1}^{\infty} (E_{1,\,\mathrm{fn}} + E_{1,\,\mathrm{bn}}). \tag{48}$$

The counter-e.m.f. is given by:

$$\dot{E}_1' = -\dot{E}_1\,. \tag{49}$$

(b) *Equivalent secondary impedance* The equivalent secondary impedance $\dot{Z}_{2\mathrm{e}}$ can be expressed:

$$\begin{aligned}\dot{Z}_{2\mathrm{e}} &= \dot{E}_1'/\dot{I}_1\\ &= \sum_{n=1}^{\infty} 2mhf_1\mu_0(1/n)(k_{\mathrm{wn}}k_{\mathrm{Lpn}}N_{\mathrm{ph}})^2 j(k_{\mathrm{mfn}}K_{\mathrm{Hfn}} + k_{\mathrm{mbn}}K_{\mathrm{Hbn}})\\ &= r_{2\mathrm{e}} + jx_{2\mathrm{e}},\end{aligned} \tag{50}$$

where $r_{2\mathrm{e}}$ and $x_{2\mathrm{e}}$ are calculated from

$$r_{2\mathrm{e}} = \mathrm{Re}(\dot{Z}_{2\mathrm{e}}),\ x_{2\mathrm{e}} = \mathrm{Im}(\dot{Z}_{2\mathrm{e}}). \tag{51}$$

(c) *Equivalent circuit for short-primary LIM* Figure 8.5 shows the equivalent circuit. The terminal impedance is given by:

$$\dot{Z}_{\mathrm{t}} = (r_1 + r_{2\mathrm{e}}) + j(x_1 + x_{2\mathrm{e}}) \tag{52}$$

Under constant voltage drive, the actual primary current I_1 is calculated as

$$I_1 = V_1/Z_{\mathrm{t}},\ \text{where}\ Z_{\mathrm{t}} = [(r_1 + r_{2\mathrm{e}})^2 + (x_1 + x_{2\mathrm{e}})^2]^{\frac{1}{2}} \tag{53}$$

Using the calculated value of I_1, the flux densities $B_{\mathrm{x1}}^{\mathrm{III}}$ and $B_{\mathrm{z1}}^{\mathrm{III}}$ are determined from Equations (26) and (27), respectively.

(d) *Thrust and vertical force* Using the concept of Maxwell's magnetic stresses, equations for the thrust and the vertical (normal) force are derived as follows. The total thrust of the double-sided LIM:

$$F_{\mathrm{x}} = \sum_{n=1}^{\infty} \mathrm{Re}\left(\frac{1}{\mu_0}B_{\mathrm{x1,\,fn}}^{\mathrm{III}}\,B_{\mathrm{z1,\,fn}}^{\mathrm{III}\,*} + \frac{1}{\mu_0}B_{\mathrm{x1,\,bn}}^{\mathrm{III}}\,B_{\mathrm{z1,\,bn}}^{\mathrm{III}\,*}\right)\ h(2L_1) \tag{54}$$

where * denotes the conjugate part. Here, it is noteworthy that the thrust can be

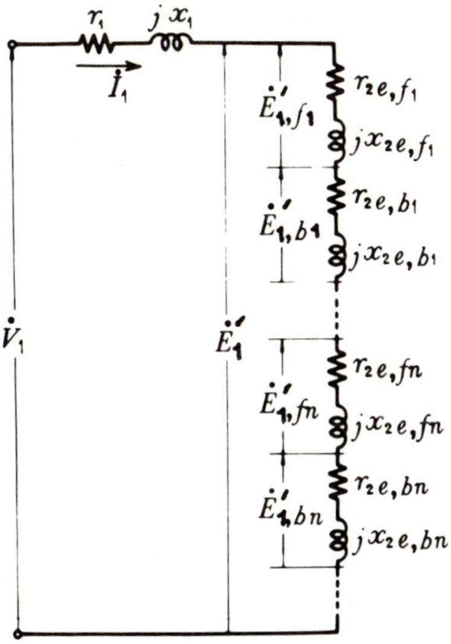

Figure 8.5 Equivalent circuit for short-primary LIM.
r_1 : primary resistance
x_1 : primary leakage reactance
V_1 : primary phase voltage

calculated simply by summation of the forces interacting between the same order space harmonic components. The vertical force acting on one side of the primary:

$$F_z^{IV} = \int_{-L/2}^{L/2} \tfrac{1}{2} \left(\frac{1}{2\mu_0} B_{z1}^{III} B_{z1}^{III*} - \frac{1}{2\mu_0} B_{x1}^{III} B_{x1}^{III*} \right) h \, dx. \tag{55}$$

(e) *Output power, power factor and efficiency* The output power P_0, power factor $\cos \varphi_1$ and efficiency η are obtained from

$$P_0 = F_x v_2 = (1 - s) F_x v_1 \tag{56}$$

where $v_1 = 2\tau f_1$, the actual synchronous speed; $s = (v_1 - v_2)/v_1$, the slip.

$$\cos \varphi_1 = (r_1 + r_{2e})/Z_t \tag{57}$$

$$\eta = P_0 / m V_1 I_1 \cos \varphi_1 \tag{58}$$

8.2.5 *Computation procedure*

(a) *Correction of the transverse end effect and the slot effect* In the preceding analysis, the transverse (y-direction) end effect was neglected. It is convenient to use an equivalent conductivity σ_e corrected by the Russell-Norsworthy coefficient instead of the intrinsic conductivity σ to allow for the transverse end effect. Furthermore, the effect of the slot on the flux distribution can be corrected by use of Carter's coefficient.

(b) *Selection criteria for L_c and n_{max}* In the computer-aided analysis of LIMs, $M\tau$ is used for L_c (the distance between the primary windings), where an integer M has to be selected of a sufficiently large value that the flux density immediately before the entry end of the short-primary converges to a small enough value to be neglected, which implies that there is no interaction between adjacent primary members. However, M should not be chosen larger than is necessary, from the point of view of required computing times.

It is advantageous to change the value of M according to both the synchronous speed v_1 and the secondary speed v_2 or the slip s.

For convergence of the series, the harmonic order n_{max}, for which the summation has to be made in a practical case, may be determined according to the following criterion. The space harmonic component with $\tau_{1n} \simeq \tau$ dominates intrinsically the performance characteristics of the LIM. Allowing for the above fact, the minimum value of n is adopted to satisfy $k_{pn} \simeq 0$ [see Equation (2)]. Then,

$$n_{max} = [(M + p + \beta - 1/mq)/\beta] \qquad (59)$$

where the square brackets indicate Gauss's symbol.

If the effect of the slot harmonic m.m.f. is considered, a larger number than n which satisfies $\tau_{1n} \simeq t_s/2 = \tau/2mq$ must be selected.

8.3 Characteristics of high-speed LIMs

8.3.1 *Examples of design parameters*

Table 8.1 shows numerical examples of high-speed double-sided LIMs. In Table 8.1 LIM 1 was originally designed for a high-speed traction vehicle, and this has been used for the theoretical investigation [5–9, 16].

LIM 2 (Figure 8.6) was designed for the TLRV of the US Department of

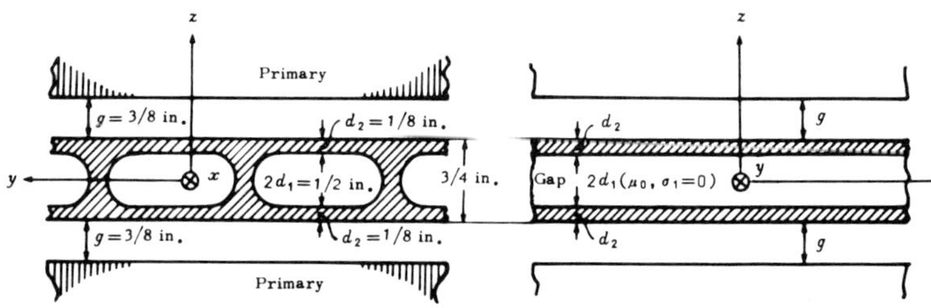

(a) Hollow aluminium reaction rail (LIM2). (b) Equivalent sandwich secondary.

Figure 8.6 Hollow aluminium reaction rail and its equivalent secondary.

Transportation [17, 18]. The secondary member of the LIM is a hollow aluminium reaction rail as shown in Figure 8.6 (a). In the analysis, this reaction rail is replaced by a sandwich structure which consists of two walls of aluminium web and an air gap in the centre region as shown in Figure 8.6 (b). An attempt is made to predict the performance characteristics of the motor by the mathematical model presented here.

LIM 3 is an experimental motor in the laboratory and the test results will be compared with the theoretical results in Section 8.5.

Table 8.1

Numerical examples of high-speed LIMs

Parameter	Symbol	Unit	LIM 1	LIM 2	LIM 3
Predicted rating					
Primary voltage	$(3)^{\frac{1}{2}} V_1$	V	1610	7136	440
Frequency	f_1	Hz	220	165	220
Primary current	I_1	A	1760	530	92
KVA	KVA	kV A	4910	6550	70
Thrust	F_X	kg		2265	
Synchronous speed	v_1	km h^{-1}	483	534.6	494.2
Cruising speed	v_2	km h^{-1}		483	483
Primary design					
Number of poles	p		6	5	4
Pole pitch	τ	mm	304.8	450	312
Stack height	h	mm	304.8	190.5	150
Number of phases	m		3	3	3
Slots/pole/phase	q		4	5	4
Slot pitch	t_S	mm	25.4	30.0	26.0
Slot width	w_S	mm	12.7	20.0	13.0
Slot depth	d_S	mm	32.8	48.0	30.0
Slots per side			82	85	58
Primary length	L	mm	2057	2520	1482
Coil pitch	β		10/12	10/15	10/12
Turns per coil	N		1	4	3
Winding connection			Two sides in series, star-connected.		
Series turns/phase	N_{ph}		48	200	96
Primary resistance	r_1	Ω	0.0175 at 75°C	0.288 at 150°C	0.104 at 75°C
Primary leakage reactance	x_1	Ω	0.2287 at 220 Hz	2.24 at 165 Hz	0.940 at 220 Hz
Mechanical clearance	g	mm	9.55	9.525	5.0
Total air gap	g_1	mm	35.0	38.1	25.0
Secondary design					
Material, Type			Al, Slab	Al, Hollow	Al, Slab
Resistivity	$1/\sigma$	Ω m	3.0×10^{-8} at 35°C	4.31×10^{-8} at 20°C	4.82×10^{-8} at 40°C
Thickness, overall	$2d$	mm	15.9	19.05	15.0
Web thickness	d_2	mm	–	3.175	–
Hole width	$2d_1$	mm	–	12.70	–
Overhang	c	mm	76.2	109	175
Height (or width)	h_2	mm	457.2	545	500

Table 8.2

Selection of L_c and n_{max} for LIM 2

s	0–0.1	0.1–0.2	0.2–0.3	0.3–0.5	0.5–0.7	0.7–1.0
L_c	300τ	250τ	200τ	150τ	100τ	50τ
n_{max}	915	765	615	465	315	165

8.3.2 *Distribution of magnetic flux density*

The space distribution of the amplitude of the x- and z-components of flux density at the primary core surface is calculated from Equations (26) and (27). Examples of the results of such calculations are shown in Figures 8.7 and 8.8. In the calculation for LIM 2, $L_c = M\tau$ and n_{max} were selected according to slip as shown in Table 8.2. Note that since there exists the relation $B^{III}_{x1,m} = \mu_0 J_1$, Figure 8.7 shows also the distribution of surface current density j_1 given by Equation (18). It is seen in Figure 8.8 that in the low-slip region, the z-component air gap flux density retards, because of the end effect behind the exit end (in the extended primary core portion L_c is assumed to make a relatively simple analysis possible). But because $B^{III}_{x1,m}$ is zero throughout L_c, its influence on the various characteristics (except the vertical force) is very little. The calculation of the vertical force F^{IV}_z, integrating numerically only its space distribution within the actual primary length as Equation (55), is found to be in fairly good agreement with experimental results as discussed later for LIM 3.

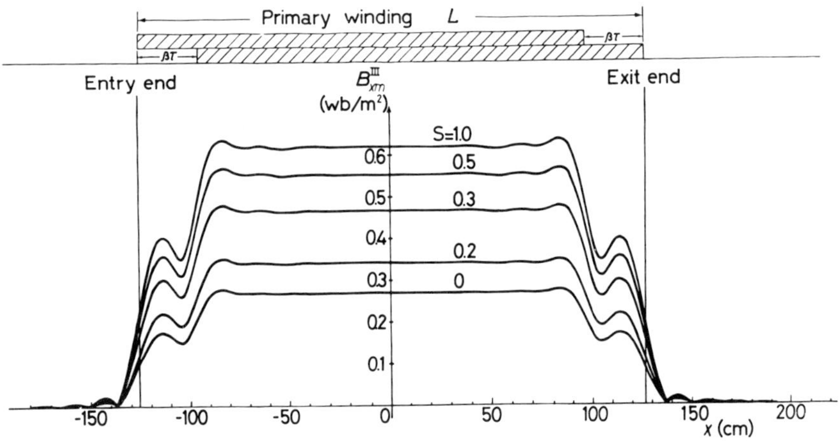

Figure 8.7 Space distribution of the x-component flux density for LIM 2.

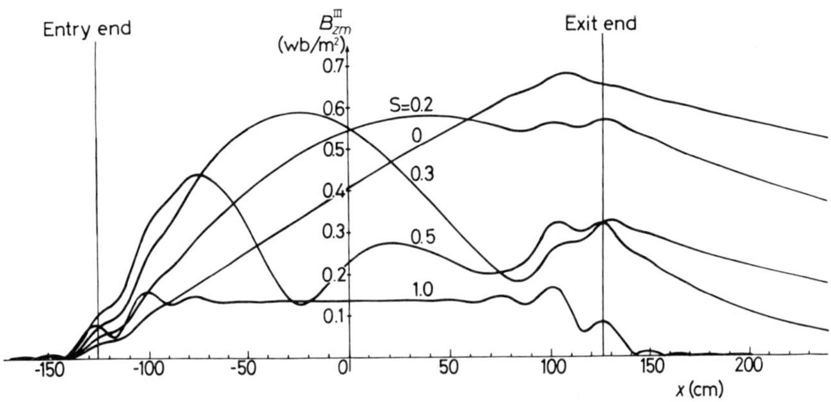

Figure 8.8 Space distribution of the z-component flux density for LIM 2.

Figures 8.9 and 8.10 show the space distribution of the thrust and the vertical force, respectively.

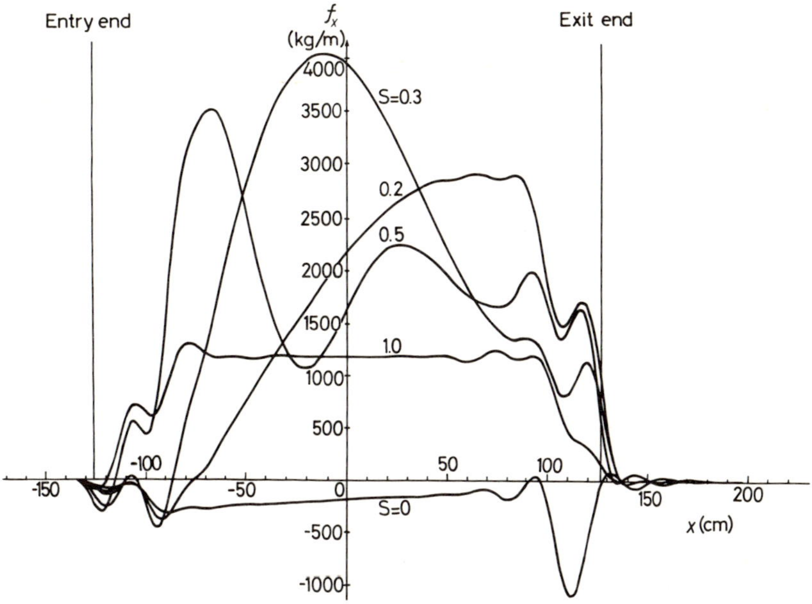

Figure 8.9 Space distribution of the thrust for LIM 2.

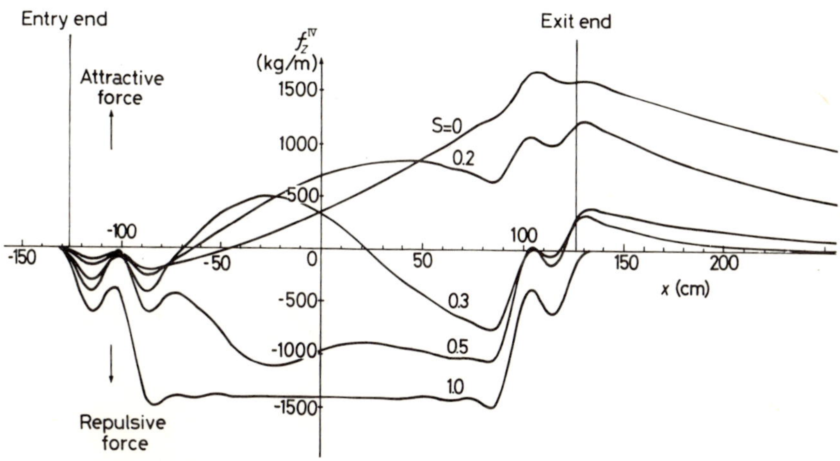

Figure 8.10 Space distribution of the vertical force acting on the primary for LIM 2.

8.3.3 *Constant voltage operation characteristics*

Figure 8.11 (a) shows the very pronounced reduction in thrust because of the end effect in the high speed region of s =0 –0.2 (483–386 km h^{-1}). Figure 8.11 (b) shows the vertical (normal) force acting on one side of the primary. In a high-speed

(a)

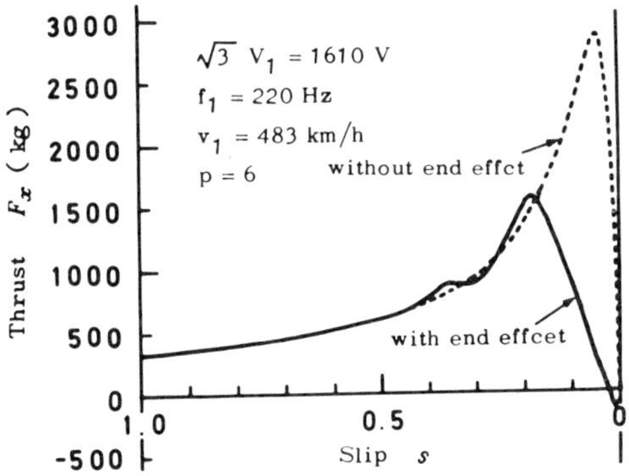

(b)

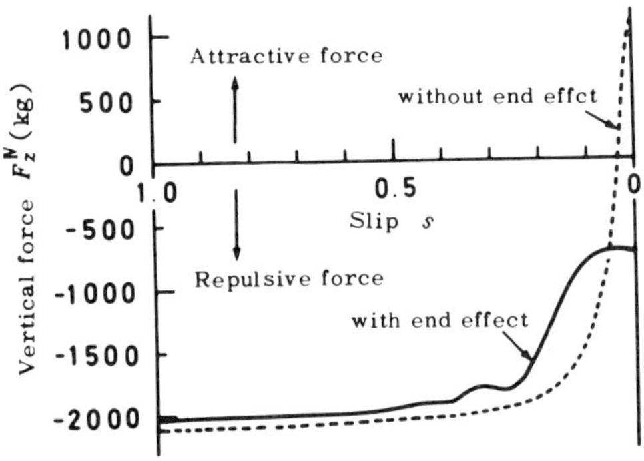

(c)

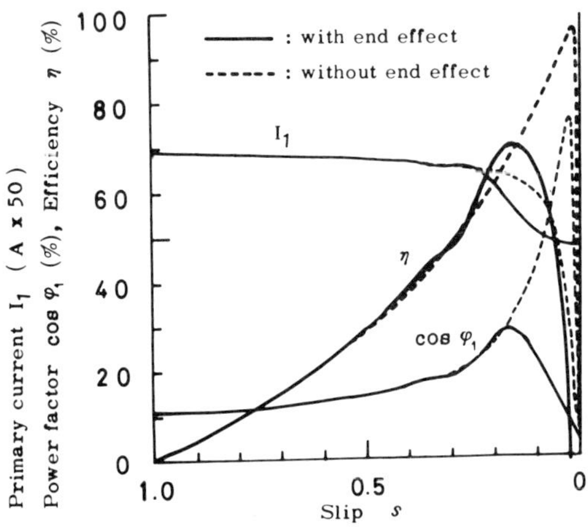

Figure 8.11 Computed characteristics for LIM 1.
(a) Thrust – slip curve.
(b) Vertical force – slip curve.
(c) Primary current, power factor and efficiency – slip curves.

LIM with a relatively thick and non-magnetic conducting plate such as LIM 1, a repulsive force is produced all over the motoring region from starting to synchronous speed (at which an attractive force is produced without the end effect). Figure 8.11 (c) shows the primary current, power factor and efficiency. The primary current at $s = 0$ decreases only to about 70% of the starting current because an additional current due to the end effect is to be supplied quite apart from the conventional magnetising current. Efficiency is not reduced to the same extent, but the power factor is very poor.

Figure 8.12 shows the calculated performance for LIM 2 under constant voltage

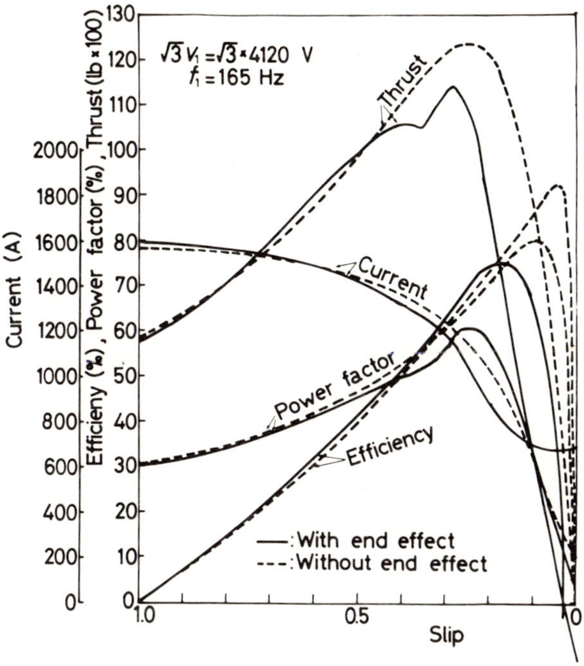

Figure 8.12 Computed characteristics for LIM 2.

drive. As the secondary effective resistance may be especially large due to the thin wall of the hollow aluminium reaction rail, the performance characteristics are improved to some extent as compared with Figure 8.11. However, it is still true that the end effect demands a primary current, at the rated slip of 0.097, of 680 A which is about 28% excess of the rated value of 530 A. The rated slip should therefore be selected to be larger than the normally designed value. Further it should be added that, on comparison of Figure 8.12 described above and Figure 8.11 in reference[17], large differences in the calculated results were observed.

8.4 Effects of variation of design parameters

The effects of varying some of the design parameters on the LIM performance are numerically investigated below.

8.4.1 *Thickness of secondary*

In this section, the characteristics of LIM 1 are investigated, assuming that a constant primary voltage is impressed to obtain the rated primary current I_{1n} = 1760 A at the rated slip s_n = 0.1 and that the supply frequency f_1 is fixed at 220 Hz.

Figure 8.13 shows the variation in thrust-slip curves for various thicknesses of the secondary aluminium plate for a constant mechanical clearance g = 9.55 mm. Here the reference curve indicated by 'ref.' gives the characteristics for a single aluminium plate of the thickness $2d$ = 15.9 mm and the overhang length $c = \tau/4$ = 76.2 mm. This prototype LIM was designed to satisfy the mechanical strength and current density of the secondary, neglecting the end effect.

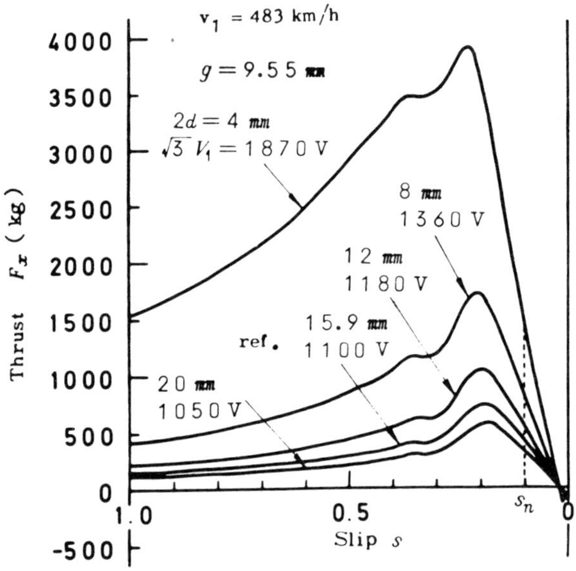

Figure 8.13 Thrust – slip curves of LIM 1 for various thicknesses of secondary plate.

Figure 8.14 summarises the values of the performance parameters at the rated slip. The performance is increased by decreasing the thickness of aluminium plate. This means that the end effect is suppressed mainly by increasing the effective resistance of the secondary. However, the minimum value of $2d$ may be determined by the allowable current density.

8.4.2 *Overhang length of secondary*

Figure 8.15 shows thrust-slip curves for different overhang lengths c of the secondary aluminium plate $2d$ = 15.9 and 4 mm. Using the Russell-Norsworthy coefficient, the effective resistivity of the secondary ρ_e increases with decreases in c. Consequently, the thrust increases as c decreases.

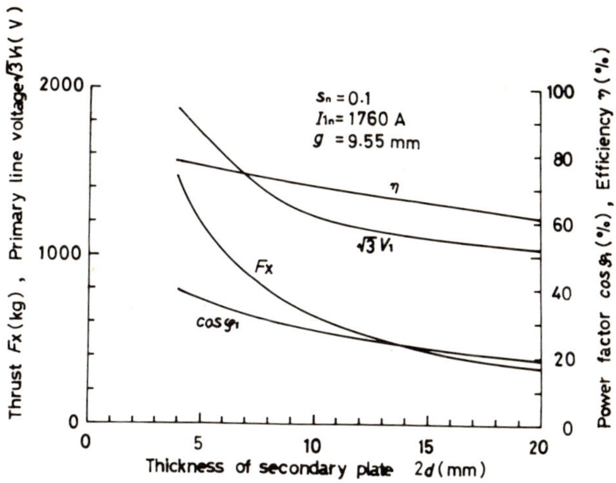

Figure 8.14 Effects of variation of thickness of secondary plate (LIM 1).

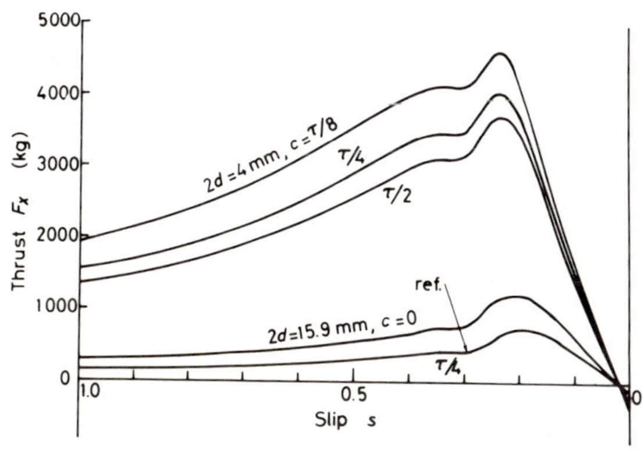

Figure 8.15 Thrust – slip curves of LIM 1 for various overhang lengths.

8.4.3 *Conductivity and permeability of secondary*

In this section, the effects of varying the conductivity and permeability of the secondary material are investigated for examples such as the copper-iron alloys with different iron contents γ and iron alone. The experimental equations used here for their permeability μ and resistivity ρ are

$$\rho = 0.084 \times 10^{-8}\gamma + \rho_{Cu}, \rho_{Cu} = 2.1 \times 10^{-8} : \text{resistivity of copper} \tag{60}$$

$$\mu = \mu_0 \mathrm{e} \exp(0.0621\gamma), \mu_0 : \text{free space permeability} \tag{61}$$

where $\rho = \rho_{cu}, \mu = \mu_0$ for $\gamma = 0$, and $5\rho_{cu}$, $500\mu_0$ for $\gamma = 100\%$.

Figure 8.16 (a) shows how the thrust F_x and vertical force F_z^{IV} vary with the different iron contents in alloys of copper and iron. Figure 8.16 (b) shows the thrust per kV A, primary line voltage, power factor and efficiency at the rated slip. As shown in Figure 8.16 (a) and (b), F_X of the LIM with the iron secondary shows a value about six times as large as that of the reference case, but $(3)^{\frac{1}{2}} V_1$ is increased about 4.5 times. The thrust per kV A of the 80% copper-iron alloy is 1.43 times larger than that of the reference case. The smaller the vertical force F_Z^{IV}, the more favourable for practical use. In the iron secondary F_Z^{IV} is about twenty times larger than that of the reference case.

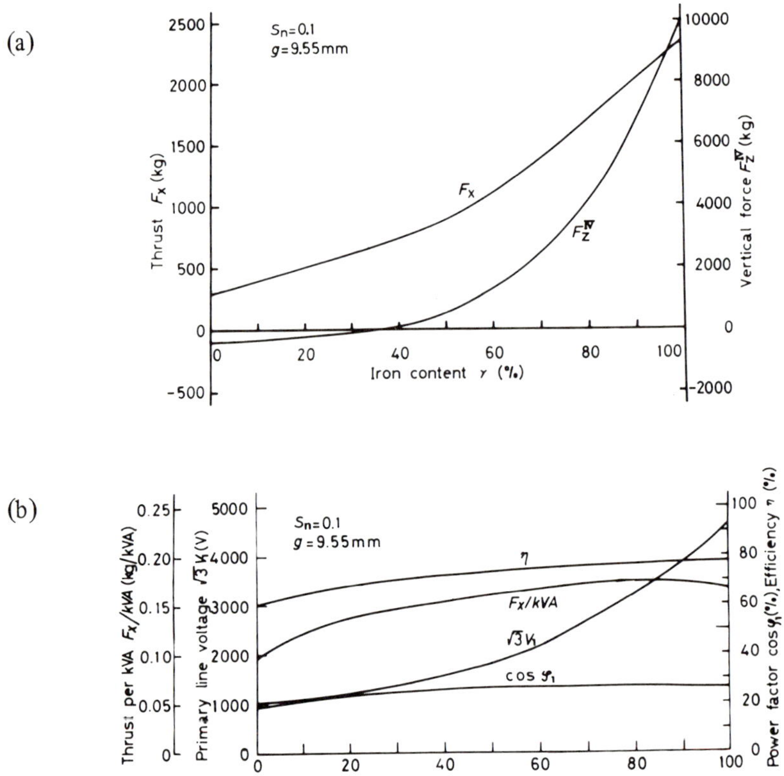

Figure 8.16 Effects of varying iron content in Cu-Fe alloy as the secondary material.
(a) Thrust and vertical force.
(b) Primary voltage, thrust per lVA, power factor and efficiency.

8.4.4 *Sandwich structure of secondary*

Two types of the sandwich, composite secondaries as shown in Figure 8.17 are considered as examples.

Type A is the structure obtained by attaching a relatively thin aluminium sheet to both sides of a thick iron plate. Case 1 and case 2 are a conventional N-S type and a N-N type, respectively. In type B with concrete as the thicker material, case 3 and case 4 represent N-S type and N-N type, respectively. Thus a thin aluminium sheet is reinforced for practical use. Especially case 2 and case 4 which correspond to two single-sided LIMs with and without back iron, respectively.

Figure 8.18 shows the thrust-slip curves for the various cases of a composite sandwich secondary, including that for the iron secondary. As shown in Figure 8.18, the rated thrust F_x of case 1 is about three times larger than that of the reference case. Case 2 shows almost the same characteristics as case 1 It is found

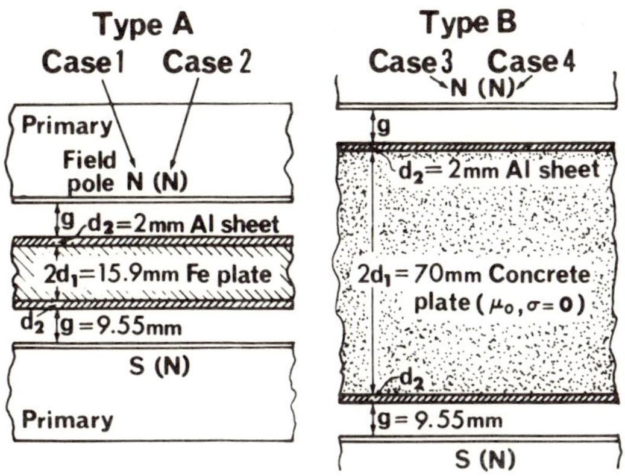

Figure 8.17 Examples of the composite sandwich secondary.

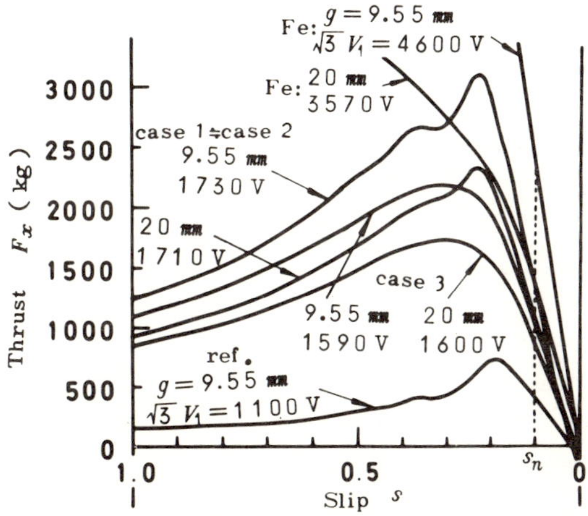

Figure 8.18 Thrust – slip curves for various cases of the sandwich secondary.

that F_x of case 3 is about twice as large as that of the reference case in spite of the large air gap introduced by the concrete plate $2d_1 = 70$ mm. This means that a relatively large air gap does not necessarily degrade the performance of high-speed LIMs. Of course, case 4 is not a practical one.

The characteristics at the rated slip for various cases of the composite sandwich secondary are summarised in Figure 8.19. Figure 8.19 (a) shows that both F_X and F_Z^{IV} at the rated slip vary with the configuration of the sandwich structure. F_Z^{IV} for cases 1, 2 and 3 show an attractive force and are nearly equal in magnitude to the repulsive force of the reference case. By comparison, in Figure 8.19 (b), the thrust per KVA for cases 1 and 2 are 1.89 times, and the corresponding value for case 3 is 1.63 times larger than that of the reference case.

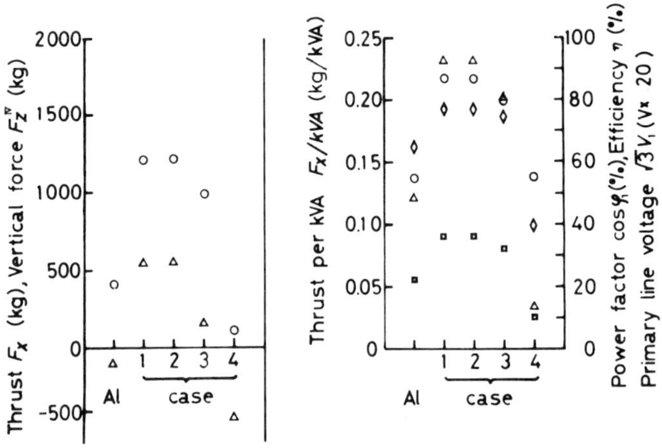

(a) o F_X
Δ F_Z^{IV}

(b) o$3^{\frac{1}{2}} V_1$, □ cos φ_1
ΔF_X/kVA, ◇η

Figure 8.19 Comparison of characteristics for various cases of the sandwich secondary (LIM 2).

In the case of the iron secondary, if a displacement from the symmetrical position is made larger, a more than proportionally larger attractive force acts on the secondary all over the motoring region. This means essentially highly unstable motion and the use of an iron secondary is very dangerous and not advisable for high-speed vehicle applications for which safety is of extreme importance. By modifying case 1 or case 2 by the use of 2 mm thick aluminium sheets on both sides of the same iron thickness, the attractive force may be reduced to several per cent of its value even in the high speed region of $s = 0–0.3$.

8.4.5 *Hollow aluminium reaction rail*

All of the reaction rails for the LIMRV, the PTACV and the TLRV at the US DOT's High Speed Ground Transport Centre are hollow aluminium extrusions. The fact that their wall thicknesses are small, unintentionally helps to suppress the end effects.

Here the effects of varying the hole width of the hollow reaction rail on the LIM performance are considered.

Figure 8.20 shows the above influences in respect of LIM 2, where it is assumed that the supply voltage and frequency are the rated values, and the wall thickness and the mechanical clearance are fixed at designed values. In Figure 8.20, the curves for $2d_1 = 12.7$ mm show the characteristics with the actual rail and the curves for $2d_1 = 0$

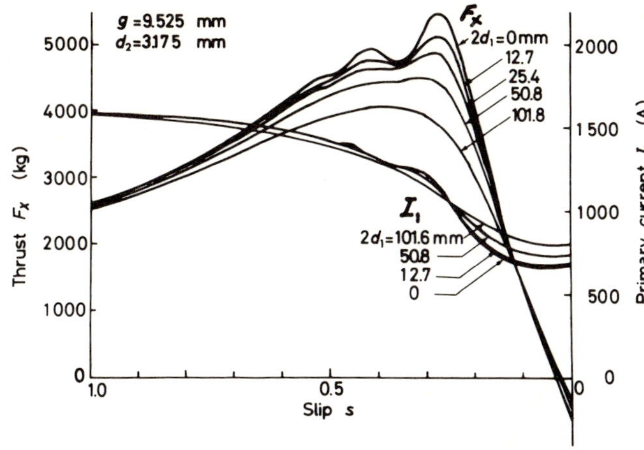

Figure 8.20 Thrust and primary current – slip curves of LIM 2 for various hole widths of the hollow reaction rail with constant mechanical clearance.

display the characteristics with a single aluminium plate 6.35 mm thick. It is seen from the figure that the thrust and primary current change little with large variation of the hole width.

Figure 8.21 summarises the effects of varying the hole width on the performance characteristics at a slip of 0.15.

From Figures 8.20 and 8.21, it is concluded that even if the hole width increases to several times the originally designed value, the quality of the characteristics of the LIM is only slightly reduced. Therefore, if it is necessary to modify the rail design to increase its lateral stiffness, an increase in rail thickness with no increase in wall thickness is a practical possibility.

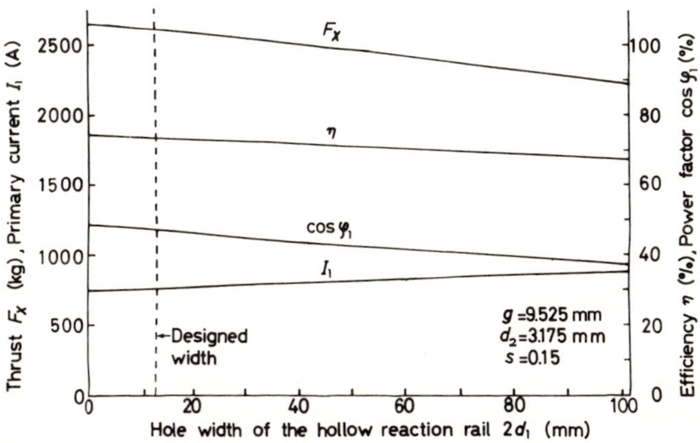

Figure 8.21 Effects of varying hole width on the LIM 2 performance parameters at slip 0.15.

8.4.6 *Mechanical clearance*

It is clear that the total air gap should be as small as mechanically possible because a large air gap requires a large magnetising current.

In the case of LIMs for high-speed ground transportation, however, it is desirable from the viewpoint of the safety of the exercise that the mechanical clearance is fairly large. The permissible mechanical clearance in relation to the LIM performance is therefore now examined.

Figure 8.22 shows the characteristics of LIM 2 for various mechanical clearances

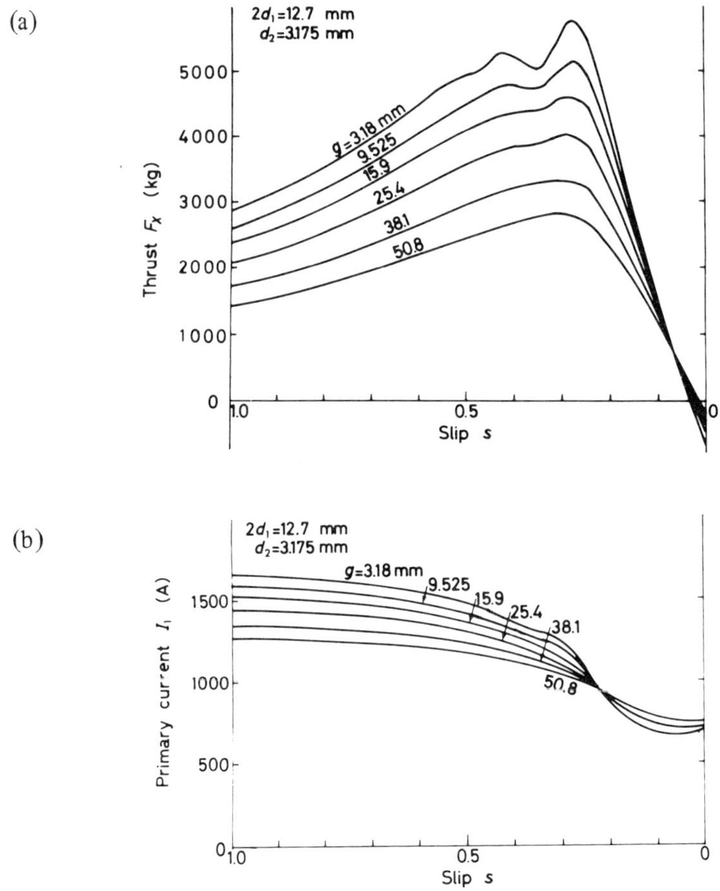

Figure 8.22 Characteristics of LIM 2 for various mechanical clearances at the rated supply voltage and frequency. (a) Thrust – slip curves. (b) Primary current – slip curves.

at the rated V_1 = 4120 V and f_1 = 165 Hz. Figure 8.23 displays the performance parameters at slip of 0.15 as a function of mechanical clearance.

It can be seen from Figure 8.22 (b) that the primary current decreases in the region of s greater than 0.22 and increases slightly in the region of s smaller than 0.22

as the mechanical clearance increases. It is to be noted from Figure 8.23 that the thrust at $s = 0.15$ with $2d_1 = 12.7$ mm only reduces to about 88% at $g = 25.4$ mm and about 68% at $g = 50.8$ mm in comparison with that at the designed $g = 9.525$ mm, respectively.

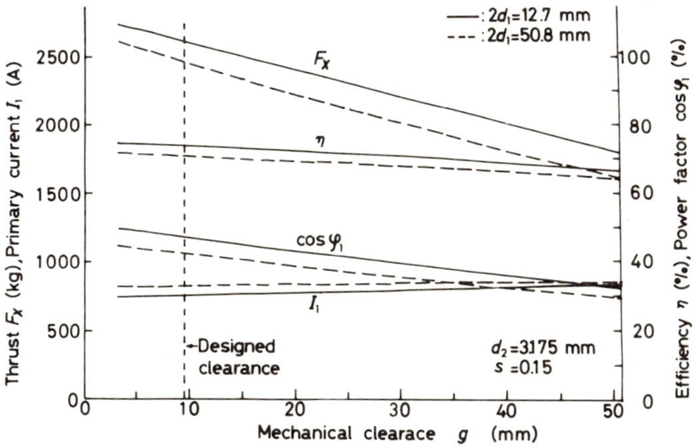

Figure 8.23. Effects of varying mechanical clearance on the LIM 2 performance parameters at slip 0.15.

Thus, we can draw an interesting conclusion that increasing the mechanical clearance does not, within limits, considerably affect the increase of the primary current in the region of slip for the normal operation of the high-speed LIM. In other words, the end effect rather than the air gap length exerts a profound influence on the magnetising current at high speeds. This fact permits an enlargement of the mechanical clearance without largely degrading the LIM performance. Thus, severe safety problems posed by the double-sided motor straddling a vertical reaction rail may be solved.

8.4.7 *Number of poles*

Here the effects of the number of poles on the characteristics of LIM 2 are considered. As a criterion for the comparison of performance, a value of rated slip s_n is selected at which the thrust F_x is two-thirds as large as the maximum thrust $F_{x,\,max}$, and then a constant primary voltage is supplied so as to obtain the rated current I_{1n} (= 530 A) at s_n.

Figure 8.24 shows the thrust per unit length and primary current versus slip characteristics for various pole numbers. Effects of varying the number of poles on the rated characteristics are illustrated by Figure 8.25. It can be seen from these figures that by increasing the number of poles, the end effect is essentially reduced and the LIM performance is remarkably improved. Thus, the number of poles should be as large as possible. There is no essential difference whether the number of poles is even or odd so long as the infinite primary core length is assumed.

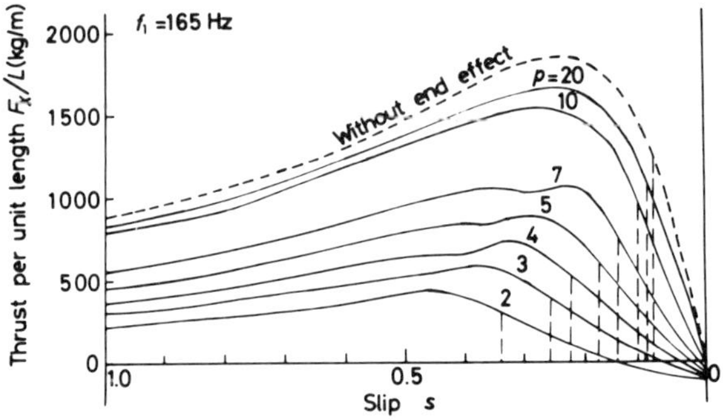

Figure 8.24 Thrust per unit length – slip curves of LIM 2 for various pole numbers.

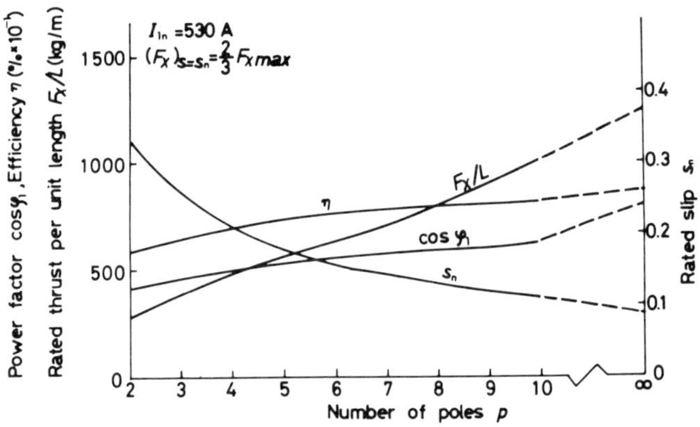

Figure 8.25 Effects of varying number of poles on the rated characteristics (LIM 2).

8.5 Comparison with experimental results

In order to check the validity of the theoretical results, experiments were performed on a laboratory test facility for a high-speed LIM. Each primary core of LIM 3 consisted of a sector shape (80.8° arc) stack 150 mm high and 1550 mm long with a 1100 mm centre line radius. A rotating aluminium disc 15 mm thick and 2700 mm in diameter was used as the secondary member. A.c. supply to the LIM was supplied from the 70 kV A thyristor inverter which provided a variable voltage from 40 to 500 V and a variable frequency from 20 to 250 Hz. The thrust and vertical force were measured by the corresponding load cells which were appropriately located on each primary side, respectively.

Figure 8.26 shows a comparison of the experimental results with computed results. The measured curves for thrust, vertical force and primary current versus slip were given directly on an X-Y recorder. The broken curves show the theoretical values for a sinusoidal wave of applied voltage. On the other hand, the dot-dash-curves show the computed results with consideration of both the space harmonics due to the end effects as well as the time harmonics due to the inverter fed voltage[19]. It is found that the computed values show an extremely close agreement with the measured values; especially, the dot-dash-curves and the measured curves for primary current or thrust, and in the low slip region they overlap each other completely.

Figure 8.27 illustrates a comparison between the computed and measured curves

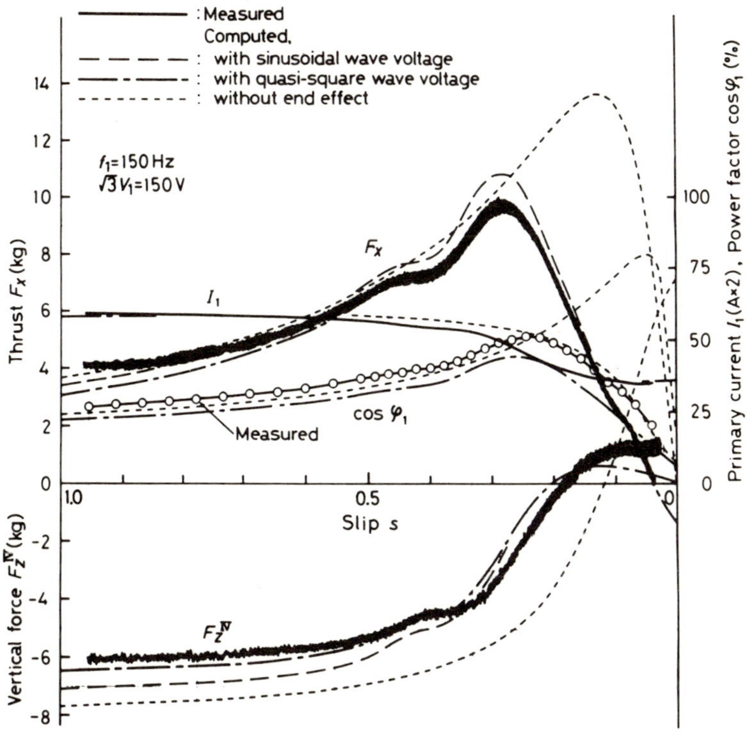

Figure 8.26 Comparison between calculation and measurement for the laboratory machine (LIM 3) driven by thyristor inverter.

of thrust, primary current and vertical force for various frequencies and a constant value of volts/Hz. As for thrust and primary current, the agreement can be said to be sufficiently satisfactory. Both the computed and measured primary current-slip curves at 90, 120 and 150 Hz coincide perfectly. The computed values for vertical force, except at 150 Hz, do not agree precisely with the experimental results.

8.6 Conclusions

The performance equations for LIMs with either an even or odd number of poles are derived directly from Maxwell's equations, using a space harmonic technique to take into account the end effect and the effects of a distributed winding. These equations are expressed in terms of the terminal voltage and all the actual design parameters, and are highly suitable for the LIM performance prediction and design calculations.

The effects of various design parameters on the LIM performance under constant voltage drive are specifically discussed concretely. In order to suppress the end effect and improve the LIM performance, the secondary effective resistance and the number

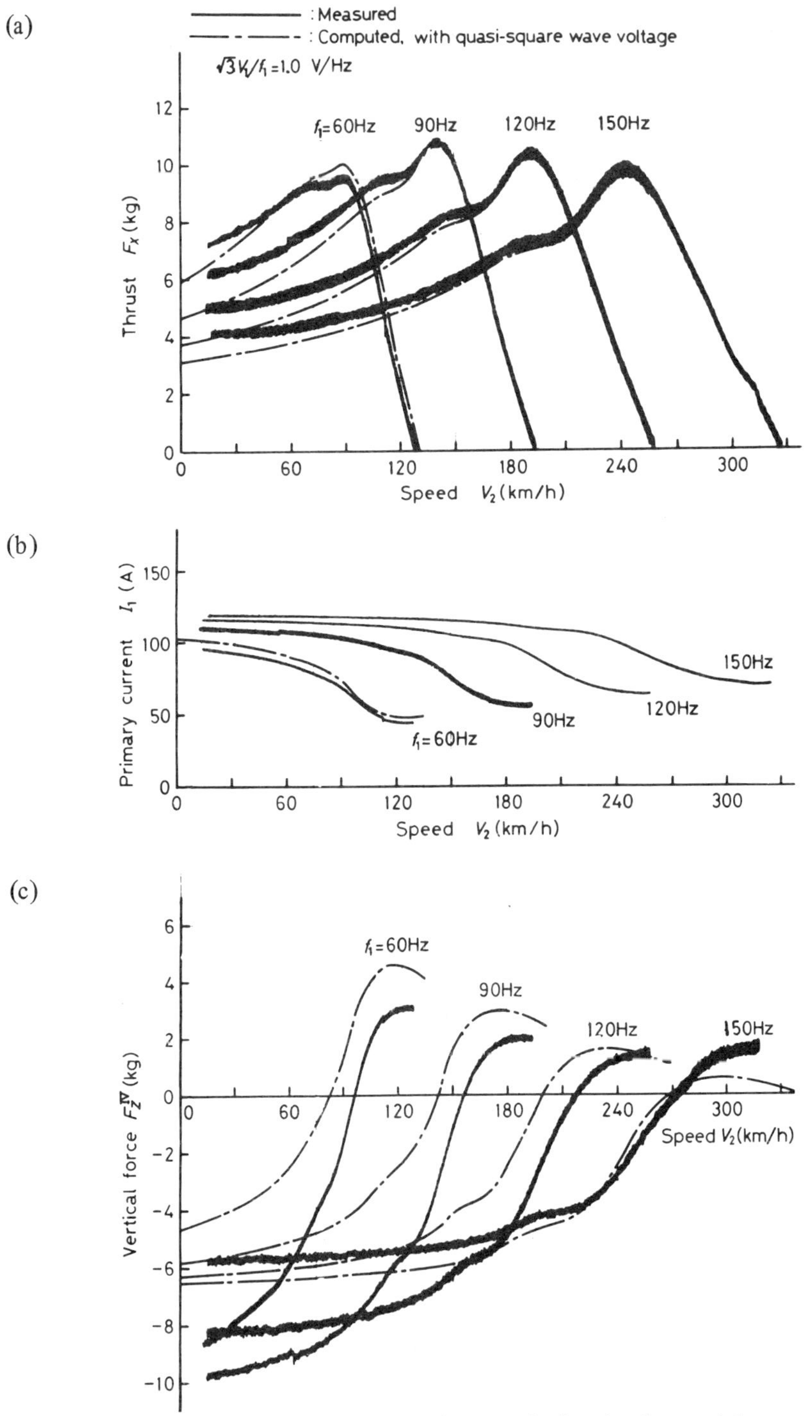

Figure 8.27 Comparison between computed and measured values for characteristic curves at different supply frequencies and constant volts/Hz operation (LIM 3).
(a) Thrust – speed curves. (b) Primary current – speed curves. (c) Vertical force – speed curves.

of poles (or primary length) should be as large as possible. It is often necessary to use a material for the secondary plate having a resistivity several times higher than that of aluminium. The composite sandwich secondary with a thin aluminium sheet on both sides of an iron or concrete plate, or a hollow aluminium reaction rail is a practical solution. It is found that even if the hole width of a hollow reaction rail increases to several times the normal designed value, the LIM performance is scarcely reduced, and an enlargement of the mechanical clearance by about 50 mm may be tolerable without serious deterioration in LIM performance. Thus serious safety problems posed by the double-sided LIM may be solved.

Experiment proved the validity of the theory presented here. The theoretical results provide a guide to the high-speed LIM designs and the performance equations will be utilised in future computer-aided design.

Acknowledgments

The authors wish to express their thanks to the members of their group, especially N. Fujii who carried out much of the experimental work and M. Ikeguchi for the computer calculation for LIM 2. This research was supported partially by scientific research grants from the Ministry of Education of Japan.

References

[1] Laithwaite, E. R. and Barwell, F. T. (1969) 'Applications of linear induction motors to high-speed transport systems', *Proc. IEE*, **116**, (5), pp. 713–724

[2] Yamamura, S. (1972) *Theory of Linear Induction Motors*, John Wiley & Sons, New York

[3] Nasar, S. A. and del Cid, L. (1973) 'Certain approaches to the analysis of single-sided linear induction motors', *Proc. IEE*, **120**, (4), pp. 477–483

[4] Kliman, G. B. and Elliott, D. G. (1973) Linear induction motor experiments in comparison with mesh/matrix analysis', IEEE Conf. Paper c73 129-4, Winter Power Meeting

[5] Nonaka, S. and Yoshida, K. (1972) 'Harmonic analysis of linear induction motors', Report RM-72-3 of the Studying Committee on Rotating Machinery

[6] Nonaka, S. and Yoshida, K. (1973) 'Space harmonic analysis of linear induction motors', *Trans. IEE Japan*, **93-B**, (3), pp. 89–96; *Elect. Eng. Japan*, **93**, (2), pp. 42–50

[7] Nonaka, S. and Yoshida, K. (1973) 'Studies of vertical forces of double-sided linear induction motors', *Trans. IEE Japan*, **93-B**, (10), pp. 471–478

[8] Nonaka, S. and Yoshida, K. (1973) 'Characteristics of double-sided linear induction motors with the secondary conductor vertically displaced from a symmetrical position', *Trans. IEE Japan*, 93-B, (11), pp. 543–550; *Elec. Eng. Japan*, 93, (6), pp. 38–46

[9] Nonaka, S. and Yoshida, K. (1974) 'The characteristics of high-speed linear induction motors analysed using a space harmonic technique', IEE Conference Publication No. 120, pp. 179–184

[10] North, G. G. (1973) 'Harmonic analysis of a short stator linear induction machine using a transformation technique', IEEE Conf. Paper c72 061-5, 1972, Winter Power Meeting; *IEEE Trans.*, 1973, (9–10), **PAS-92**, pp. 1733–1743

[11] Oberretl, K. (1973) 'Dreidimensionale berechnung des linearmotors mit berücksichtigung der endeffekte und der wicklungsverteilung', *Arch. f. Elektrotechn.*, **55**, (4), pp. 181–190

[12] Nonaka, S. and Yoshida, K. (1973) 'On the characteristics of high-speed linear induction motors', Report RM-73-9 of the Studying Committee on Rotating Machinery

[13] Nonaka, S., Yoshida, K. and Ikeguchi, M. (1974) 'Space harmonic analysis of linear induction motors with an odd number of poles on each side', *Kyushu Branch Conv. Four Insts. Elect. Eng. Japan*, No. 309

[14] Nonaka, S. and Yoshida, K. (1970) 'Equivalent circuits for double-sided linear motors', *J. IEE Japan*, **90**, (5), pp. 890–899; *Elect. Eng. Japan*, **90**, (3), pp. 32–41

[15] Nonaka, S. and Yoshida, K. (1970) 'Analysis of double-sided linear motors', *J. I.E.E. Japan*, **90**, (5), pp. 880–889; *Elect. Eng. Japan*, **90**, (3), pp. 21–31

[16] Nonaka, S. and Yoshida, K. (1971) 'Characteristics of linear motors with sandwich compound conducting plate', *J. IEE Japan*,91, (2), pp. 355–364; *Elect. Eng. Japan*, **91**, (1), pp. 183–193

[17] Dannan, J. H., Day, R. N. and Kalman, G. P. (1973) 'A linear-induction-motor propulsion system for high-speed ground vehicles', *Proc. IEEE*, **61**, (5), pp. 621–630

[18] 'Development and manufacture of a linear induction motor propulsion system for a tracked air cushion research vehicle', U.S. Department of Transportation, NTIS Document FRA-RT-72-35

[19] Nonaka, S., Hayashi, K. and Yoshida, K. (1973) 'Thrust characteristics of high-speed linear motors driven by the three-phase inverter', *Technol. Rep. Kyushu Univ.*, **46**, (2), pp. 102–108

9

Three-Dimensional Analysis of the Linear Motor

K. OBERRETL

9.1 Introduction

Interest in the linear motor has been growing continuously, particularly in connection with projects involving high-speed railway trains. Since the power consumption of such high-speed trains at speeds of 400 to 500 km h^{-1} is approximately 10 to 20 MVA, an exact analysis of such motors is of great practical importance.

Previous theories [1–15] have either been one- or two-dimensional, or have neglected to take into account longitudinal or transverse edge effects and do not contain the m.m.f. harmonics of the winding. In contrast, the theory given below is three-dimensional and takes into account all important design parameters. To this end, an analytical method is used because it represents the easiest way to incorporate the multitude of parameters.

9.2 Mathematical model

Figure 9.1 shows the coordinate system which will be used. Figure 9.2 shows the mathematical model in the y, z-plane. In order to simplify the calculation of the transverse edge effect, an infinite array of adjacent primary members is considered, in

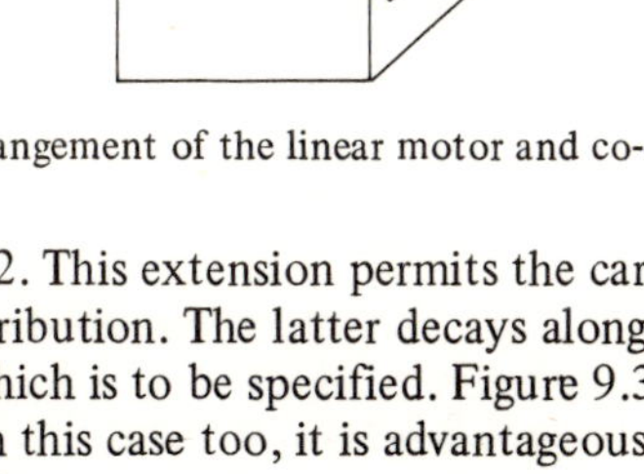

Figure 9.1 Arrangement of the linear motor and co-ordinate system.

accordance with Figure 9.2. This extension permits the carrying out of a Fourier analysis of the m.m.f. distribution. The latter decays along the winding overhang according to a function which is to be specified. Figure 9.3 shows the mathematical model in the y, x-plane. In this case too, it is advantageous to assume an infinite array of

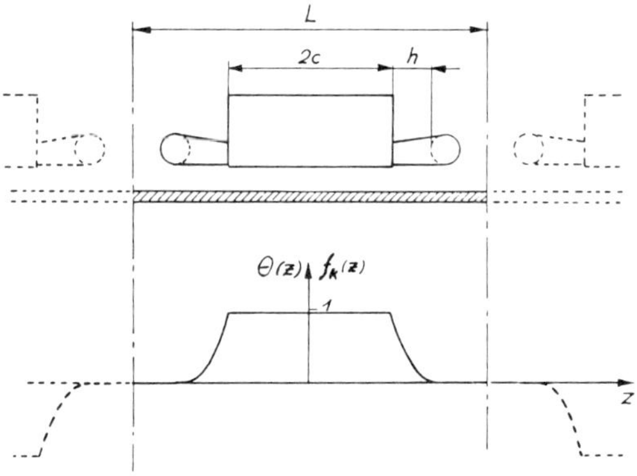

Figure 9.2 Mathematical model and m.m.f. distribution in the z-direction.

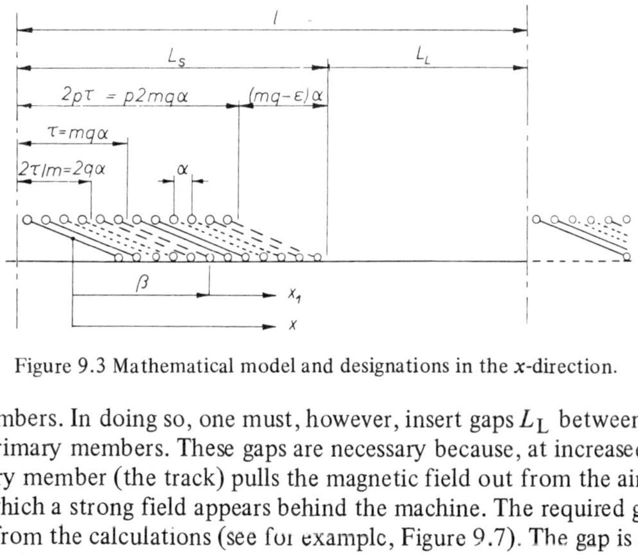

Figure 9.3 Mathematical model and designations in the x-direction.

primary members. In doing so, one must, however, insert gaps L_L between the individual primary members. These gaps are necessary because, at increased speeds, the secondary member (the track) pulls the magnetic field out from the air gap as a result of which a strong field appears behind the machine. The required gap length is obtained from the calculations (see for example, Figure 9.7). The gap is chosen to be not larger than necessary, since otherwise long computation times result. Various types of three-phase double-layer winding are feasible with linear motors. The so called 'compact' double-layer winding with half-filled end slots has a low longitudinal edge effect and is in common use. (For the so called 'extended' winding see Section 9.9.) The following equations are given for the 'compact' winding.

Using the dimensions indicated in Figure 9.3 the expression for the slot pitch α is:

$$p2mq\alpha + (mq-\epsilon)\,\alpha + L_L = l, \quad l = L_s + L_L$$
$$\alpha = \frac{L_s}{L_s + L_L}\,\frac{2\pi}{mq\,(2p+1)-\epsilon} \text{ in radians} \tag{1}$$

In the following treatment L_S, L_L, l, x_1, x_2, z, etc. are understood to be in units of length, α and x in units of radians.

Negative values for the gap length L_L may also be used. For $L_L = -(mq-\epsilon)\,\alpha l/2\pi$

the windings overlap in such a way that an infinitely long machine results. In this case the longitudinal edge effect disappears as in the case of a rotating motor. In order to avoid indeterminate expressions in the winding factor $^{v}k_{\mathrm{w}}$ (of the form 0/0) it is recommended that the gap length L_{L} be chosen 0.1 to 0.5% smaller than that stated above. For the same reason, the gap length must never be an integral multiple of the length L_{S} of the machine.

It is also worthwhile mentioning that the theory given below applies both to double-sided as well as single-sided winding arrangements, if, in the case of the single-sided arrangement, the single-sided track is backed with a return core for the magnetic circuit. The values computed below for the induced voltage, E, the electrical power P, the force F, etc. are always understood to apply to a single side.

9.3 The m.m.f. distribution

A symmetric current system is considered, bearing in mind that the supply of electrical power may be from a frequency converter. In doing so, asymmetric voltages appear at the terminals as a result of the longitudinal edge effect. One may, of course, also proceed in reverse order. In the case of multipolar machines, the asymmetrics of the terminal voltages are very small. Thus, in the case of the 16-pole motor discussed in Section 9.6, for example, the induced voltages for the various phases differ only by ± 2.6% under nominal operating conditions, and by even less at larger values of slip.

9.3.1 *Distribution in the x-direction*

A compact double-layer winding is considered with half-filled end slots. The winding consists of $1, 2, \ldots k \ldots m$ phases, $1, 2, \ldots \kappa, \ldots q$ coils in each coil group, and $1, 2, \ldots, \rho, \ldots 2p$ coil groups per phase.

The current of the kth phase of the ρth coil group, written in complex form, is[23]:

$$I_{1,k,\rho} = I_1 (2)^{\frac{1}{2}} \exp\left\{\mathrm{j}\left[\omega t - (k-1)\frac{\pi}{m} - (\rho - 1)\pi\right]\right\} \tag{2}$$

The instantaneous value is obtained, for example, by taking real parts. Thus, the current in the phase $k = 1$ and coil group $\rho = 1$ assumes its maximum value at $t = 0$. For the νth harmonic of the m.m.f. distribution of the first coil group of the first phase, according to reference[17] is:

$$\begin{aligned} {}^{\nu}\Theta &= \frac{N}{\pi\nu}\,{}^{\nu}k_s I_1 (2)^{\frac{1}{2}} \exp \mathrm{j}(\omega t - \nu x) \\ \nu &= \pm 1, \pm 2, \pm 3, \ldots \\ {}^{\nu}k_s &= \sin \nu(mq - \epsilon)\,\alpha/2 \end{aligned} \tag{3}$$

where N denotes the number of turns in each coil, $\omega = 2\pi f$ the angular supply frequency, and ϵ the reduction in width of the coils in units of slots. For the m.m.f. of the κth coil of the $\rho = 1, 2, 3, \ldots 2p$th coil group of the kth phase one has:

$$\begin{aligned} {}^{\nu}\Theta_{\kappa,\rho,k} = \frac{N}{\pi\nu}\,{}^{\nu}k_s I_1 (2)^{\frac{1}{2}} \exp \Bigg\{\mathrm{j}\Bigg[& \omega t - \nu x - (k-1)\frac{\pi}{m} \\ & + \nu(\kappa - 1)\,\alpha + \nu(\rho - 1)\,mq\alpha + \nu(k-1)\,q\alpha - (\rho - 1)\,\pi\Bigg]\Bigg\} \end{aligned}$$

Summation over κ, ρ and k yields:

$$ {}^{\nu}\Theta = \sum_{\kappa=1}^{q} \sum_{\rho=1}^{2p} \sum_{k=1}^{m} {}^{\nu}\Theta_{\kappa,\,\rho,\,k} $$

$$ = 2\frac{N}{\pi\nu} I_1 (2)^{\frac{1}{2}} ({}^{\nu}k''_{\omega}\,{}^{\nu}k'_{\omega}) \exp\left\{ \mathrm{j}\left(\omega t - \nu x + \nu\beta + \frac{\pi}{2m}\right)\right\} \tag{4} $$

$$ {}^{\nu}k''_{\omega} = \frac{1}{2} \frac{\sin p\,(\nu m q \alpha)}{\sin\frac{1}{2}\,(\nu m q\alpha - \pi)} \frac{\sin \nu q\alpha/2}{\sin \nu\alpha/2} \sin\left(\nu \frac{mq - \epsilon}{2}\alpha\right) \tag{5} $$

$$ {}^{\nu}k'_{\omega} = \frac{\sin\frac{1}{2}\,(\nu m q\alpha - \pi)}{\sin\frac{1}{2}\,(\nu q\alpha - \pi/m)},\ {}^{\nu}k_{\omega} = {}^{\nu}k''_{\omega}\,{}^{\nu}k'_{\omega} \tag{6} $$

$$ \beta = (pmq - \tfrac{1}{2})\alpha,\quad p = \text{integer} $$

$$ \nu = \pm 1, \pm 2, \pm 3, \ldots $$

The angle β can be eliminated if a new co-ordinate x_1, is introduced:

$$ x_1 = x\frac{l}{2\pi} - \frac{l}{2\pi}\beta \tag{7} $$

The origin of x_1 is shown in Figure 9.3. x_1 is given in units of length.

9.3.2 *Distribution in the z-direction*

If it is assumed that the m.m.f. decays, for example, sinusoidally along the winding overhang, then a Fourier analysis of the m.m.f. distribution shown in Figure 9.2 yields, in this case:

$$ f_k(z) = \frac{4}{\pi} \sum_{n}^{\infty} {}^{n}y \cos n\frac{\pi}{L} z,\quad n = +1, +3, +5, \ldots $$

$$ {}^{n}y = \frac{\left(\frac{L}{2h}\right)^2}{\left(\frac{L}{2h}\right)^2 - n^2} \left[\frac{1}{n}\left(\sin \pi n\frac{c+h}{L}\right) - \frac{2h}{L}\cos\frac{\pi n c}{L}\right] \tag{8} $$

It was shown in reference [5] by means of field measurements that a sinusoidally decaying m.m.f. distribution along the winding overhang is in good agreement with the distribution of the leakage field. In principle, of course, any other distribution can be substituted, provided that corresponding results of field measurements are available.

9.3.3 *Resulting m.m.f. distribution, current distribution*

$$ \Theta(x_1, z, t) = 2\frac{N}{\pi} I_1 (2)^{\frac{1}{2}} \frac{4}{\pi} \sum_{n}^{\infty} \sum_{\nu}^{\infty} \frac{{}^{\nu}k_{\omega}}{\nu}\,{}^{n}y \left(\cos n\frac{\pi}{L} z\right) \exp \mathrm{j}\left(\omega t - \nu x_1 \frac{2\pi}{l} + \frac{\pi}{2m}\right). \tag{9} $$

From this the following current distribution in the z-direction is obtained:

$$\alpha_{1,z} = \frac{\partial \Theta}{\partial x_1} = \frac{1}{\mathrm{j}} \frac{16}{\pi} \frac{N}{l} I_1 (2)^{\frac{1}{2}} \sum_n \sum_\nu {}^{\nu}k_\omega \, {}^{n}y \left(\cos n \frac{\pi}{L} z \right)$$

$$\exp \mathrm{j} \left(\omega t - \nu x_1 \frac{2\pi}{l} + \frac{\pi}{2m} \right)$$

From the continuity condition the following current distribution in the x-direction (in the winding overhang) is obtained:

$$\alpha_{1,x} = - \int \frac{\partial \alpha_{1,z}}{\partial z} \, \mathrm{d}x_1$$

$$= \frac{8}{\pi} \frac{N}{L} I_1 (2)^{\frac{1}{2}} \sum_n \sum_\nu {}^{\nu}k_\omega \, {}^{n}y \left(\sin n \frac{\pi}{L} z \right) n \frac{1}{\nu} \exp \mathrm{j} \left(\omega t - \nu x_1 \frac{2\pi}{l} + \frac{\pi}{2m} \right).$$

The resulting m.m.f. distribution in the x-direction is represented for 5 different instants of time in Figure 9.4 and the reduced magnitude of the end poles is shown.

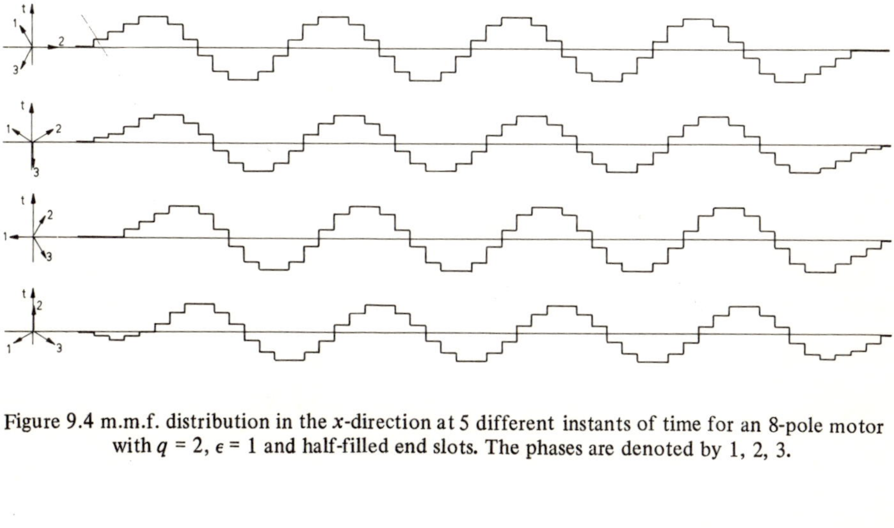

Figure 9.4 m.m.f. distribution in the x-direction at 5 different instants of time for an 8-pole motor with $q = 2$, $\epsilon = 1$ and half-filled end slots. The phases are denoted by 1, 2, 3.

9.4 Solution of the field equations

The solution is carried out analytically using the method of Bernoulli-Fourier. The origin of the coordinate system is placed on the surface of the track, and all quantities relating to the primary member are given with reference to the track by means of:

$$x_1 = x_2 + 2\tau f(1-s)t \qquad (10)$$

$$= x_2 + \frac{\mathrm{mq}\alpha}{\pi}\frac{l}{2\pi}\,\omega(1-s)t$$

as indicated in Figure 9.5.

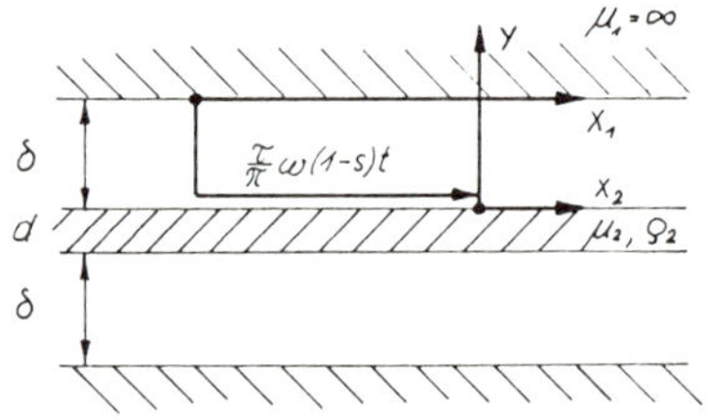

Figure 9.5 Relationship between the primary and secondary co-ordinate system.

9.4.1 *The air gap*

In the determination of the m.m.f. distribution it was assumed that the component of the current density in the primary member $J_{1,\,y} = 0$. This holds exactly for the most essential part, the slotted section. Since the secondary member, the track, is very thin relative both to the pole pitch and to the width of the track, it is permissible also, in this case, to set $J_{2,\,y} = 0$. Thus, because $\partial A/\partial t = -\rho J$ the corresponding components of the magnetic vector potential become:

$$A_{2,\,y} = A_{\delta,\,y} = 0. \qquad (11)$$

The following equations hold:

$$\frac{\partial^2 A_{\delta,\,z}}{\partial x_1^2} + \frac{\partial^2 A_{\delta,\,z}}{\partial y^2} + \frac{\partial^2 A_{\delta,\,z}}{\partial z^2} = 0$$

$$\frac{\partial^2 A_{\delta,\,x_1}}{\partial x_1^2} + \frac{\partial^2 A_{\delta,\,x_1}}{\partial y^2} + \frac{\partial^2 A_{\delta,\,x_1}}{\partial z^2} = 0 \qquad (12)$$

$$\operatorname{div} A = 0 : \frac{\partial A_{\delta,\,x_1}}{\partial x_1} + \frac{\partial A_{\delta,\,z}}{\partial z} = 0$$

On the basis of the boundary condition Equation (9), the following expression in the form of a product suggests itself as a particular solution for the n, νth harmonic:

$${}^{n,\nu}A_{\delta,z} = X_1(x_1)\,Y_1(y)\,Z_1(z)$$

$$X_1 = C_1 \exp \mathrm{j}(\omega t - \nu 2\pi x_1/l)$$

$$Z_1 = C_2\,{}^{n}y\,(\cos n\pi z/L)$$

Using the method of separation of variables:

$$Y_1 = C_3 \left(\cosh \lambda y + \frac{C_4}{C_3} \sinh \lambda y\right), \ \lambda^2 = \left(\nu \frac{2\pi}{l}\right)^2 + \left(n \frac{\pi}{L}\right)^2$$

and the solution is:

$$^{n,\nu}A_{\delta,z} = C_{10} \exp \mathrm{j} \left(\omega t - \nu \frac{2\pi}{l} x_1\right) {}^{n}y \left(\cos n \frac{\pi}{L} z\right)(\cosh \lambda y + C_{12} \sinh \lambda y), \quad (13)$$

$$C_{10} = C_1 C_2 C_3$$

The other component is obtained from Equation (12):

$$^{n,\nu}A_{\delta,x_1} = -\int \frac{\partial^{n,\nu}A_{\delta,z}}{\partial z} \mathrm{d}x_1 \quad (14)$$

In order to be able to match this solution to the boundary condition for $y = \delta$, it is necessary to sum all particular solutions:

$$A_{\delta,z} = \sum_{n}^{\infty} \sum_{\nu}^{\infty} C_{10} \exp \mathrm{j} \left(\omega t - \nu \frac{2\pi}{l} x_1\right) {}^{n}y \left(\cos n \frac{\pi}{L} z\right) (\cosh \lambda y + C_{12} \sinh \lambda y) \quad (15)$$

$$A_{\delta,x_1} = \mathrm{j} \frac{\pi}{L} \sum_{n}^{\infty} \sum_{\nu}^{\infty} \frac{l}{2\pi\nu} {}^{n}y \, C_{10} \, n \left(\sin n \frac{\pi}{L} z\right) \exp \mathrm{j} \left(\omega t - \nu \frac{2\pi}{l} x_1\right) (\cosh \lambda y + C_{12} \sinh \lambda y) \quad (16)$$

9.4.2 *The track*

The following differential equations hold:

$$\frac{\partial^2 A_{2,z}}{\partial x_2^2} + \frac{\partial^2 A_{2,z}}{\partial y^2} + \frac{\partial^2 A_{2,z}}{\partial z^2} = \gamma_2 \mu_2 \frac{\partial A_{2,z}}{\partial t}$$

$$\frac{\partial^2 A_{2,x_2}}{\partial x_2^2} + \frac{\partial^2 A_{2,x_2}}{\partial y^2} + \frac{\partial^2 A_{2,x_2}}{\partial z^2} = \gamma_2 \mu_2 \frac{\partial A_{2,x_2}}{\partial t}, \quad (17)$$

$$\mathrm{div}\, A = 0 : \frac{\partial A_{2,x_2}}{\partial x_2} + \frac{\partial A_{2,z}}{\partial z} = 0$$

The following expression in the form of a product again suggests itself as a particular solution for the n, νth harmonic:

$$^{n,\nu}A_{2,z} = X_2(x_2)\, Y_2(y)\, Z_2(z)\, T(t)$$

From Equations (13) and (10) the vector potential at the surface of the track is:

$$^{n,\nu}A_{\delta,z} = C_{10} \exp \mathrm{j} \left({}^{\nu}s\omega t - \nu 2\pi x_2/l\right) {}^{n}y \left(\cos n\pi z/L\right) (\cosh \lambda y + C_{12} \sinh \lambda y) \quad (18)$$

$$^{\nu}s = 1 - \nu(1 - s)\, mq\alpha/\pi \quad (19)$$

Taking account of Equation (18) it is expedient to assume solutions of the following form:

$$X_2(x_2) = C_5 \exp j\left(-\nu \frac{2\pi}{l} x_2\right)$$

$$Z_2(z) = C_6 \left(\cos n\frac{\pi}{L} z\right)^{n} y$$

$$T(t) = C_7 \exp j\, {}^{\nu}s\omega t$$

The method of separation of variables thus yields for the particular solution:

$$ {}^{n,\,\nu}A_{2,\,z} = C_{13}\ (\cosh \xi y + C_{u} \sinh \xi y)\, {}^{n}y \left(\cos n\frac{\pi}{L} z\right) \exp j\left({}^{\nu}s\omega t - \nu\frac{2\pi}{l} x_2\right)$$

$$\xi = \left[\left(\nu\frac{2\pi}{l}\right)^2 + \left(n\frac{\pi}{L}\right)^2 + j\, {}^{\nu}s\omega\gamma_2\mu_2\right]^{\frac{1}{2}} \tag{20}$$

The other component ${}^{n,\,\nu}A_{2,\,x_2} = -\int(\partial^{n,\,\nu}A_{2,z}/\partial z)\, dx_2$ is obtained from Equation (17). Regarding the boundary conditions, the solution is obtained from the sum:

$$A_{2,\,z} = \sum_{n}^{\infty}\sum_{\nu}^{\infty} C_{13} \exp j\left({}^{\nu}s\omega t - \nu\frac{2\pi}{l}x_2\right)\, {}^{n}y \left(\cos n\frac{\pi}{L} z\right)(\cosh \xi y + C_{11} \sinh \xi y) \tag{21}$$

$$A_{2,\,x_2} = j\frac{\pi}{L}\sum_{n}^{\infty}\sum_{\nu}^{\infty}\frac{l}{\nu 2\pi} C_{13}\, {}^{n}y \left(\sin n\frac{\pi}{L} z\right) n \exp j\left({}^{\nu}s\omega t - \nu\frac{2\pi}{l}x_2\right)(\cosh \xi y + C_{11} \sinh \xi y) \tag{22}$$

9.4.3 *Determination of the constants* $C_{10}, C_{11}, C_{12}, C_{13}$

From the symmetry condition at the centre of the track:

$$y = -\frac{d}{2},\ H_{2,\,x_2} = 0 \ \text{ or } \ \frac{1}{\mu_2}\frac{\partial A_{2,\,z}}{\partial y} = 0$$

and,

$$C_{11} = \tanh \xi\frac{d}{2} \tag{23}$$

The symmetry condition $H_{2,\,z} = 0$ leads to the same expression for C_{11}. From the continuity of the tangential component of H at the surface of the track:

$$y = 0,\ H_{\delta,x_2} = H_{2,\,x_2} \ \text{ or } \ \frac{1}{\mu_0}\frac{\partial A_{\delta,\,z}}{\partial y} = \frac{1}{\mu_0\mu_{r2}}\frac{\partial A_{2,\,z}}{\partial y}$$

and hence,

$$C_{12} = \xi C_{11}/(\lambda\mu_{r2}) \tag{24}$$

The condition $H_{\delta,\,z} = H_{2,\,z}$ leads to the same expression for C_{12}. From continuity of the tangential component of E at the surface of the track:

$$y = 0,\ E_{\delta,\,z} = E_{2,\,z}$$

or, because,

$$\boldsymbol{E} = -\partial \boldsymbol{A}/\partial t,\ A_{\delta,\,z} = A_{2,\,z}$$

it follows that,

$$C_{10} = C_{13} \tag{25}$$

The continuity of the normal component of B at the surface of the track also leads to Equation (25). Now, C_{10} remains to be determined. At the surface of the primary member, with $\mu_1 = \infty$:

$$y = \delta,\ H_{\delta,\,x_1} = \frac{\partial\Theta}{\partial x_1} \text{ or } \frac{1}{\mu_0}\frac{\partial A_{\delta,\,z}}{\partial y} = \frac{\partial\Theta}{\partial x_1}.$$

From this C_{10} may be obtained thus:

$$C_{10} = \frac{\mu_0 16 N I_1 (2)^{\frac{1}{2}}\ {}^{\nu}k_{\omega}/(\pi l)}{j\lambda\,(\sinh\lambda\delta + C_{12}\cosh\lambda\delta)} \exp\,[\mathrm{j}\pi/(2m)] \tag{26}$$

The corresponding condition for $H_{\delta,\,z}$ leads to the same expression for C_{10}.

9.4.4 *Fields in the air gap, fields in the track, current density in the track*

From the vector potential, which is now known, making use of $\boldsymbol{B} = \text{rot}\,\boldsymbol{A}$:

$$B_y = \frac{\partial A_x}{\partial z} - \frac{\partial A_z}{\partial x},\ B_x = \frac{\partial A_z}{\partial y},\ B_z = -\frac{\partial A_x}{\partial y}$$

and the fields may be calculated as follows:

$$B_{\delta,\,y} = \mathrm{j}\sum_n\sum_\nu \frac{l}{\nu 2\pi}\lambda^2 C_{10}\,{}^{n}y\left(\cos n\frac{\pi}{L}z\right)\exp\mathrm{j}\left(\omega t - \nu\frac{2\pi}{l}x_1\right)(\cosh\lambda y + C_{12}\sinh\lambda y) \tag{27}$$

$$B_{\delta,\,x_1} = \sum_n\sum_\nu C_{10}\exp\mathrm{j}\left(\omega t - \nu\frac{2\pi}{l}x_1\right){}^{n}y\left(\cos n\frac{\pi}{L}z\right)\lambda\,(\sinh\lambda y + C_{12}\cosh\lambda y) \tag{28}$$

$$B_{\delta,\,z} = -\mathrm{j}\frac{\pi}{L}\sum_n\sum_\nu \frac{1}{2\pi\nu}{}^{n}y\, n\, C_{10}\left(\sin n\frac{\pi}{L}z\right)\exp\mathrm{j}\left(\omega t - \nu\frac{2\pi}{l}x_1\right)\lambda(\sinh\lambda y + C_{12}\cosh\lambda y) \tag{29}$$

$$B_{2,\,y} = \mathrm{j}\sum_n\sum_\nu \frac{l}{2\pi\nu}\lambda^2 C_{13}\,{}^{n}y\left(\cos n\frac{\pi}{L}z\right)\exp\mathrm{j}\left({}^{\nu}s\omega t - \nu\frac{2\pi}{l}x_2\right)(\cosh\xi y + C_{11}\sinh\xi y) \tag{30}$$

$$B_{2,x_2} = \sum_n \sum_\nu C_{13} \exp \mathrm{j} \left({}^\nu s\omega t - \nu \frac{2\pi}{l} x_2 \right) {}^n y \left(\cos n \frac{\pi}{L} z \right) \xi (\sinh \xi y + C_{11} \cosh \xi y) \tag{31}$$

$$B_{2,z} = -\mathrm{j} \frac{\pi}{L} \sum_n \sum_\nu \frac{l}{2\pi\nu} C_{13} {}^n y\, n \left(\sin n \frac{\pi}{L} z \right) \exp \mathrm{j} \left({}^\nu s\omega t - \nu \frac{2\pi}{l} x_2 \right) \xi (\sinh \xi y + C_{11} \cosh \xi y) \tag{32}$$

For the electric field in the air gap, using $\boldsymbol{E} = -\partial \boldsymbol{A}/\partial t$, it follows that:

$$E_{\delta,x_1} = \frac{\omega\pi}{L} \sum_n \sum_\nu \frac{l}{2\pi\nu} {}^n y n C_{10} \left(\sin n \frac{\pi}{L} z \right) \exp \mathrm{j} \left(\omega t - \nu \frac{2\pi}{l} x_1 \right) (\cosh \lambda y + C_{12} \sinh \lambda y) \tag{33}$$

$$E_{\delta,z} = -\mathrm{j}\omega \sum_n \sum_\nu C_{10} {}^n y \left(\cos n \frac{\pi}{L} z \right) \exp \mathrm{j} \left(\omega t - \nu \frac{2\pi}{l} x_1 \right) (\cosh \lambda y + C_{12} \sinh \lambda y) \tag{34}$$

For the current density in the track, using $\boldsymbol{J} = E/\rho$, then:

$$J_{2,x_2} = \frac{\pi}{\rho_2 L} \sum_n \sum_\nu {}^\nu s\omega \frac{l}{2\pi\nu} C_{13} {}^n y\, n \left(\sin n \frac{\pi}{L} z \right) \exp \mathrm{j} \left({}^\nu s\omega t - \nu \frac{2\pi}{l} x_2 \right) (\cosh \xi y + C_{11} \sinh \xi y) \tag{35}$$

$$J_{2,z} = -\frac{\mathrm{j}}{\rho_2} \sum_n \sum_\nu {}^\nu s\omega C_{13} \exp \mathrm{j} \left({}^\nu s\omega t - \nu \frac{2\pi}{l} x_2 \right) {}^n y \left(\cos n \frac{\pi}{L} z \right) (\cosh \xi y + C_{11} \sinh \xi y) \tag{36}$$

9.4.5 *Magnetic flux, induced voltage*

The flux enclosed by the κth coil of the ρth coil group of the kth phase is;

$$\Phi_{\kappa,\rho,k} = \int_{z=-L/2}^{+L/2} \int_{x=\left[-(mq-\epsilon)\frac{\alpha}{2}+\gamma\right]\frac{l}{2\pi}}^{\left[(mq-\epsilon)\frac{\alpha}{2}+\gamma\right]\frac{l}{2\pi}} (B_{\delta,y})_{y=\delta} f_k(z)\, \mathrm{d}z \mathrm{d}x,$$

$$\gamma = (\rho - 1)\, mq\alpha + (k-1)\, q\alpha + (\kappa - 1)\alpha$$

$$\rho = +1, +2, +3, \ldots 2p.$$

The flux-linkage function $f_k(z)$ depends on the distribution of the winding and corresponds exactly with Equation (8) or Figure 9.2.

By carrying out the integration using Equations (27) and (7):

$$\Phi_{\kappa,\rho,k} = \mathrm{j} \frac{4L}{\pi} \sum_n \sum_\nu \left(\frac{l}{2\pi\nu} \right)^2 \left(\sin \nu \frac{mq-\epsilon}{2} \alpha \right) \lambda^2 C_{10} {}^n y^2$$

$$\exp \mathrm{j} (\omega t - \nu\gamma + \nu\beta) (\cosh \lambda\delta + C_{12} \sinh \lambda\delta)$$

Summation over the coils and coil groups yields the following expression for the flux of the kth phase:

$$\Phi_k = \sum_{\kappa=1}^{q} \left[\sum_{\rho=1,3,5\ldots}^{(2p-1)} \Phi_{\kappa,\rho,k} - \sum_{\rho=2,4,6\ldots}^{2p} \Phi_{\kappa,\rho,k} \right]$$

$$= \sum_{n}^{\infty} \sum_{\nu}^{\infty} \frac{8L}{\left(\nu \frac{2\pi}{l}\right)^2 \pi} {}^{\nu}k''_{\omega} C_{10} \lambda^2 (\cosh \lambda\delta + C_{12} \sinh \lambda\delta)^n y^2$$

$$\exp\left\{ j \left[\omega t + \nu q \alpha \frac{1}{2} (m + 1 - 2k) \right] \right\} \tag{37}$$

The induced voltage is:

$$E_k = -N \frac{d\Phi_k}{dt}$$

$$= -j\omega N \sum_{n} \sum_{\nu} \frac{8L}{\left(\nu \frac{2\pi}{l}\right)^2 \pi} {}^{\nu}k''_{\omega} C_{10} \lambda^2 (\cosh \lambda\delta + C_{12} \sinh \lambda\delta) \exp j \Big[\omega t$$

$$+ \nu q \frac{\alpha}{2} (m + 1 - 2k) \Big]^n y^2 \tag{38}$$

In order to calculate the voltage at the terminals, only the slot leakage (in accordance with Equation (86a) of reference [17], for example) and the ohmic voltage drop of the primary winding remain to be added vectorially to the induced voltage E_k given by Equation (38), since the harmonic-, the tooth-tip, and (part of) the overhang leakage have already been taken into account. In the case of motors with large winding overhang in the y-direction, the part of the winding overhang leakage field which is not linked with the track remains to be added.

The determination of the end winding leakage flux is a difficult task even for conventional rotating machines, (see reference [28]). The linear motor theory already given contains part of the end winding leakage. The rest of it, depending on the special shape of the end windings and pressure plates (magnetic, unmagnetic) can be taken into account by the inductance L_{S1} according to:

$$U_k(2)^{\frac{1}{2}} = R_1 I_{1,k}(2)^{\frac{1}{2}} + j\omega (L_{N1} + L_{S1}) I_{1k}(2)^{\frac{1}{2}} - E_k \tag{39}$$

$$L_{S1} = \mu_0 l_{S1} 2p4q^2 N^2 \lambda_{S1}.$$

With rotating machines, the end winding leakage permeance amounts to about $\lambda_{S1} \approx 0.3$, as is known. Therefore, values between $\lambda_{S1} = 0 \ldots 0.3$ are to be expected with linear motors. For the motor considered in Section 8.4, $\lambda_{S1} = 0.1$ has proved to be adequate. l_{S1} denotes the average length of one-coil at one end of the motor.

9.4.6 *Power in the air gap, track losses*

Using Equations (38) and (2) the power transmitted through the air gap may be obtained:

$$P_{1,\delta} = \sum_{k=1}^{m} \mathrm{Re}\left(\frac{-E_k}{(2)^{\frac{1}{2}}} \frac{I_{1,k}^*}{(2)^{\frac{1}{2}}}\right)$$

$$= \mathrm{Re}\left[\frac{-1}{2} I_1 (2)^{\frac{1}{2}} \omega N \sum_n \sum_\nu \frac{8L}{\left(\nu \frac{2\pi}{l}\right)^2 \pi} {}^{\nu}k_\omega C_{10} \lambda^2\right.$$

$$\left.(\cosh \lambda\delta + C_{12} \sinh \lambda\delta)\, {}^{n}y^2 \exp -\mathrm{j}\frac{\pi}{m2}\right] \tag{40}$$

The track losses of the linear motor are obtained from:

$$P_{v2} = P_{1,\delta} - F_x v \tag{41}$$

where the traction force F_x is to be substituted in accordance with Equation (42).

9.5 Traction force, vertical force

The forces can be calculated advantageously by way of Maxwell's fictitious surface stresses $\boldsymbol{T}$. To do so, it is only necessary to know the fields in the air gap. According to reference[18], the expression for the force vector is:

$$\boldsymbol{F} = \oint\!\!\oint \boldsymbol{T}\, \mathrm{d}a,\ \boldsymbol{T} = \Phi\boldsymbol{n},\ \boldsymbol{n} = \boldsymbol{j},$$

$$\boldsymbol{T} \equiv \boldsymbol{T}_y = \boldsymbol{i}T_{yx} + \boldsymbol{j}T_{yy} + \boldsymbol{k}T_{yz}.$$

The surface of the track over the period l is chosen as the surface of integration. It can be shown that, for reasons of symmetry, only the surface with $\boldsymbol{n} = \boldsymbol{j}$ yields a contribution, where $\boldsymbol{n}$ is the outward pointing normal to the surface. The traction force is therefore obtained from:

$$F_x = \int_0^l \int_{-L/2}^{L/2} T_{yx}\, \mathrm{d}x_1\, \mathrm{d}z$$

$$T_{yx} = \frac{\mathrm{Re}(B_{\delta,y})\, \mathrm{Re}(B_{\delta,x_1})}{\mu_0},\ y = 0$$

and the vertical force (force in the y-direction) is given by:

$$F_y = \int_0^l \int_{-L/2}^{L/2} T_{yy}\, \mathrm{d}x_1\, \mathrm{d}z$$

$$T_{yy} = \frac{[\mathrm{Re}(B_{\delta,y})]^2 - [\mathrm{Re}(B_{\delta,z})]^2 - [\mathrm{Re}(B_{\delta,x_1})]^2}{2\mu_0},\ y = 0$$

After a long calculation, integration yields, using Equations (27) to (29), the relatively simple and easily manageable result for the time-averaged value of the force:

$$F_x = \frac{LI^2}{8\pi\mu_0} \sum_n^{\infty} \sum_\nu^{\infty} \frac{1}{\nu} \lambda^3 {}^n y^2 |C_{10}|^2 \, \mathrm{Im}\,(C_{12}) \tag{42}$$

$$F_y = \frac{LI^2}{32\pi^2\mu_0} \sum_n^{\infty} \sum_\nu^{\infty} \frac{\lambda^4}{\nu^2} {}^n y^2 |C_{10}|^2 (1 - |C_{12}|^2) \tag{43}$$

9.6 Numerical example and field plot

As an example, for a motor for a high-speed traction vehicle, the following specifications are defined: length L_S = 5.25 m, width $2c$ = 0.30 m, number of poles $2p$ = 16, air gap δ = 0.021 m (including the Carter coefficient), track d = 0.01 m, L = 0.56 m, μ_{r2} = 1, ρ_2 = 0.16 x 10^{-6} Ωm, supply frequency f = 200 Hz, number of phases m = 3, number of slots per pole and phase q = 3, reduction of coil width ϵ = 3, number of turns N = 1, winding overhang h = 0.05 m, primary current $I_1(2)^{\frac{1}{2}}$ = 2000$(2)^{\frac{1}{2}}$ A, synchronous velocity v = 126 m s^{-1}, pole pitch τ = 0.315 m, slot pitch τ/mq = 0.035 m (150 slots).

In this case a gap length of L_L = 10 to 20 m proves sufficient; for L_L = –0.210 m (–0.209) the machine can be considered infinitely long, henceforth referred to as 'without longitudinal edge effect'. The upper boundary for the harmonic numbers ν and n are specified in each of the individual diagrams.

Figure 9.6 shows a section of the 3-dimensional field distribution in the air gap

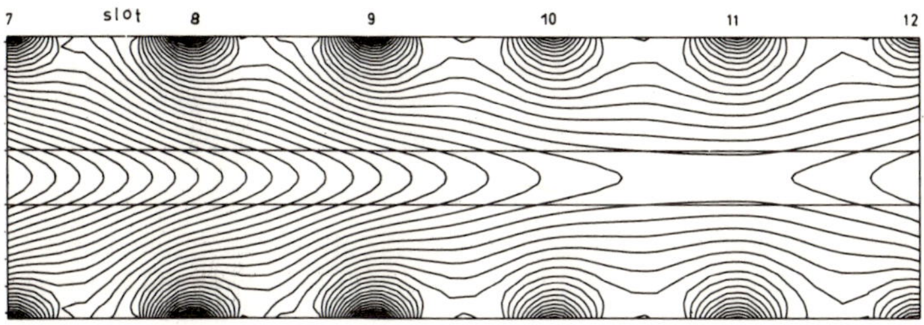

Figure 9.6 Section from the 3-dimensional field distribution in the air gap and the track at approximately nominal speed. Computed for time t = o and z = o, s = 4%, ν_{max} = 650, n = 1. The track runs from left to right. The section shows approximately half a pole at the front of the machine between x_1 = –2.5025 . . . –2.3275 m from a total of 16 poles, 150 slots.

and track. The representation of the field distribution by means of field lines, as in Figure 9.6, is possible only for z = 0, since only in this case is one component of $\boldsymbol{B}$ equal to zero (namely B_z). In all other planes only the points of intersection of the field with the plane of the page can be represented, which is not very illustrative. From this field plot, an extremely interesting result is obtained, that is the fact that the massive track initially refuses to allow the field to penetrate and that it pushes some of the field lines into a direction almost parallel to the track. Only in the middle of the machine do the field lines run in a reasonably straight line across the air gap, provided that the machine is suitably dimensioned.

Figure 9.7 shows the field distribution in the air gap in the longitudinal direction in respect of the important *y*-component. It is recognised that the field reaches its full value only in the last third of the machine (despite suitable dimensioning of the machine). If the electrical conductivity of the track were to be higher, the magnetic time constant of the track would also increase and result in an even slower rate of increase of the field (with a further corresponding loss of traction). Also remarkable is the 'magnetic tail' which occurs behind the machine.

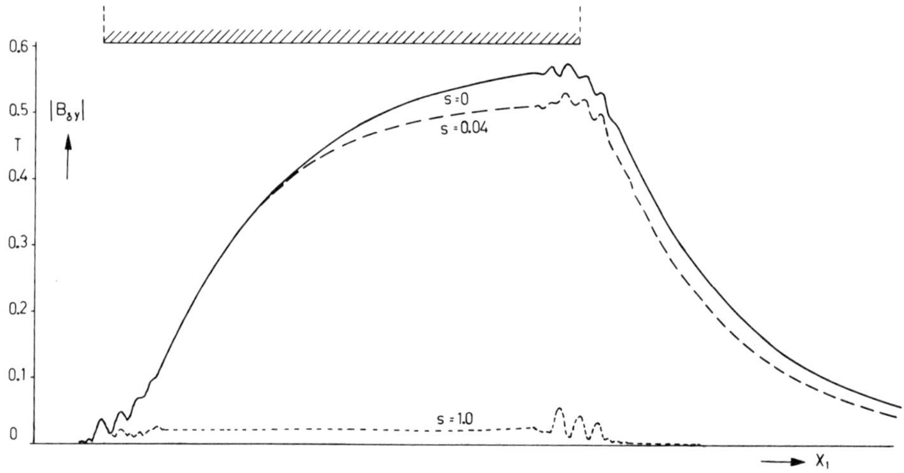

Figure 9.7 Field distribution at the surface of the primary member in the longitudinal direction with the slip s as parameter. The maximum value in time of the y-component is shown. (Computed for $z = o, y = \delta, v_{max} = 90, n_{max} = 5, L_L = 10$ m).

Figure 9.8 represents the field distribution in the air gap along the width of the machine. The increase in flux at the edges at approximately nominal speed and on no load is striking and is important for the calculation of the degree of saturation of the iron.

Figure 9.9 shows the variation of the eddy current density in the longitudinal direction at the centre line of the track. The peaks in current density at the entrance and exit of the machine are quite striking. The fluctuations in eddy current density are due to the alternating fields which occur at the entrance and the exit.

Figure 9.10 shows the variation of the traction force for different track materials. The important conclusion follows that, for a minimum thickness of 10 mm of the track, aluminium is worse than bronze in the entire range, as far as producing traction force is concerned. In the case of bronze, the loss in tractive force as a result of the longitudinal edge effect is still tolerable. In the case of aluminium, the large loss in tractive force is, as was already pointed out, due to the slower rate of increase of the field as a result of the larger magnetic time constant, i.e. it is due to the larger longitudinal edge effect.

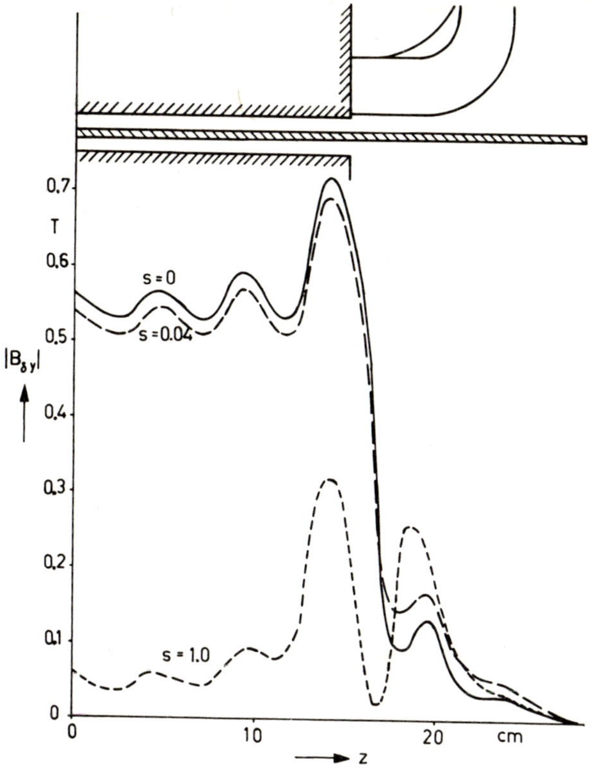

Figure 9.8 Field distribution at the surface of the primary member in the transverse direction with the slip s as parameter. The maximum value in time of the y-component is shown. (Computed for $x_1 = 0, y = \delta, \nu_{max} = 90, n_{max} = 23, L_L = 10$ m).

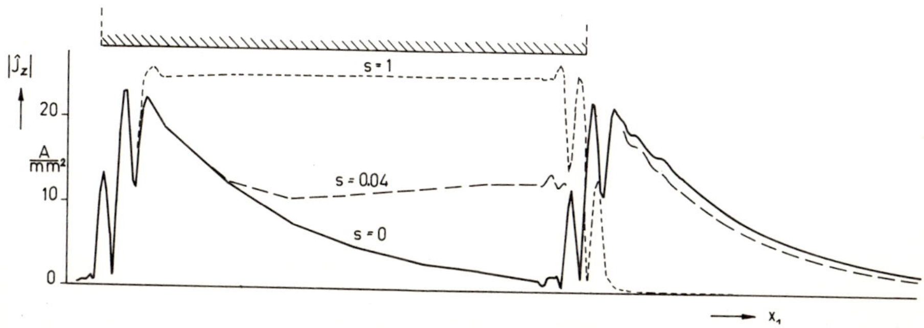

Figure 9.9 Eddy current distribution in the longitudinal direction at the centre line of the track with slip s as parameter. Here it is the maximum value in time of the z component that is shown. (Computed for $z = 0, y = -d/2, \nu_{max} = 90$ and $270, n_{max} = 5, L_L = 10$ m).

Figure 9.11 shows the variation in vertical force F_y, which is produced in the motor and which must be absorbed by the structural members. In the case of the single-sided linear motor, this force may be used for the levitation of the vehicle.

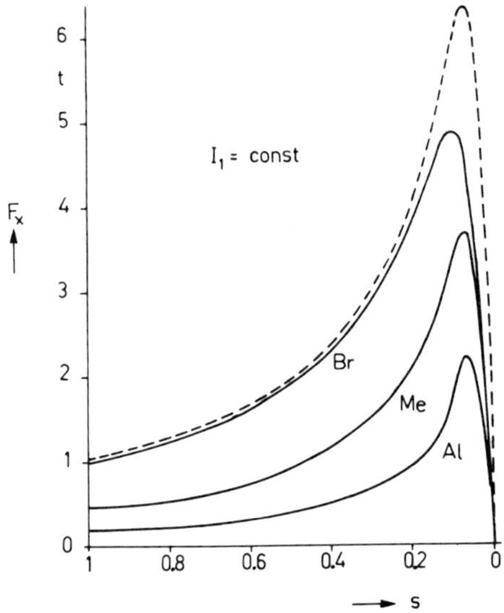

Figure 9.10 Traction force F_x as a function of the slip s for tracks made of bronze ($\nu_{max} = 90$, $n_{max} = 5$, $L_L = 10$ m) for brass (Me) and aluminium ($\rho_2 = 0.286 \times 10^{-7}$ Ωm) of a thickness $d = 0.01$ m. Computed for constant primary current of $I_1\, 2^{\frac{1}{2}} = 200 \times 2^{\frac{1}{2}}$ A. The dotted line applies to the bronze track without longitudinal edge effects ($L_L = -0.2090$ m).

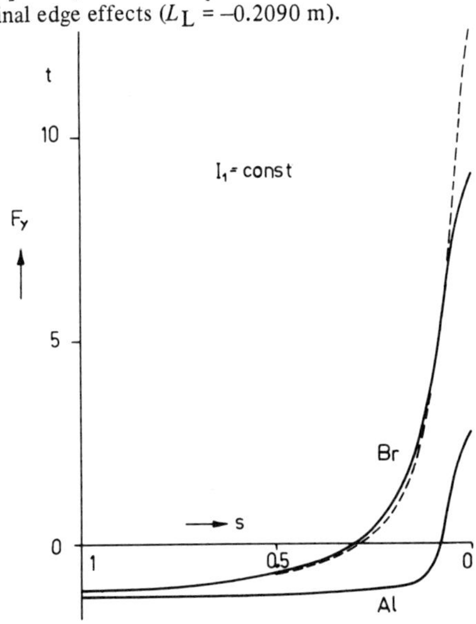

Figure 9.11 Vertical force F_y as a function of the slip s for different tracks of thickness $d = 0.01$ m.
——— Bronze ($\nu_{max} = 90$, $n_{max} = 5$, $L_L = 10$ m), Al = aluminium
– – – – Bronze without longitudinal edge effect. Br = bronze. Computed for constant primary current of $I_1\, 2^{\frac{1}{2}} = 2000 \times 2^{\frac{1}{2}}$ A.

Note especially the change of direction of the force at approximately 3/4 of the synchronous speed. At small values of slip the vertical force is attractive (small eddy currents). At large values of slip it is repulsive (large eddy currents). The vertical force can, according to Figure 9.11, reach approximately twice the value of the traction force (Figure 9.10).

9.7 Convergence

The convergence of the series is very satisfactory. In the z-direction about $n = 7$ suffices. In the x-direction the sum must be taken up to the first pair of slot harmonics with open slots:

$$\nu = \pm (Z_1 + p) \frac{L_S + L_L}{L_S} 1.1, \text{ where } Z_1 \text{ is the slot number,}$$

and up to the second pair of slot harmonics with semi-open slots:

$$\nu = \pm (2Z_1 + p) \frac{L_S + L_L}{L_S} 1.1$$

A detailed treatment of the influence of the slot openings is given in reference [19].

9.8 Comparison between tests and calculation

9.8.1 *The American high-speed linear motor of Pueblo, Colorado*

Several linear motor theories have been published in recent years. However, very few verifiable comparisons between tests and calculations have been made. The linear motor of the American LIMRV at Pueblo, Colorado (which is under the responsibility of the Department of Transportation) now offers the possibility of verifying theories on a full-scale model. Design data and test results are known from references [24] and [25].

(1) *Input data and convergence* The double-sided linear motor has the following nominal values: power output 1865 kW at 400 km h^{-1}, traction force 16.68 kN, 173 Hz, 2000A, 3600 kVA, 1040 V (line to line).

Design data: Number of poles $2p = 10$, length of laminations $L_S = 3.79$ m, width of laminations 0.254 m, mechanical clearance 0.011 . . . 0.022 m, thickness of aluminium rail 0.016 m. Since the rail is hollow, an effective thickness of 0.0072 m is given with a specific conductivity of $\gamma = 24 \times 10^6$ mho m m^{-2} (aluminium alloy).

The end windings are of special shape and are not identical on both sides of the machine, as may be seen from Figure 9.12. For the length h, which is the distance between end winding and lamination according to Figure 9.2, an average value of $h = 0.05$ m has been taken. The influence of h on the results is not very pronounced, i.e. for $h = 0.02$ m the traction force is about 5% lower.

For the additional end winding leakage permeance in Equation (39) which depends on the special shape of the end coils according to Figure 9.12, a value of $\lambda_{S1} = 0.1$ has proved to be adequate (see also Section 9.4.5).

The space L_L, which is necessary between successive primaries, is most easily obtained by calculating the traction force at zero slip, $s = 0$. It has been proved that

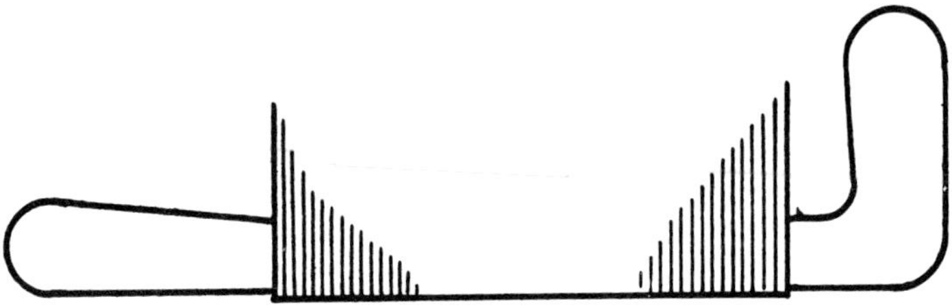

Figure 9.12 End winding arrangement.

$s = 0$ implies the worst case. For slip values $s > 0$, the necessary space L_L is always smaller. Figure 9.13 shows the calculated traction force F_x as a function of the space L_L at a frequency $f_1 = 156$ Hz and at a speed of $v = 400$ km h^{-1}. Obviously, a space of $L_L = 30$ m is sufficient at this velocity. Higher values of L_L would not. change the results but only increase the computing time, since the harmonic order ν_{max} would be increased also.

The space L_L is approximately proportional to the velocity. Therefore, length $L_L = 5$ m is sufficient for $v = 67$ km h^{-1}.

The motor has open slots. Therefore, the harmonics ν have been summed up to the first pair of slot harmonics, as described in Section 9.7. The computing time (on IBM 370/158) was 20 s per slip value (i.e. about 7 dollars, for $\nu_{max} = 430$, $n_{max} = 7$). The influence of the upper limit on the terminal voltage U_{ph} and the induced voltage

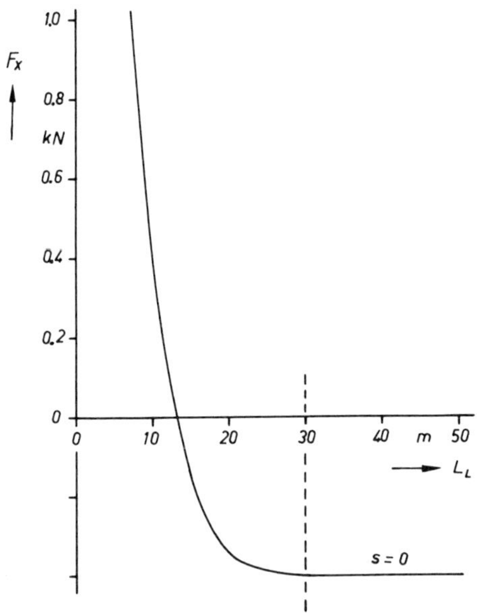

Figure 9.13 Determination of the space L_L. Traction force F_x as a function of the interspace L_L at slip s = o and frequency f_1 = 156 Hz.

E_{ph} is shown in Figure 9.14. It can be seen that only the first pair of slot harmonics with $\nu \approx 370$ and the second pair with $\nu \approx 740$ contribute significantly to the voltages (of course, when $q = 2$ the influence would be greater).

In the z direction, a summation up to $n_{max} = 7$ is sufficient, as far as integrated quantities such as force, voltage, etc., are concerned. The calculation of flux density needs higher values for n_{max}. From the calculations made it can be concluded that the convergence of the series is very satisfactory.

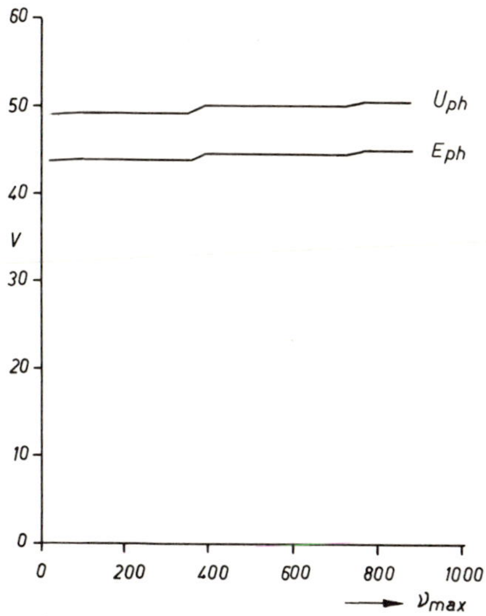

Figure 9.14 Terminal voltage U_{ph} and induced voltage E_{ph} as a function of ν_{max}. Frequency f_1 = 26 Hz, space $L_L = 5.0$ m, $n_{max} = 7$.

All the other input data necessary for the calculation are obtained easily from the design data:

$$L = 0.53 \text{ m}, 2c = 0.254, q = 5, \epsilon = 5, N = 1, m = 3,$$

$$R_{1ph} = 0.0102\ \Omega, \lambda_{S1} l_{S1} = 0.0465 \text{ m}, 2p4q\lambda_{N_1} \equiv \Lambda_{1,1} = 192.$$

(2) *Results and comments* Figure 9.15 shows the traction force F_x as a function of the frequency at standstill and constant current $I_1 = 1500$ A. Figure 9.16 gives the traction force F_x as a function of the slip s at constant voltage $U_{ph} = 50$ V and constant frequency $f_1 = 26$ Hz. The synchronous speed is $v_{syn} = 67$ km h^{-1}. Figure 9.17 displays the current of the primary and Figure 9.18 the efficiency and power factor as functions of the slip s. The calculations agree very well with the measurements. Recently, measurements at high speeds have been carried out as well. Figure 9.19 shows the traction force F_x at constant current $I_1 = 2000$ A and at a frequency of $f_1 = 156$ Hz. Here, the synchronous speed is $v_{syn} = 400$ km h^{-1}.

Measurements, especially of traction force F_x and slip s, at such high velocities are very difficult to make and are of limited accuracy. Improved tests with new equipment

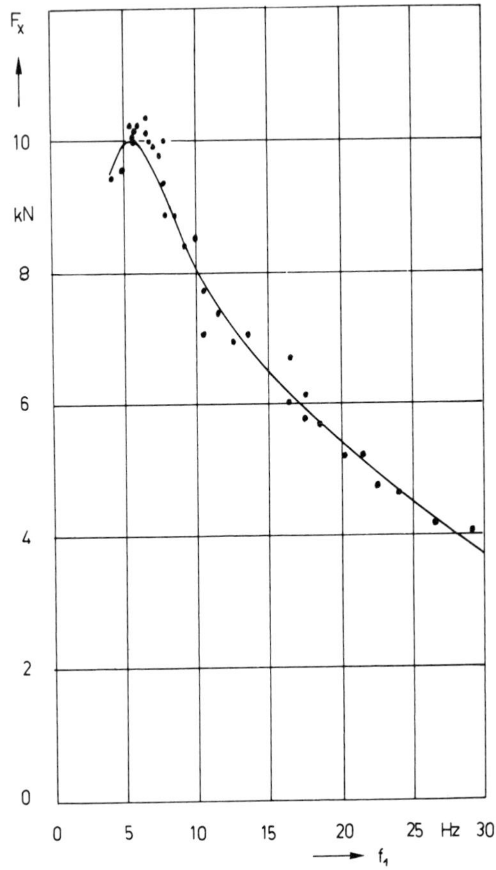

Figure 9.15 Traction force F_x as a function of the frequency f_1 at standstill for constant current I_1 = 1500 A. Distance primary-to-primary 0.0603 m ($2\frac{3}{8}$ inches). .... Test points. ---- Calculation.

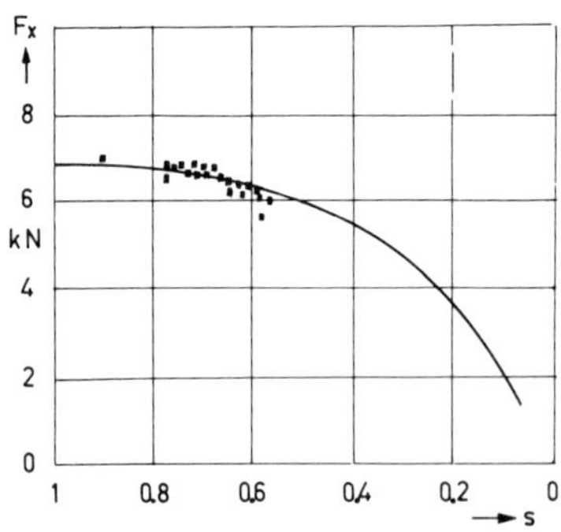

Figure 9.16 Traction force F_x as a function of slip s at constant frequency f_1 = 26 Hz and constant terminal voltage U_{ph} = 50 V. Synchronous speed v_{syn} = 67 km h^{-1}, distance primary-to-primary 0.0381 m ($1\frac{1}{2}$ inches). .... Test points. —— Calculations (with ν_{max} = 430, n_{max} = 7, L_L = 5 m).

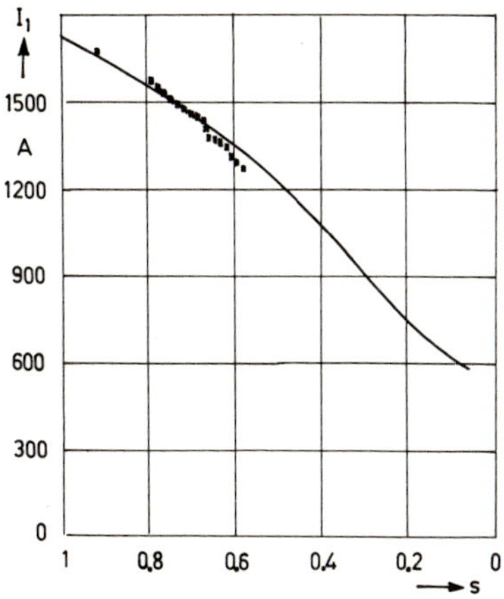

Figure 9.17 Current I_1 as a function of slip s. Motor as in Figure 9.16. ▪ ▪ ▪ ▪ Test points. ——— Calculations.

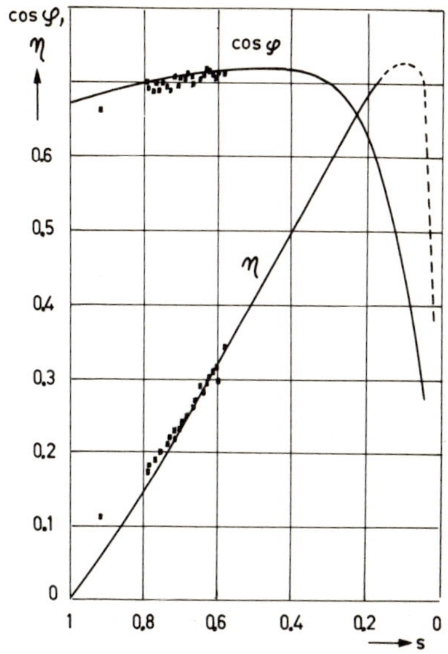

Figure 9.18 Power factor cos φ and efficiency η as a function of slip. Motor as in Figure 9.16. ▪ ▪ ▪ ▪ Test points. ——— Calculation.

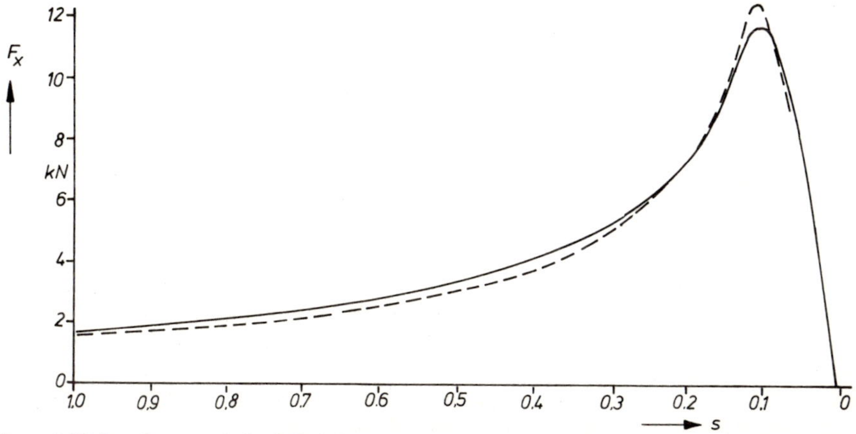

Figure 9.32 (*see footnote below*) Traction force as a function of slip at a frequency of f_1 = 157.1 Hz and constant current I_1 = 2000 A. Synchronous speed v_{syn} = 402 km h^{-1}, primary to primary 0.0381 m ($1\frac{1}{2}$ inches). ---- Tests. ——— Calculations.

are planned. The comparison of Figure 9.19 with the calculations given in Figure 9.20 shows good agreement in the high-speed region. In the low speed region (i.e. for high values of slip s) the force measured is obviously too low, since extrapolation of the curve for $s \geqq 0.78$ leads to negative forces, which is quite impossible. The measurements given in Figure 9.19 must not therefore be considered final. Tests near the pull-out force should also be made.*

* Note added in proof: Improved tests (see Figure 9.32 above) show good agreement with the calculations for all speeds.

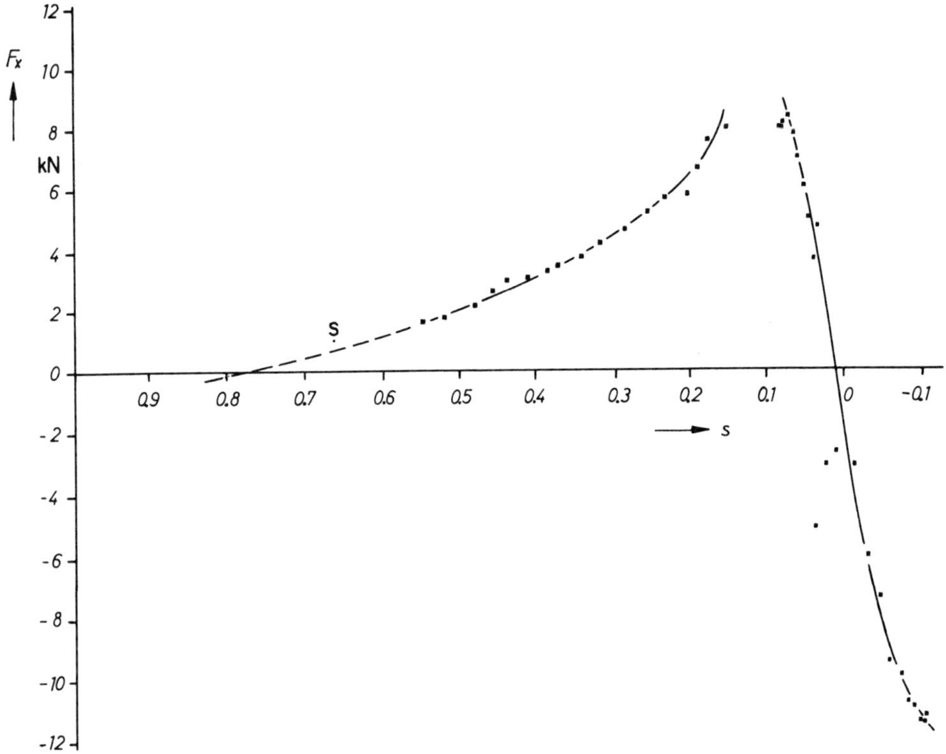

Figure 9.19 Measured traction force as a function of a slip s at a frequency of f_1 = 156 Hz and constant current I_1 = 2000A. Synchronous speed v_{syn} = 400 km h^{-1}, primary-to-primary 0.0476 m ($1\frac{7}{8}$ inches). Measurement not final. ■ ■ ■ Test points.

The no-load slip s_0 is very low, i.e. about 1%, as can be seen from tests and calculations according to Figure 9.19 and Figure 9.20. This is quite remarkable at this high synchronous speed. Similar experience has been obtained with other linear motors, provided that they had double-layer windings. With single-layer windings, values of no-load slip s_0 = 10 . . . 20% may be obtained, since with these windings the m.m.f. at the exit end of the motor is not zero, see reference[30].

9.8.2 *Investigations on a six-pole rotating model with a stator segment*

Further comparisons were made on a six-pole linear motor with a rotating copper reaction rail at supply frequencies of 50 Hz and 240 Hz. With this model velocities up to 115 km h^{-1} were obtained. The influence of the air gap harmonics was remarkable at high velocities. From measurements and calculations, it was found that in this motor all teeth with numbers 1, 3, 5, . . . have a notably higher flux density than teeth with numbers 2, 4, 6 . . . Further experiments indicated that the influence of additional laminations at the entry or exit edge was negligible.

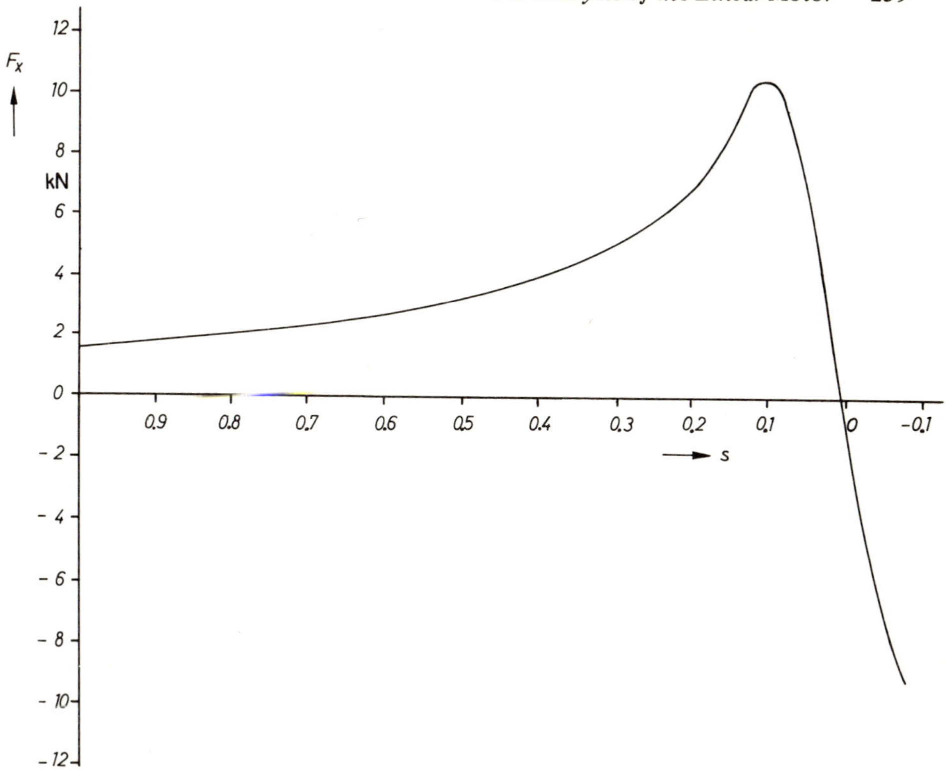

Figure 9.20 Calculated traction force F_x as a function of slip s = (v_{max}· 1620 n_{max} = 7, L_L = 30 m). Otherwise as in Figure 9.19.

(1) *Design data* The stator segment of the double-sided motor had a length of 0.528 m, a width of 0.10 m and a full pitch 'compact' double-layer winding with $q = 2$ slots per pole and phase. Other data are: $N = 12$ turns per coil, slot pitch 0.011 m, axial extension of the end coils $h = 0.03$ m, Carter's coefficient 1.09, number of stator slots $Z_1 = 48$, slot openings semi-closed. Because the winding with $2p = 6$ necessitated only 42 slots, 6 slots were empty. The empty slots were placed at the entry or exit edge in order to show the influence of additional laminations and in consequence the validity of the corresponding assumption in the mathematical model. The rotor consisted of a copper annulus (not a disc) of 2.0 m diameter, thickness 0.0673 m, width 0.23 m, air gap 0.020 m, resistivity $\rho_2 = 0.206 \times 10^{-7}\ \Omega$m.

(2) *Test equipment* The torque was measured with an oscillating dynamometer. The tooth fluxes were obtained from search coils on top of the slot wedges. The pitch of the search coil was equal to the slot pitch.

(3) *Results and comment* The differences in force, currents, fluxes, losses etc. due to additional laminations at the entry or exit edge were very small, i.e. almost within

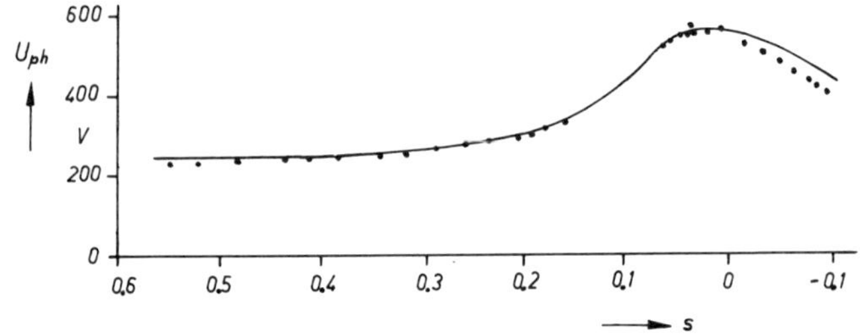

Figure 9.21 Terminal voltage as in function of slip *s*. Motor as in Figure 9.19. ..... Test points.
——— Calculation.

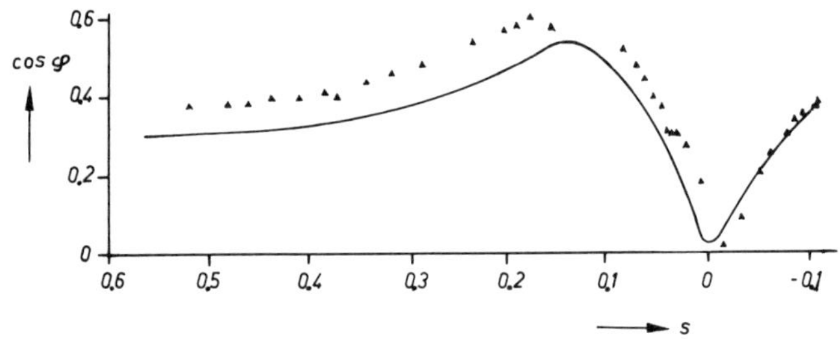

Figure 9.22 Power factor cos φ as a function of slip *s*. Motor as in Figure 9.19. ▲▲▲ test points.
——— Calculation.

the accuracy of the measuring equipment, even for 240 Hz (115 km h^{-1}). The empty slots at the entry edge had no influence at all, whereas with the empty slots at the exit edge the motor performance was slightly improved. All measurements shown in the following figures were obtained with the empty slots at the entry edge.

Figure 9.23 shows the calculated and measured thrust at 50 Hz; Figure 9.24 shows the same quantities at 240 Hz (with a corresponding synchronous speed of 115 km h^{-1}). It can be seen from Figure 9.24 that the air gap harmonics have a considerable influence, although they have never been mentioned in the relevant literature on linear motors. The upper limit of the sum was ν_{max} = 200 and ν_{max} = 500, thus taking the second pair of slot harmonics into account in both cases, according to the rule given in Section 9.7.

Figure 9.25 gives efficiency as a function of the speed. Figure 9.26 displays the distribution of the tooth-fluxes within the motor. Obviously, teeth with odd numbers have a remarkably higher flux than teeth with even numbers. This is because the winding has two slots per pole and phase (q = 2) which leads to a higher flux in the inner coil. The calculated fluxes for the teeth with number 32 and 33 show a difference of 20%, see Table 9.1, and are in agreement with the measurements. The calculation was rather lengthy because it necessitates integration in the directions of both x and z. It is therefore much easier to calculate the y-component of the air gap field at $z = 0$ for discrete x values, as shown in Figure 9.27.

Figure 9.26 and 9.27 show that the air gap field reaches its nominal value near to

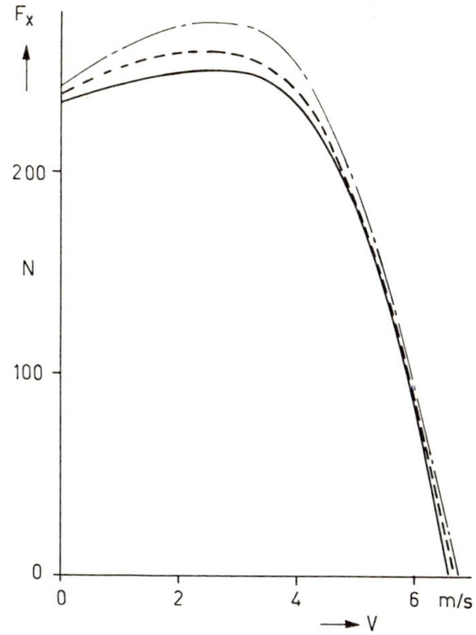

Figure 9.23 Thrust F_x as a function of speed 50 Hz. —·— Calculated with ν_{max} = 13 (fundamentals only), n_{max} = 7, L_L = 0.5 m. – – – – Calculated with ν_{max} = 200 (harmonics up to the second pair of slot harmonics included). n_{max} = 7, L_L = 0.5. ——— Measured.

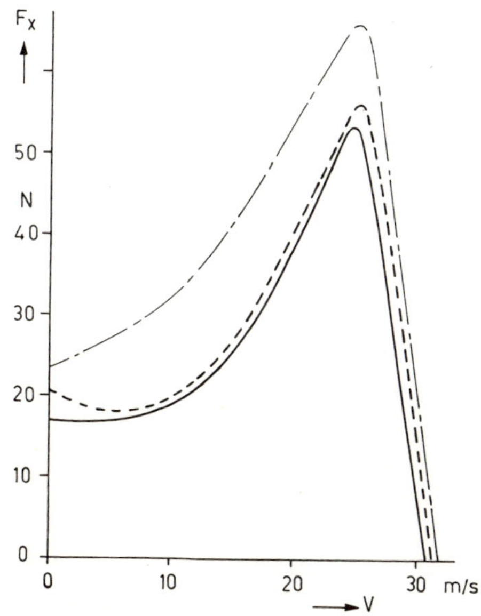

Figure 9.24 Thrust F_x as a function of speed at 240 Hz. ———–——— Calculated with ν_{max} = 30 (fundamentals only), n_{max} = 7, L_L = 2.0 m. – – – Calculated with ν_{max} = 510 (harmonics up to the second pair of slots harmonics included), n_{max} = 7, L_L = 2.0 m. ——— Measured.

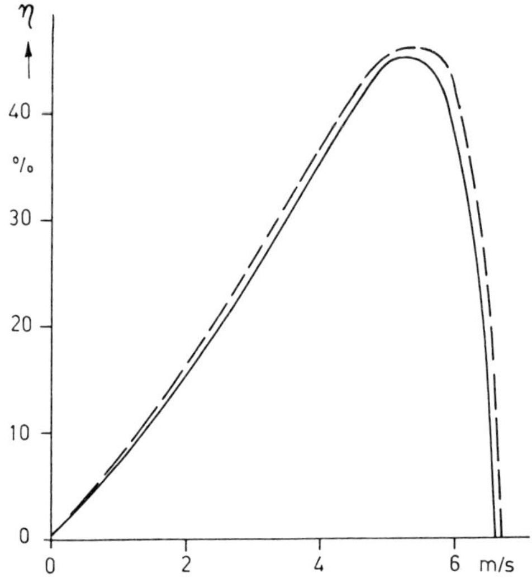

Figure 9.25 Efficiency as a function of speed at 50 Hz. ——— Calculated with $\nu_{max} = 200$, $n_{max} = 7$. ——— Measured.

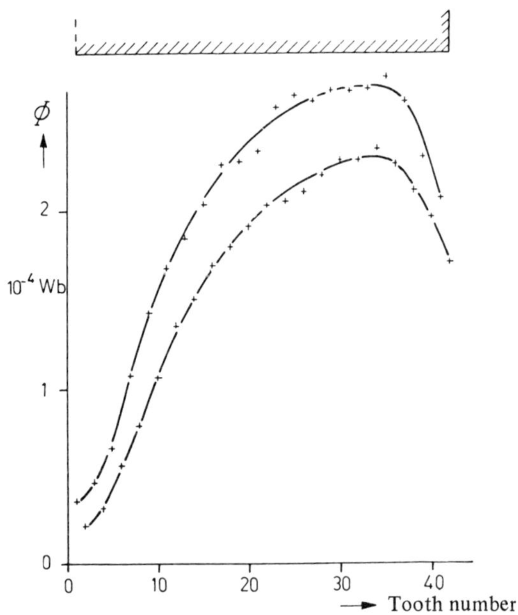

Figure 9.26 Tooth fluxes in the motor (mean values in time). Upper curve: teeth with odd numbers. Lower curve: teeth with even numbers. Measured at slip $s = 0$, $I_1 = 13.7$ A, $f_1 = 50$ Hz.

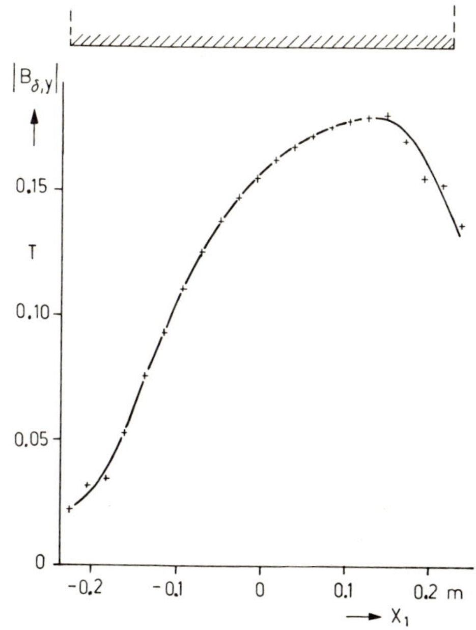

Figure 9.27 Amplitude of the air gap flux density $|B_{\delta,\ y}|$ as a function of length x_1 at the surface of the primary. Calculated for $s = 0$, $z = 0$, $\nu_{max} = 200$, $n_{max} = 7$, $f_1 = 50$ Hz, $I_1 = 13.7$ A.

the exit edge of the motor. At the entry edge, the air gap field is cancelled almost completely by currents in the reaction rail.

Figure 9.28 gives the calculated air gap field in the x-direction at the surface of

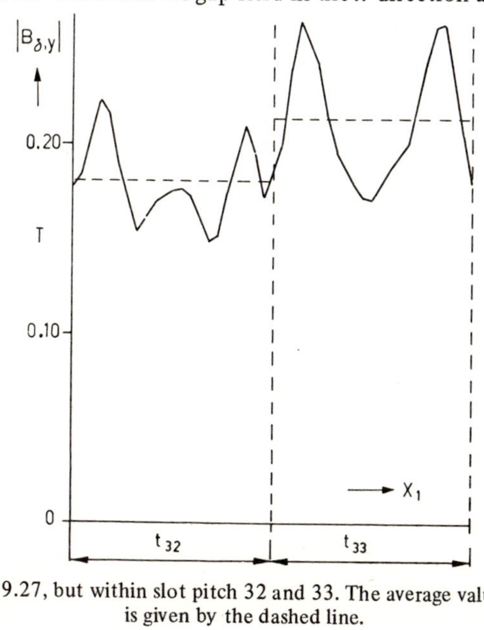

Figure 9.28 As Figure 9.27, but within slot pitch 32 and 33. The average value of the flux density is given by the dashed line.

teeth 32 and 33 at $z = 0$. Because ν was summed up to ± 200 the second pair of slot harmonics may be recognised in Figure 9.28. The mean flux densities of teeth 32 and 33 are 0.181 T and 0.214 T respectively (at $z = 0$).

Figure 9.29 shows the air gap flux density in the z-direction at $x_1 = 0.1265$ m. Table 9.2 gives the measured and calculated currents of the primary at different values of slip s (averages of the three phase currents).

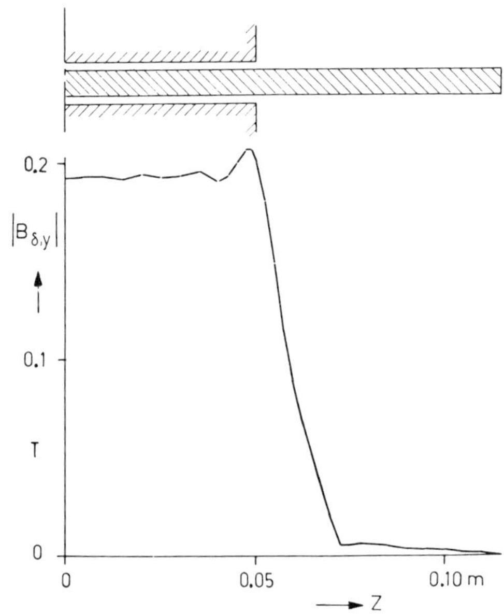

Figure 9.29 Amplitude of the air gap flux density $|B_{\delta,y}|$ as a function of z (i.e. in the transverse direction) at the surface of the primary. Calculated for slip $s = 0$, $x_1 = 0.1265$ (i.e. slot 33), $\nu_{max} = 13$, $n_{max} = 35$, $I_1 = 13.7$ A.

Table 9.1

Amplitudes of tooth fluxes for slip $s = 0$, 50 Hz, $I_1 = 13.7$ A. Calculated values for $\nu_{max} = 200$, $n_{max} = 35$.

Tooth	32	33
Calculated	2.20×10^{-4} Wb	2.60×10^{-4} Wb
Measured	2.30×10^{-4} Wb	2.72×10^{-4} Wb

Table 9.2

Measured and calculated currents I_1 for 50 Hz and 240 Hz at different values of slip *s*.

Supply	s = 1.0	0.8	0.6	0.4	0.3	0.2	0.1	0
50 Hz, 115 V, Y, coil groups in series, measured (Amperes)	26.0	24.8	22.7	19.6		15.8	14.5	13.7
Calculated (Amperes)	26.5							13.7
240 Hz, 250 V, Y, coil groups in series, measured (Amperes)	16.9	16.7	16.4	15.8	15.0	13.3	10.7	10.1
Calculated (Amperes)	17.2							9.9

9.9 Special double-layer windings for linear motors

Various types of three-phase double-layer winding are feasible with linear motors. A so called 'extended' double-layer winding has been proposed in references [26] and [27]. The arrangement of the winding is shown in Figure 9.30 and contrasted with

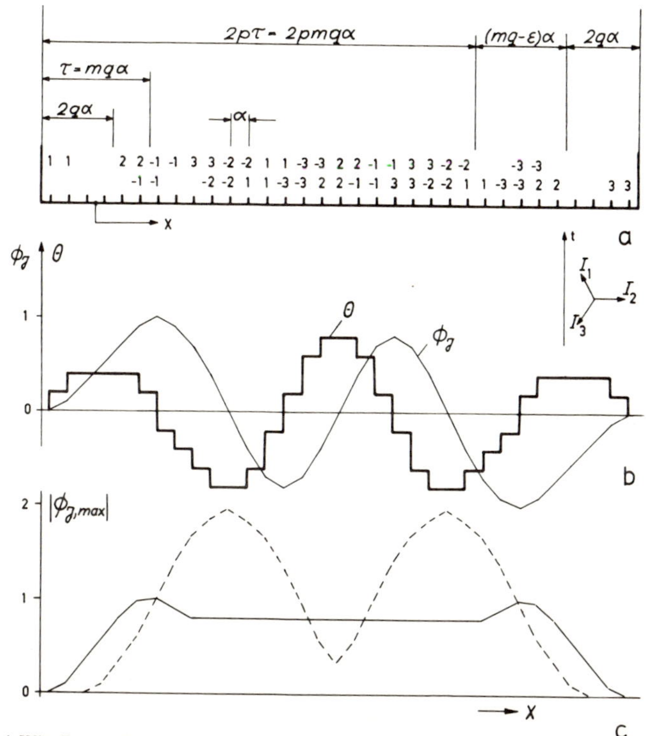

Figure 9.30 (a) Winding and notation

(b) M.m.f. distribution and core flux ϕ_J for an extended double-layer winding with 4 poles

(c) Amplitude of the core flux $|\phi_{J,\,max}|$ for the compact (– – – – –) and extended (———) winding. The rotating machine has $|\phi_{J,\,max}| = 0.81$.

the conventional (compact) winding of Figure 9.31. It can be seen from these figures that, based on the assumption of no iron saturation and of negligible reaction effect of the secondary, the core flux density is much lower for the extended winding. Furthermore, the longitudinal edge effect is also lower with the extended winding, i.e. the efficiency is somewhat higher (see references [26] and [27]).

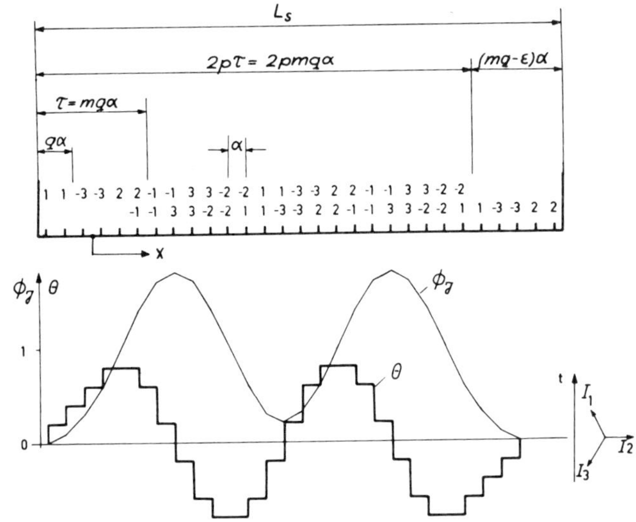

Figure 9.31 M.m.f. distribution θ and core flux $\phi_{J,\ max}$ for a compact double-layer winding with 4 poles. Phase numbers are denoted by 1, 2 and 3.

References

[1] Okhremenko, N. M. (1964) 'Magnetfeld der flachen Induktionspumpe', *Elektrischestwo*, pp. 18–26. In Russian

[2] Laithwaite, E. R. (1966) *Induction Machines for Special Purposes*, Butterworth, London

[2a] Lee, C. H. (1967) 'Study of linear induction motor and its feasibility for high speed ground transportation', U.S. Department of Transportation Report No. 67

[3] Poloujadoff, M. and Sabonnadière, J. C. (1968) 'Utilisation d'une méthode de partition dans la résolution de certaines équations aux derivée partielles dont le domaine comparte une bande infinie', *C. R. Acad. Sci. Paris,* **266**, 230–233, 272–275, 1412–1415

[4] Budig, K., Timmel, H. and Dittrich, W. (1968) 'Auslegung von Linearmotoren als Antrieborgane für Fahrzeuge', *Wiss. Z. d. T. H. Karl-Marx-Stadt,* **10**, 321–334

[5] Preston, T. W. and Reece, A. B. J. (1969) 'Transverse edge effects in linear induction motors', *Proc. IEE*, **116**, 973–979

[6] Waltke, G. (1969) 'Uber das Betriebsverhalten des Wanderfeld-MHD-Wandlers und des asynchronen Linearmotors, Dissertation TU-Braunschweig

[7] Nasar, S. A. (1969) 'Electromagnetic fields and forces in a linear induction motor, taking into account edge effects', *Proc. IEE*, **116**, 605–609

[8] Bolton, H. (1969) 'Transverse edge effect in sheet-rotor induction motors', *Proc. IEE*, **116**, 725–731

[9] Wang, T. C. (1969) Linear Induction Motor for High Speed Ground Transportation. *AICE* pp. 964–974

[10] Ooi, B. T. and White, D. C. (1970) 'Traction and normal forces in the linear induction motor', *IEEE Trans. Power Appar. Systs.*, **89**, 638–645

[11] Schulz, T. (1970) 'Langsendeffekte von induktiven Dreiphasen-MHD-Wandlern mit offenem magnetischem Kreis und ihre Simulation an Segmentstator-Asynchronmaschinen', *Wiss. Ber. AEG-Telefunken*, **43**, 139–152

[12] Weh, H., Braess, H. and Mosebach, H. (1971) 'Die rechnerische Behandlung asynchroner linearer Wandler', *Energy convers.*, **11**, 25–37

[13] Gutt, H. J. (1971) 'Wanderfeldmotoren im Vergleich zu Drehfeldmaschinen herkömmlicher Bauart', *Elektrotech. Z. Ausg. A*, **92**, 342–347

[14] Yamamura, S., Ito, H., and Ishikawa, Y. (1971) 'Einfluss. des Randeffektes auf die Eigenschaften des Linearmotors', *J. Inst. Electr. Eng. Japan,* 91, 145–153. In Japanese.

[15] Jufer, M. (1971) 'Determination des caracteristiques specifiques du motor lineaire', *Rev. Gen. Electr.*, **80**, 105–113

[16] Angst, G. (1962) 'Polyphase induction motor with solid rotor', *Trans. IEEE*, **80**, part III, 902–909

[17] Oberretl, K. (1965) 'Die Oberfeldtheorie des Käfigsmotors unter Berücksichtigung der durch die Ankerrückwirkung verursachten Statoroberströme and der parallelen Wicklungszweige', *Archiv. Elektrotech.*, **49**, 343–364

[18] Oberettl, K. (1963) 'Die Ermittlung von magnet. Feldern, Wirbelströmen und Kraften in komplizierten Fällen durch Simulation an Gittermodellen', *Archiv. Elektrotech.*, **48**, 297–313

[19] Oberretl, K. (1970) Das zweidimensionale Luftspaltfeld einer Drehstromwicklung mit offen Nuten', *Archiv. Elektrotech.*, **53**, 371–381

[20] Oberretl, K. (1965) 'Uber den Einfluss von parallelen Wicklungszweigen, Dreieckschaltung, Spulensehnung, Nutschlitzbreite und Nutenschrägung auf das Drehmoment von Käfigläufermotoren,. *Elektrotech. Z. Ausg. A*, **86**, 619–627

[21] Nürnberg, W. (1963) *Die Asynchronmaschine* Springer, Berlin, Göttingen, Heidelberg.

[22] Rummich, E. (1972) 'Der Sektormotor und seine theoret'. Behandlung *Bull. SEV.*, **63**, 1485–1490

[23] Oberretl, K. (1973) 'Three-dimensional analysis of the linear motor, taking into account edge effects and the distribution of the winding', *Archiv. Elektrotech.*, **55**, 181–190

[24] Engineering staff: Linear induction motor research. AiResearch Manufacturing Company. DOT-Report 71-7094 Vol. III (1971)

[25] Chirgwin, K. M. (1974) 'Test results from the U.S. linear induction motor research vehicle programme', IEE-Conference Publ. No. 120 pp. 236–243

[26] Oberretl, K. (1974) 'Linear motor with compact or extended double layer winding', *Archiv. Elektrotech.*, **56**, 55–58

[27] Oberretl, K. (1974) 'Linear motors with special double-layer windings', Conference on Linear Electric Machines, London October 1974, IEE-Publication No. 120

[28] Oberretl, K. (1963) 'Leakage fields, eddy current losses, temperature rise, forces and parasitic sheet-fusion in the end zone of turbo-generators', *Elektrotech. Maschinenbau*, **80**, 539–550. English translation available on request.

[29] Dannan, J. H., Day, R. N., and Kalman, G. P. (1973) 'Linear induction motor propulsion system for high-speed ground vehicles', *Proc. IEEE*, **61**, No. 5, 621–630

[30] Oberretl, K. (1974) 'Single sided linear motor with cage in the secondary', *Archiv. Elektrotech.*, **56**, 305–319

[31] Gagliano, A., Kant, M. and Mouillet, A. (1972) 'Contribution à l'étude du champ magnétique et des courants induits dans une bande conductrice mince de dimensions finies', *Rev. Phys. Appl.*, 7, 253–260

10

The URBA System

M. BARTHALON

10.1 A fast growing problem: urban transportation

Urban transportation has recently become a major issue due to social and technical evolution, to urban growth and to the nuisances of existing transportation systems. Innovation seems to be the only way out.

10.1.1 *Recent evolution*

The last few years have seen growing signs of commuters' discontent or even rate, for example the troubles in Paris' St Lazare Station and even riots (in Japan). The customer's reactions are more dangerous and more political than those shown for other shortcomings of modern society because urban transportation involves environment, quality of life and the physical and psychological fatigue of the working class.

From a marketing and commercial point of view it is clear that since the year 1900 when buses and the Metro were born, there has been a sudden increase of transportation needs in both quantity and quality, for various reasons: the percentage of urbanised population jumped from 25% to 75% as they moved to the suburbs (Paris surface transport was multiplied by 16), housing rigidity and job instability increased trips in number and length, whilst there was an increase in both the standard of living and of the claims of commuters and trackside dwellers; streets became saturated by the automobile.

In 1971, studies for the 6th French plan showed that to avoid deterioration of city transportation, it was necessary to invest 25 billion francs for the Paris region only. In fact, less than half of the plan has been accomplished because railways both open and underground are too costly, the streets's immediate underground is now encumbered by pipe and cable networks, and automobile traffic prevents open cuttings and underground work involves a high percentage of labour with fast-growing salaries.

The Paris example is fascinating: the RER (Regional express subway) has absorbed most of the investment money of RATP for 15 years and its transport capacity is less than two years of traffic growth. As a result the transportation situation is getting worse. Transportation is always installed too late: airports are a good example of this.

More recently the energy crisis has highlighted the need to dissuade commuters from the automobile: but railway lines are still heavy, noisy and costly and are well suited for centres or suburban trunk lines only. For less dense suburbs and middle-sized towns a railway network is not dense enough to attract customers. Buses are well adapted for end-of-journey distribution. However, they are too slow for secondary axes and rocades. Further their capacity is too low (3000 per hour).

The need for an intermediate capacity and cost system was indicated by URBA, 8 years ago and has since been generally recognised. The recent proposal to use street cars to solve the problem is nothing but an admission of failure to find a solution. The history of science and technology does not offer *one* example where a return to an original system has been successful. The reasons which led to the abandonment of street cars are now more valid than ever.

It follows that there is a large, urgent and unsatisfied need for light distribution systems for close and distant suburbs which should be of low cost, somewhere between those of the bus and the subway.

10.1.2 *Nuisances and shortcomings of existing transportation*

Governments have achieved very little in recent years in the field of transportation because of a number of basic conflicts between users and the internal nuisances, between operators and trackside dwellers, between collective and individual transports, between economy and quality of life, between technical progress and exploitation reliability.

Noise and vibration accounts for 75% of town dwellers' fatigue because of its influence on their nervous systems. In recent years the effect of induced vibration due to railcars on building structures such as Colisée in Rome, Dôme in Milan and others has been discovered. Pollution from transportation accounts for 45% of chemical air pollution. This figure is fast increasing with the success of controls of other pollution sources which are easier to curb. The burnt gases endanger life by promoting bronchitis and cancer; they attack facades and the structures of buildings.

Urban transport accidents account for 50% of the total accident casualties. Many victims remain handicapped for life. Together with time loss in transportation, street accidents account for an important drop in national income (5 to 10%).

The waste of urban territory by autoroutes has cast doubt as to their usefulness; in Los Angeles more than 25% of the city's area is occupied by highways. They are made even more objectionable by the fact that they usually destroy the few remaining forests and open-grounds.

Captive customers are in hopeless situation as public transportation presents an insufficient network and deteriorates further.

10.1.3 *Problems of urban growth and structure*

More recently, urban development has shifted towards medium sized cities, new towns and renovation of suburban centres, which results in modules of 30 to 100 000 inhabitants. Their connection together or to the old centre requires a system capacity of 3000 to 10 000 passengers per hour. Distances are rather large; there is therefore a need for a fast, low cost per mile and nuisance-free system which tires neither commuters nor trackside-dwellers.

Furthermore, a city is a constantly changing system. A low cost system, installed early enough could guide and control a city's growth. It should adapt itself to the city's further development. Its track should be light so as to constitute a network of flexible and additive meshes which could be easily changed from time to time.

The present accumulation of unsolved problems regarding transportation is a typical deadlock, not the first to be encountered throughout society's evolution. It stems from the fact that cities are trying to solve 1975's transport problems with the technical solutions of 1900 when every aspect of the 1900 situation has dramatically changed. No existing industrial business could afford such a widely open gap between market evolution and product evolution.

10.1.4 *Only one escape – innovation*

Throughout man's history, whenever he was faced with an apparently impossible situation, the only possible escape was to use his brain to invent and develop new solutions for the new problems. Innovation is the only key to the future in cities, indeed as it is everywhere else.

The I.R.T., the official French transportation research institute has reckoned and publicised the fact that subways and buses alone will not meet the challenge. It has established the need and a detailed specification for a new distribution system, with low cost investment. The trouble is that existing monopolies have, because of their large size, a remarkable resistance to change and innovation.

A second problem is that more than 400 new systems have been proposed by inventors, promoters or industry, most of them not market orientated, badly designed, undeveloped, potentially unreliable and virtually valueless.

10.1.5 *How to conceive and develop a successful innovation*

Let us abandon for the time being the idiosyncrasies of the transportation world in order to see how industrial innovators proceed: they first discover a need which should become a market after the 15 years necessary for development; within the need they locate a specific gap, they analyse the operating qualities necessary to fill it and they establish a detailed specification. Then they design a product structure meeting the specification and capable of producing a feasible and reliable machine. That structure then becomes the guide-line of a coherent development programme which will strive to accumulate the maximum amount of experience and know-how along that same line and to clear uncertainties and unknowns. This avoids money-wasting sidetracks which lead to an excessive variety of solutions with the only result of disturbing financiers, customers and Governments.

Throughout such a systematic approach modern electronics allows the engineer to develop the invention on test rigs and benches as well as on computers. He will not waste his money on building costly experimental vehicles and tracks. At the end of ten years of such work, he will need only one experimental track/unit to *check* the predicted results on only one vehicle which should operate successfully right from the start, as do aircraft. How does this translate in the transportation world?

Figure 10.1 shows the basic structure of URBA. This has remained unchanged from the original, which was the supporting backbone of subsequent development programmes on which improvements were based.

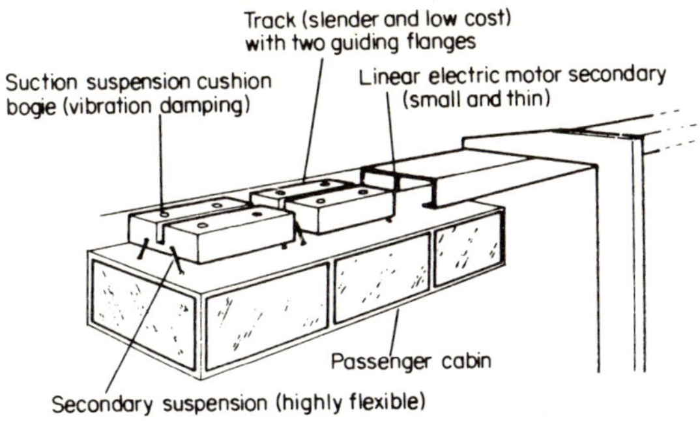

Figure 10.1 URBA's basic suction suspended structure.

10.2 Making the basic decisions

10.2.1 *Forecasting the user's requirements*

The first and most important step is to imagine how the towns of the twenty first century will develop and to try to determine what the transportation needs will be at that time. At first sight it appears that each time the population is offered a personalised and flexible transportation means such as the automobile, it chooses it immediately as a basic system. '*A fortiori*', such a reaction will be worldwide at the end of this century, when the standard of living will have increased. Furthermore constraints will be added, mainly as far as nuisances are concerned: noise, fumes, visual intrusion, etc. *All prospective views in the transportation field must therefore aim at a more attractive system than the automobile.*

It is useful to analyse the reason for this almost automatic choice made by the population in favour of such transportation means: first of all, the main reason is quite often the saving in actual transportation time, particularly each time the existing links are poor, or when door-to-door liaisons imply too large a number of interchanges and too large a time or distance between the stations (departure and arrival) and the destination points. However, inside this transportation time, the duration that the user resents as the most disagreeable one is the waiting time, rather than the time elapsed when the vehicle is in motion or when he is sitting in his automobile and listening to the radio. This is partly due to the discomfort of this waiting time: queueing, pushing, freezing, getting wet, being in a draught, etc.

The appeal of the automobile contains the notion of the 'personal sphere' described by the psychologist. The human being resents all interferences within a sphere of 1 metre around him, but greatly appreciates a freedom of behaviour, of choice and of interest in and from this sphere. For example, even if the journey duration is longer, he will prefer the bus which will allow him to sit, which will offer a landscape or an amusement and which drives nearer to his home.

Finally, and more generally speaking, it must be noted that the automobile, due to its flexibility, fits better into the year-by-year evolution of towns.

Journey time problem in the future is well illustrated by Table 10.1. Note the importance of the interchanges and the waiting times in travel using the present Metro system: one quarter only of the duration is used for the actual running time of the train.

Figure 10.2 URBA's installation on the city highway. This photomontage gives evidence of the small visual intrusion and the aesthetic value of the slender elevated track of URBA. The cost of the track is relatively low compared to that for other transportation systems.

Table 10.1 also shows the two possible shortcomings in the PRT principle. It replaces the waiting and stop times in the several stations along the line by a waiting time in the departure station while vehicles which are not programmed for the arrival station are passing by. Furthermore, it leaves unchanged the wasted time outside the system (50% of the total).

To sum up, the study of needs at the end of the century shows evidence that the evolution which starts now will lead to a transportation system comprising individual duo-mode vehicles, travelling individually at low speed on urban street networks, and integrated into a system at high speed. But we shall also see that it is a long road from proposal to reality.

Table 10.1

Average displacement time, minutes

	1972 Metro as a reference	URBA first generation (Figure 10.2)	PRT*	URBA bus mode SDM	Individual duo-mode
Home to departure station Arrival station to work	25	25	25	14	14.5
Travel station to station					
Constant speed	12.5	10	10	10	10
Waiting time, acceleration and braking time, stops	12.5	6	3	6	0.5
Total	50	41	38	30	21.5

*personalised rapid transit

10.2.2 *Establishing a coherent specification and development programme for a transportation system*

In the highly competitive situation which prevails presently in the development of new transportation means (more than 400 projects), the successful team will be the one which studies transportation means as a complete, integrated system. However, quite often this study is only made 'on paper' and in a very theoretical fashion. The study on the type of solutions which are proposed should be made in depth, on the chance that it will lead to a solution which works, on the internal and external coherence of the electromechanical system.

Such coherence must be good not only for the immediate application within the next few years, but also for the next fifty years. This leads to a study of a whole family of successive systems, of which the basic operating constraints should be examined in order to find their consequences on system and components design. A serious study of a family of systems must also identify the technical and technological difficulties, the fundamental problems at component level which impose the critical path of each stage and which will have to be solved one by one. Finally, the solutions to these problems will have to be incorporated progressively in the successive systems of the family.

When the specification of the final system has been roughly set up, as a function of this evolution, it is necessary to consider today's technical solutions, their state of development, their potential evolution and to appraise them in the light of the technical and technological problems common to all systems which do not comprise trains of coupled vehicles.

All this cannot be successful without a methodical approach, making use of all modern tools of research and development and the most elaborate methods of creativity, evaluation, industrialisation and customer acceptance.

This requires a prospective view. Therefore, a viable system family can only be conceived by projecting oneself in the long term future, and then working backward by studying as a master pattern of the family, the system immediately feasible, but retaining the ability to evolve towards the next systems of the family.

This whole process is time-consuming and quite complex. It could only be totally mastered by starting very early in 1966, five years before the urgency and importance of urban transport needs had begun to be realised by the industry. The conception in 1966/1968 of the several URBA systems presently developed follows the above general rules and Section 10.5 relates how this was achieved.

10.2.3 *Consequences on evolution of transport systems*

The above facts throw some light on the forthcoming evolution of transport systems. A first generation will see the development of new solutions at the level of system and vehicle components: propulsion, suspension and sophisticated controls with correlated improvements in reliability, in installation and exploitation costs, in station-to-station travelling time and in traffic capacity for a given investment.

A second generation will make good use of these breakthroughs at system level to build what could be called simplified public duo-mode systems (SDM) involving large capacity vehicles (30 to 40 seats) assembled in trains on the main trunk and operating like buses off the line from intermediate or end stations. Time-distance between trains remains reasonably within presently forecasted possibilities (45 to 60 seconds).

The third generation has become known as Personal Rapid Transit Systems which offer small vehicles going non-stop station to station. A number of fascinating demonstrations have been organised for the Press and Officials on a small scale in the period 1972–1975. The only trouble is that most of these systems, when built for commercial operation, have been disappointing to say the least and have, in at least one instance, led Governments to abandon PRT.

The last generation could divide the travel time by three and can be called individual duo-mode (I.D.M.). The 'bubbles' or personal spheres could be carried by pallets at high speed on the main line and dropped for autonomous operation at widely distant stations. However, such a system requires first the development of five basic ingredients:

. 100% fail safe and weather proof, braking and passive switching
. vehicles with 0.99999 minimum reliability (which means 0.998 at system's level)
. sophisticated control systems for 100% safe and reliable automated operation with 5 to 15 seconds vehicle intervals
. a rendez-vous system, weather and fool-proof, to insert station outgoing vehicles in the flow of incoming main line vehicles
. a radically lower cost for the infrastructure (track, controls, computers, etc..)

10.3 Basic problems of personal rapid transit systems and duo-mode

Let us now review in more detail the five main problems of PRTs, switching, reliability, automatics, rendez-vous and cost, to facilitate better forecasting of their future and to help shape a satisfactory basic design.

10.3.1 *Reliable braking and passive switching in all circumstances*

All duo-mode or personal rapid transit systems include small or large vehicles driven at a few seconds interval and able to leave the main track and to return to it. This involves an important safety problem, which is illustrated by what will happen on an icy day on a downward slope of 5 to 10% when one of the vehicles suddenly comes to a stop. A simple calculation shows that a very slight difference in the deceleration rates between the vehicles leads to dramatic collisions. This risk is still greater when it must be controlled in the case of sudden power failure.

A similar safety problem arises for vehicles on the passive switch under the same weather conditions and electrical power failure, with one aggravating factor: the requirement of stopping without hitting the core of the switch or the vehicle located on the main track. One can see the risks incurred in the event of superposition of two or more of the preceding cases.

From all this, one can conclude that no compromise is possible on braking and passive switching safety; safety must be absolute and independent of operating conditions, track or vehicle situation and power supply. This cannot be achieved if traction or (and) braking is achieved through a wheel with erratic tangential friction on a track or a rail.

Protection methods against shocks, inspired from the automobile, (such as a padding or shields on the vehicle, the track, the switches, etc.) are, for the time being, the only answer to these safety problems. But, by their nature, these safety methods are a major impediment to a fast acceptance of PRTs by the public bodies in charge of transportation, especially because, as in the motorway accidents, PRT accidents will cause chain collisions in which several tens of vehicles will be involved.

Only linear motors and wheel-less suspension with direct braking action on the track can achieve satisfactory braking whatever the weather or power supply. Safety is higher in the case of systems suspended from above because snow and fallen objects could lie on the track in the other systems.

10.3.2 *Reliability of controls and vehicles*

One of the most obvious characteristics of sophisticated systems is the large number of interdependent elements.

It is known in probability calculations that the reliability F of a family of n components of reliability ($f = 1 - \epsilon$) is:

$$F = (1 - \epsilon)^n \simeq 1 - n\epsilon$$

where ϵ is the component failure rate.

The origin of this large number lies primarily in the automated controls and in the computers and is a well-known problem to the specialist in electronics. These systems involve a huge number N of interdependent components which are (as has been shown during the experiments of ARAMIS, BART, Morgantown and Dallas) submitted to the contingencies of atmospheric conditions of erratic radio interferences and of operation of a tracked system.

Their reliability $F = 1 - N\epsilon$ has a tendency to be very poor in spite of the small number of electronic or electrical component failures, because N reaches an order of magnitude of 10^6 to 10^8. One can imagine the problems these systems will meet by watching what happens on a motorway, in spite of the fact that the human brain is able to take account, in real time, the various and numerous pieces of information such as location and speed of several preceding vehicles, the running condition of the vehicle, wind, rain, profile of the road, etc. This is why the first automated mini-metro on tyres, the 'skybus', has not been readily accepted and why the first completely automated large system suffered from serious difficulties and from accidents, in spite of the 150-second interval between trains. This proves the soundness of the very cautious policies followed by the operators of large existing urban transportation networks in the automation of their systems. Recent studies in France have in fact shown the saving achieved by eliminating the drivers was more than balanced by the first cost, the breakdowns and the increased maintenance force.

Finally, it justifies the interesting attempts made in several countries to eliminate a large part of the components involved in the automatics by measuring the distance and the relative speed between two vehicles. However, there is still no practical evidence of the actual effectiveness of these attempts.

The complexity factor is also involved at the level of vehicles themselves. Within an interval of a few seconds, all vehicles of a system are interdependent and are numerous. A PRT system of 5 s interval between vehicles and a total length of 10 km involves 400 vehicles.

If $(1-\epsilon_1)$ is the reliability of one vehicle, the reliability of the system is: $1-400\epsilon_1$. One can see that ϵ_1 must be very small to get an acceptable value of F. Now, each vehicle has a reliability $1-\epsilon_1 = 1-n\epsilon_2$, n being the number of inter-dependent parts within the vehicle and ϵ_2 the failure rate of each part.

How do wheeled vehicles and contactless vehicles compare in this regard?

The number of interdependent elements n is multiplied by a factor of 4 to 6 on a wheeled vehicle in comparison with an air cushion linear motor vehicle*. Let us take a factor of 5. The failure rate ϵ_2 of a part is a function of three factors. The base factor ϵ_3 is a function of the level of technology which we assume the same for both types of vehicle. It is multiplied by a coefficient K_1 when fatigue phenomena are involved due to the wheel/track contacts and induced vibrations.

For wheeled vehicles it is multiplied by a coefficient K_2 because of random stresses (shocks, slidings, vibrations) preventing computer simulation of the actual operating stress and of simulated fatigue testing. Most of these difficulties are eliminated in wheel-less vehicles.

Therefore:

$$f = 1 - n_2 K_1 K_2$$

On wheeled vehicles K_1K_2 is of the order of 8. This value is suggested by the following experimental facts:

- less maintenance is needed on tyred metro vehicles;
- there is a need for 5 years of experiment in actual operation of a new metro bogie or when using light alloys in bogie design;

*In wheeled vehicles there are numerous axles, bearings, gears, rotors, lubrication circuits and wheels which are interconnected. They can overheat and therefore change size, as they all driven one by the other. An ACV lift fan is independent of the other elements (the difference being between elements in series and elements in parallel).

. aircraft maintenance has been divided by 6 since jets were used;
. the failure rate has been divided by 8 in changing from the piston engine to the gas turbine;
. the power transmitted by the rear axle in automobiles has multiplied by 3 since hydraulic torque convertors were used, eliminating the torsional stress from the drive;
. the oil pipe breakage has been divided by 10 on diesel engines when they do not vibrate, etc.

The ratio between failure rates on the wheeled vehicle and air cushion vehicle is therefore:

$$R = \frac{n_1}{n_2} \frac{K_1}{1/K_2} \approx 5 \times 8 = 40$$

where n_1 is the number of parts of a wheeled vehicle and n_2 is the number of parts of an air cushion, linear electric motor vehicle.

If a wheeled vehicle has a reliability of 0.998, an air cushion vehicle can reach 0.99995.

10.3.3 *Safety of operation*

The above result is most important, as in personal transit systems, reliability and safety are tied together because locking or sudden braking leads almost unavoidably to an accident.

This explains why some PRTs have followed the same path as URBA: an air cushion and a linear motor. This is also why some experts are beginning to state that there will be no personal rapid transit systems without an air cushion and a linear motor.

It is also easy to understand why after beginning with systems comprising small four-seater vehicles and a large number of stations, promoters of these systems are coming to proposals for vehicles of 20 to 30 seats with relatively limited use on small circuits for downtown business districts, comprising just a few stations and a maximum of 50 vehicles.

These decisions are in fact a direct consequence of the formula $F = (1 - N\epsilon)$ as they divide by a factor of 5 the number of vehicles and by a factor of the same magnitude the number of automatic elements.

Finally, personal rapid transit systems' problems could be summarised by two sentences used in the mechanical industry 'As soon as a part exists, there is a chance that it will break and this will inevitably happen one day!' or 'Only parts which do not exist never fail!'. These statements derive from the experience of engineers who design, develop and operate mechanisms. When a design is satisfactory, the smallest change in a gap, surface finish, etc. is enough to cause a failure in operation. This is even more true, and on a larger scale, for a change in dimensions or in design, especially if the parts are submitted to random phenomena such as are induced by the wheels.

It is therefore an illusion to think that a system does not present any risk if it has wheels. In fact, we have just seen that, unless an existing bogie is identically copied, all wheel systems, even slightly modified, will present more risks than air cushion linear motors vehicles methodically developed, because the latter are simpler. This leaves very little chance of success for highly-automated, newly-designed wheeled systems.

Finally, all this foretells many difficulties at the start of operations and a poor reliability of all complex hybrid systems comprising a large number of parts, which, when they are combined, add all their potential of mechanical failures to those of wheel systems, cable driven systems, conveyors, friction mechanisms, slipping or variable speed mechanical drives, etc.

The more a system aims at numerous and complex functions, the more the elements and the components of this system must be simple and reliable.

10.3.4 *The rendez-vous problem*

All systems considered for the end of this century must allow a programmed diversion of individual vehicles from the main flow of vehicles at the entrance of the station and a programmed re-injection of these vehicles into the main flow when they have left the station. In both cases, passenger accelerations must be maintained at an acceptable value for comfort reasons, and the injection of the vehicle into the very narrow slot must be achieved successfully.

All this can be safely achieved only if a propulsion system presents a precision of the order of 1/1000, in the programming of the movement of the vehicle, its location, its speed and its acceleration. It is believed that no industrial propulsion system is able to achieve this precision with a wheeled vehicle or with a linear induction motor for which there is no strict space-time relationship due to slipping of the motor and (or) wheels. Of course this problem could be solved using feed-back loops but this would increase price and decrease reliability.

The compagnie d'Energétique Linéaire is beginning to test, at half scale, a system potentially safer incorporating a synchronous electromagnetic motor, the magnetic masses forming the necessary space scale without requiring feed-back loops.

The rendez-vous problem being solved, the problem remains of having special suspension and propulsion devices for each mode. Logically, this problem can only be solved by using a 'pallet' system at high speed which appears to be the only one to meet the required reliability and economic interests.

In such a system one of the problems is that of instantaneously coupling a pallet and a vehicle that have a speed difference between them. This can be solved with an air-layer-guided pallet (less shock on vehicle) in suspension (which is self-stabilising in response to the impulse due to the speed difference and eliminates interference in the same plane between the high speed structure and the launching ramp structure).

Such a system is shown in Figure 10.3. Inside an inverted U shape track an air 'bogie' is suspended from the track inner lower surface by a vacuum which is made stable by the inverted position of the cushion skirt (above the track flange). This principle adds to the air cushion advantages all the advantages of suspended vehicles (Table 10.7).

When applied to palletised duo-mode operation it lends to an efficient pick-up arrangement at rendez-vous time. The small mass of pallet, its independance of the track, a gliding vacuum cup, the self-stable suspended position, the specialisation of vehicles upper face for guided mode and lower face for street mode make pick up operation shock-less and oscillation free.

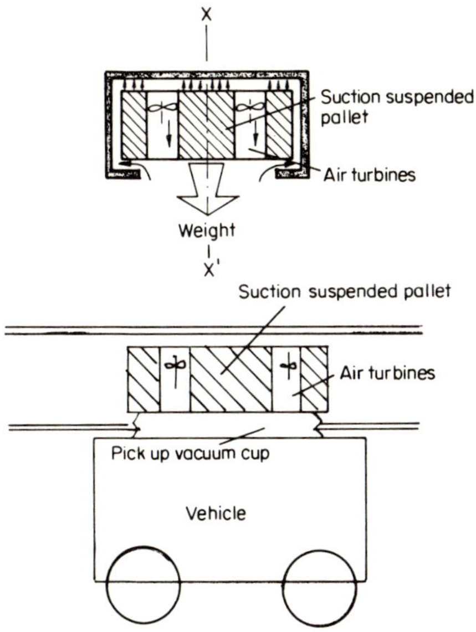

Figure 10.3 The principle of URBA's suction suspension and duo-mode pallet application.

10.3.5 *The investments*

The delays in starting up the operation of public transport lines are usually due to the inadequacy of available finance. Now the sophisticated systems under discussion aggravate this situation as their investments per transported passenger are several times greater than the investment for present public transport systems (Table 10.2) and it is probable that this table is optimistic, since 2.5 km of the new personal rapid transit systems under experiment will cost $ 65 000 000.

Under the conditions described, there is not the slightest chance of the communities accepting this cost. The only escape perhaps is to use investment cost-saving components such as linear motors and air cushions (see 10.4.2 and 10.4.3).

Table 10.2

Investment comparison of the Metro with various URBA systems

	Metro 1972 (as reference)	URBA 1st Generation	Supported PRT with wheels	URBA bus duo-mode	URBA personalised duo-mode
Investment per km for 10 000 passengers/hour (Million F)	30/40	10/15	70/100?	20/30?	50/60

10.4 Where to innovate?

10.4.1 *A coherent line of attack: design of a system family progressively evolving and self-compatible, especially at the structural level*

Considering the problems described above, one could argue that the chances of developing a system of this type are very slim. However, history, and especially that of transportation techniques, has shown that the ability of man to solve apparently insoluble problems has always been under-estimated. It is therefore reasonable to start with this optimism if, and only if, all the weapons of high technology are used and if progress is made cautiously, step by step.

As the structures of a transport system are amortised very slowly all new transportation means which do not integrate future evolution in the initial specification would fairly rapidly become obsolete.

The track is the main part of the structure and of the investment. It is difficult to modify and must stay the same throughout the progressive evolution towards higher-performance systems.

Vehicle design must be of module type for psychological (as a personal bubble automobile), for industrial and for technical reasons. The development, at each successive step, will be limited to these modules, and it will be possible to cut the vehicles into slices made of a smaller number of modules when the reliability of one module and its automatics increase.

The switching system must provide an immediately reliable technological solution and must be capable of being successively modified during a progressive evolution towards a totally safe passive switch, avoiding all mechanical moving parts which could stall or break.

It seems that guiding with sub-atmospheric pressure is a principle which allows it a vehicle to follow a curve from its internal surface without any risk of having the guiding forces suddenly stopped in case of power failure.

As already shown, it seems impossible to ensure the high reliability which will be required when PRT or duo-mode systems are used, without eliminating slipping, shocks, vibrations, complex mechanisms, large numbers of parts and inertia, which are inevitable consequences of the use of wheels.

Similarly it seems impossible to obtain the braking safety, the short stopping distances and precision required by a short interval and passive switches, as long as they depend upon such uncontrolled factors as tangent contacts between wheel and track surfaces, wear characteristics and atmospheric conditions. Wheel-less linear motor operation is therefore a must.

Finally, the evolution will spread over several decades and will comprise several successive systems. Each will have to be able to be deduced from the preceding one by changes as small as possible, and consequently, the fundamental choices of the final system must have already been incorporated in the design of the first.

These facts suggest that the conjunction of air suction suspension and linear motor propulsion seems to combine the optimum internal and external, present and future coherence of a family of transport systems.

But apart from offering solutions to the five basic problems of PRT systems, air cushion and linear motors bring advantages which have not always been fully understood and measured.

10.4.2 *Air cushion – WHY?*

Much has been written for and against the air cushion in public transport vehicles. A large part of it concerned high speed operation. The view is expressed here, personally that by putting emphasis on high-speed/air cushions, inventors were put on the wrong track. They were obliged to make higher and higher speed bids as conventional railways engineers were improving their own speed limits.

The interest in the air cushion can lie elsewhere: it is more reliable than wheels (Section 10.3.2), it brings less vibrations in vehicle and track, it allows safer emergency braking (as seen in Section 10.3.1) and finally it brings a decisive saving for vehicle and elevated tracks which represent the most important part of suburban lines investment.

It is worth recalling how important is this saving and that it is the consequence of a lighter vehicle and a lighter track . Table 10.3 shows why an air cushion vehicle is lighter.

Table 10.3

Vehicle weight savings from air cushion/linear motor

Factor	Consequence on vehicle design		Weight savings
Linear motor propulsion			
Compactness	Simpler structure	=	ΔW_1
No transmission		=	ΔW_1
Small weight	Less supporting structure	=	$\Delta W_3 + \Delta W_4$
No vibrations	Lighter structure	=	ΔW_5
			ΔW
Air cushion vehicle			
No mechanical vibrations	Lighter structure		ΔW_6
Absence of heavy sound insulation	Simpler and lighter structure		ΔW_7
Distributed weight	Lighter structure		ΔW_8
			$\Delta W'$

Order of magnitude of ΔW A lower value of $\Delta W = \Delta W_1 + \Delta W_2 + \Delta W_4 + \Delta W_5$ is given by the gain obtained on the ASTROGLIDE system when wheeled propulsion was replaced by the linear motor (see Section 10.4.3). The gain was 25% of dead weight. $\Delta W = 25\%^{+0}_{+5}$ is therefore a reasonable guess. (The gain would have been higher if the vehicle structure had been modified.)

Order of magnitude of $\Delta W' + \Delta W$ An idea of $\Delta W + \Delta W'$ is given by comparing URBA with existing wheeled systems of same capacity (Table 10.4).

Table 10.4

Weight of vehicle per metre length

Airtrans	Skybus	URBA
1.25 T	1.45 T	.67 T

$$\Delta W + \Delta W' \text{ is roughly } \frac{(1.25 + 1.45) - 2 \times 0.67}{1.25 + 1.45} \approx 50\%$$

It is therefore reasonable to state that the combination of air cushion and linear motor can bring a vehicle weight saving of roughly 50% of which half is due to the motor and half to the air cushion.

Table 10.5 shows why the track of an air cushion vehicle is lighter and Figure 10.4 adds visual evidence.

Table 10.5

Track weight savings from air cushion/linear motors

Factors	Consequences of track design factors	Saving on track weight
No vibrations, shocks, gliding	Less dynamic loading and fatigue	ΔW_1
Progressive loading on bridge	Less dynamic loading	ΔW_2
Distributed load	Lower maximum stress	ΔW_3
No ballast	Less load per unit length	ΔW_4
No track	Less load per unit length	ΔW_5
50% lighter vehicle	Less load per unit length	ΔW_6
		ΔW

An order of magnitude of ΔW is given by Table 10.6.

Track weight per metre linear (double track)

Table 10.6

Airtrans	Skybus	Aerotrain
7 T	6.5 T	3.2 T

$$\Delta W = \frac{(7 + 6.5) - 2 \times 3.2}{(7 + 6.5)} \approx 55\%$$

Two more factors emerge in favour of the air cushion.

Firstly, there is a smaller cost of development. This can be better understood with reference to Section 10.3.2 which deals with the absence of random stresses and the consequent possibility of test rig and computer simulations, as well as the decrease of the number of parts needing development. URBA was 90% developed for less than £1 000 000 in less than 6 years and needed only two prototypes.

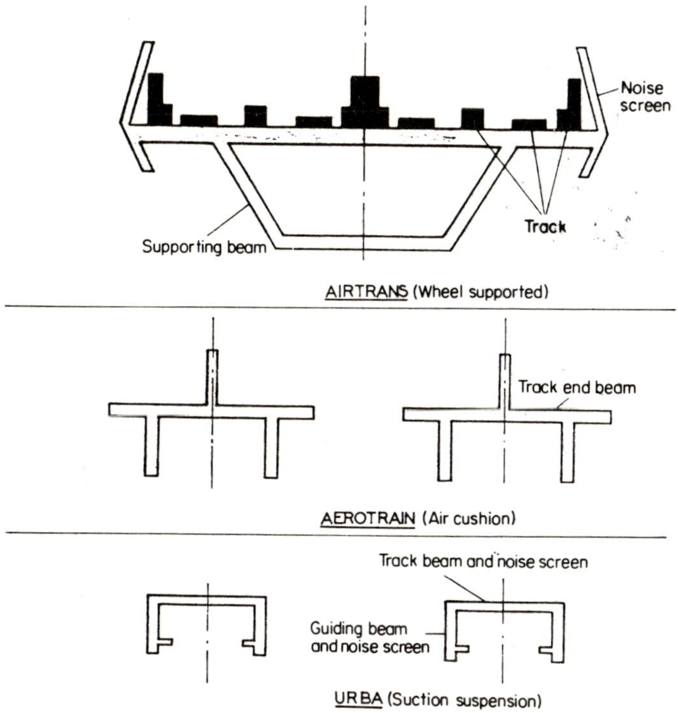

Figure 10.4 Transverse sections of tracks of three transport systems.

The second factor is the new impetus given to suspended vehicles by the knowledge that they had not yet achieved their full potential, because their wheels need to be above the track. With suction suspension, it becomes possible to suspend a vehicle from the underside of the track. This has important consequences in the simplicity and low cost of the vehicle, the rigidity and low cost of the track, noise control and weather protection.

Table 10.7 and Figure 10.4 indicate why a suspended air cushion vehicle track is even lighter than an air cushion track.

Table 10.7

Track weight saving from suspended vehicle

Factors	Consequences on track design factors	Saving on track weight
Lighter vehicle	Less load	$\Delta W'_1$
Centre of gravity below suspension	Less torsion load (–45%),	$\Delta W'_2$
	lighter supports, smaller foundations,	$\Delta W'_3$
	narrower track (–45%)	$\Delta W'_4$
Snow and ice free	No snow overload factor,	$\Delta W'_5$
	no heating pipes for de-icing	$\Delta W'_6$
Load applied under beam top	Less bending of beam's side	$\Delta W'_7$
Sides of track constitute noise screens	No extra weight for noise screens	$\Delta W'_8$
		$\Delta W'$

An order of magnitude of $\Delta W' = \Sigma \Delta W'_n$ is given by Table 10.8.

Table 10.8

Track weight per metre linear

	Aerotrain	URBA
Track weight/metre	3.2 T	2.2 T

The gain $\Delta W'$ is roughly 33%.

10.4.3 *Linear motors – why and how*

According to the systems and components analysis described above, the URBA 8 vehicle introduced in March 1968, combined three basic features: suction air suspension, suspended position and linear electric motor in a system designed for low speed operation. It was the first linear electric propulsion, passenger carrying vehicle ever operated regularly. For 6 years, it has made daily demonstrations carrying 30 000 visitors.

Since that time, many transportation systems have tried to follow the ways opened by URBA. They were more or less successful in as much as they were more or less similar to URBA.

Linear motor a must in air cushion vehicle propulsion When searching for the most suitable propulsion system in urban transportation air cushion vehicles, one must refer to the basic aims and claims of such vehicles.

They should maximise savings of dead weight so that on the one hand vehicle and track installation costs are smaller and on the other hand energy consumption due to accelerating and decelerating the vehicle is kept to a minimum.

They should eliminate the nuisance of existing systems (noise, vibrations, etc.).

They should reduce operating cost and maintenance.

It should be remembered that the wheel plays two roles: it is first a means of support and second a means of transmitting the propulsive effort to the rail. It was shown above (Section 10.3) that the use of wheels as a supporting device using steel or tyred wheels on rails or tracks is not sufficient for the new transport systems if their aim ultimately is to evolve into reliable duo-mode or PRT operation.

At present, it has been proposed to keep the wheel for propulsion (making a hybrid vehicle). Theory and practice can show it is an impractical solution. Why?

First let us recall that engineers' constant experience has always been that hybrids are rarely successful because they generally try to combine mechanisms of opposite nature which do not fit together and result in higher weight than that of a "pure" solution. Two of the best known examples are the turbo-prop aeroplanes and the tyred wheel trains. This is basically due to a lack in the logic of such basic schemes.

Secondly, in order to transmit the propulsive effort, taking into account the friction coefficient, wheels have to be applied to track or rail with a force of the order of magnitude of the vehicle's weight. A full suspension system is therefore necessary to isolate it from the vehicle and the bogie presents practically the same size and weight of the original bogie of the wheeled vehicle.

Under such conditions the hybrid vehicle keeps all the well-known shortcomings of wheeled vehicles: high weight, low acceleration, erratic braking, loss of adhesion under icing conditions, vibrations and noise, high maintenance, plus a few new problems involved in guiding the traction horizontal axis bogies on a special vertical flange.

The introduction of a linear motor by contrast brings improvements: good, reliable and safe acceleration and braking, low weight, simplicity of design and elimination of maintenance, reliability and long life, absence of vibrations, etc..

Most of these are precisely the same advantages as those of the air cushion. It seems logical therefore to use the linear motor for propulsion and the air cushion for suspension.

Once such evidence is accepted, comes the question what kind of linear motor should be used, as this generic name covers a large variety of electric motors characterised by a relative linear movement between the inductor and the secondary.

Linear motors fall into four categories: induction motors and synchronous motors, each of which may be designed for reciprocating or continuous movement.

1. *Induction motors* (asynchronous) These work from *sinusoidal* magnetic fields created in gaps by coils fed with *alternating current* (sinusoidal); the fluctuations of the field create induced current in a conducting and continuous inductor (aluminium).

The propulsion thrust results from the action (with slip) of the field on the induced current and from the displacement of the point of application. *Advantages*: simplicity, reliability, large size motors. *Disadvantages*: heating of the secondary, unsatisfactory speed control, high cost of the secondary, low power factor (in some types), guidance problems.

Synchronous motors (reluctance) These can work from a *pulsed magnetic field* created in a gap by current pulses, the primary coils, each being fed with a *succession of separate impulses.* These are often known as 'stepping motors'.

The secondary is made of a discontinuous succession of cores made of welded iron sheet, packed in order to make the core non-conducting. Each impulse magnetically attracts one core of the secondary into one gap of the inductor.

Advantages: simplicity, reliability, satisfactory and precise controls, limited cost of the secondary, high power factor. *Disadvantages*: Guidance problem, moderate unit size, tendency to instability.

3. *Reciprocating linear motors* Forward and backward movement with a limited stroke.

4. *Continuous movement linear motors* A movement with a large displacement.

Specific Power: In the cases 3 and 4, the essential conditions of specific power can be derived from the formula:

$\frac{F}{M} = \gamma$ where F is the propulsion force, M the driven mass and γ the acceleration.

Case 3 requires a huge acceleration for the forward and backward stroke within 1/50th of a second and therefore requires a *large value of F, for a small value of M.* Case 4 requires a large *tractive effort, F,* for a reduced cost of the secondary, that is to say a *small value of M.* This proves that a principle of physics which is satisfactory for Case 3 must be worth studying for Case 4.

As shown by Professor Laithwaite, there is a size limit above which induction motors present a higher specific power than synchronous reluctance motors. A French Government contract is now allowing CEL to apply dimensional analysis to size and speed effects in linear electromagnetic motors. The results should help to define more precisely the best respective zones of application of each motor type.

Meanwhile, it is interesting to recall how such motors have been developed successfully outside the transportation world.

Such achievements show that linear motors are coming into fashion and give a good idea of just how much linear motors could bring to the transportation world when their possibilities are really understood and thoroughly developed.

The attempts made in France to use linear induction motors for high speed vehicles were not successful, presumably due to the lack of know-how of French Companies involved in that field.

URBA avoided linear motor difficulties by making a few key design decisions for vehicle and track:

. aiming at moderate speeds (30/40 m.p.h.);
. not using the track's guiding flange as the motor's secondary;
. designing a suspended vehicle with secondary suspension;
. this in turn permits a multi-bogie design and therefore a multi-motor vehicle.
. all the above decisions resulting in individual motors of small dimensions

(17 kW), low inertia, with a narrow gap and small pole dimensions, better power factor, small dimensions for the secondary and therefore easier natural cooling.

Table 10.9

Induction electrodynamic motors (action field/current)		Reluctance motors (action field/core)	
Reciprocating movement	Continuous movement	Continuous movement	Alternative movement
Failure of a lot of attempts	A commercial success is developing in the application to many industrial uses along the lines proposed by Laithwaite	Technical and commercial success. Stoklec Pharmelec automated storage (Barthalon license) Potentially the largest market for small linear motors.	Success in mass production and sales. Barthalon and his licensees.
	A complete range of industrial linear motors is available from the companies below:	Traction Traklec* being developed	Black & Decker (US) Nitto Kohki (Japan) Torin (US) Balma (Japan)
	England: Lintrol linear motors		
	USA: Polynoid, Lintec linear motors		

* The Traklec electromagnetic motor has a high starting pull, a low cost secondary with no heat losses and distortion, a simple and precise control of position, speed and acceleration, allowing programmed operation and energy recovery when braking.

To sum it up: in order to be successful a tracked air cushion vehicle and track system ought to be designed for and around its linear motor instead of being designed with only the air cushion constraints in mind.

After pioneer work by Laplace, the first linear electric motor in history was presumably the one demonstrated by André Marie Ampère in Lyons (1821).

An early technological design was patented by Zehden in 1902. Later on, fundamental studies by Sthurmann (1946) and Laithwaite (1948) were a decisive turn towards a practical design. Then like any other invention, the linear induction motor had to wait until the market was ready for it. A specially spectacular and important field of application was needed to facilitate its take-off. Air cushion vehicles and more specially URBA, the first passenger carrying vehicle to use linear electric motor provided such necessary opportunity and publicity.

Success is now indicated by the number of proposed transport systems which are being successful because they partly use the same basic principles as URBA: linear motor, suspended position, moderate speed, negative air cusion. They are shown in the Table 10.10.

Table 10.10

URBA and consecutive suspended vehicles

Name of system and builder	Suspended position	Linear motor	Small size motor	Speed	Air cushion
URBA (Barthalon)	Yes	Yes	Yes	low 40 m.p.h.	Yes
ASTROGLIDE (PRT Systems)	Yes	Yes	Yes	low 35 m.p.h.	No
ASTROGLIDE 2 (PRT Systems)	Yes	Yes	Yes	low 40 m.p.h.	Yes
PENDAIR (Dennis Bliss)	Yes	Yes	Yes	low 45 m.p.h.	Yes
SIEMENS-DUWAG	Yes	Yes	Yes	low 35 m.p.h.	No

10.4.4 *Consequences of operating costs*

In most cases, cities are even more interested in operating cost than in investment because they have to pay operating costs each year and have difficulty in obtaining Government subsidies.

Table 10.11 shows the maintenance savings from use of air cushion/linear motor.

Table 10.11

Maintenance savings from air cushion/linear motor

Factors	Effect	Cost saving
Vehicle		
No vibrations	Less fatigue and loosening of parts	Δp_1 = 80% (see Section 10.3.2)
Less loaded, fewer parts	Fewer breakdowns	
Brakes with no mechanical contact	No contact damage (flats)	Δp_2
Fewer breakdowns	Less traffic losses	Δp_3
Track		
No ballast	Limited adjustment	$\Delta p'_1$
No shocks, vibrations, gliding	Less damage to track	$\Delta p'_2$
No ice or snow on track	Less maintenance	$\Delta p'_3$ specific to suspended vehicles
No foreign bodies on track	Less cleaning	$\Delta p'_4$

The total saving in maintenance cost, $\Delta p = \Sigma \Delta p_n$, is certainly difficult to assess before actual commercial operation. Some conservative evaluation has been carried on in cooperation with operators from detailed time studies of maintenance operations. The results are given in Table 10.12.

Table 10.12

Maintenance cost in centimes/seat/100 km

URBA 10 000 passenger hours/track	Conventional 10 000 passenger hours/track
72.44	151.64

The saving is at least of the order of 50%. By comparison, the maintenance gains obtained in other fields by elimination of shocks, vibrations and fatigue loading (see Section 10.3.2 above) run between 70% and 85%.

Gain in maintenance cost for air cushion/linear motors vehicles will therefore most probably be of the order of 65% ± 10%.

The recent energy crisis has increased the relative importance of energy consumption. The cost of this factor used to be small compared to maintenance and was increasing less rapidly.

Despite this, air cushion linear electric vehicles already compare favourably with conventional transports even in energy consumption.

Firstly, in urban traffic, energy is mostly used to accelerate and decelerate. Therefore, vehicle dead weight plays an important role. Weight saving results in an almost proportional energy saving (of the order of 40%). Secondly, linear induction motors have been shown to present a slightly lower consumption of electricity than conventional motors in actual operation (see for example, Astroglide results in Dallas, Texas). But the above gains are for the time being overcome by the energy cost of driving the compressors feeding air to the cushions. However, as it often happens in the first few years of development of a new technique, our experience shows that progress in that respect has been and still is rapid, due to the remarkable improvements on the sealing systems of air cushions. Energy absorbed by cushion compressors will soon be of the order of magnitude of friction and transmission losses.

10.5 URBA and its development

10.5.1 *Main technical steps between URBA and personalised duo-mode URBA*

The family of URBA systems is able to follow the main evolution which can be forecast in the transportation field during the next decades by solving the fundamental problems in a module of a minitrain URBA, without starting an endless process of development.

Having said this, it is not in the URBA promoters' mind to rush immediately towards more sophisticated systems than the URBA minitrain described in detail below. For example, the realisation of 'complete automation' with no personnel on board or in the stations will be approached cautiously, taking into consideration what has to be learnt from the present experiments.

The off-track stations can be considered immediately afterwards but only if automation has been successful.

The realisation of an URBA/PRT system, in spite of its anticipated advantages over the PRTs presently proposed, is envisaged with caution because of the problems which will probably arise.

URBA duo-mode is the final target for satisfying the needs of low and medium-speed transportation in the twenty first century. This system has already been extensively studied and partly tested; results will be published later.

However, it is already obvious from Tables 10.1 and 10.2 that a simplified version of duo-mode from the multi-system study explained above would give a door-to-door service more efficiently than a personal rapid transit system and would present less technical problems and less investment.

A review of the development of the URBA family shows very well that, at each step: (a) only one of the numerous, fundamental and absolutely necessary aspects of the personalised duo-mode system was analysed and solved; (b) all the systems will be compatible with one another, and the URBA 30 network can, as progress is made, become an automatic minitrain, a simplified duo-mode and a personalised URBA duo-mode.

To develop one of these systems too rapidly can in our opinion, lead only to uncertainty regarding the cost, the duration and the result, and there is a danger in ending up with a 'closed' system, specially designed for a given step, unable to be modernised, or to evolve towards a more sophisticated system, compatible with the preceding one.

10.5.2 *The minitrain URBA 30*

The development of URBA 30 is the main step of the global strategy of URBA. It corresponds also to an important and even more immediate market.

URBA 30 is a suspended automatic minitrain for medium size towns (between 200 000 and 900 000 inhabitants) such as Rouen, Bordeaux, Nantes, etc. It comprises 30-seater vehicles, coupled in trains of two, four, and six vehicles. Vehicles have a driver console for safety but in principle are automatically driven.

The vehicles are suspended through a wide-base suspension to six air-bogies, guided and sustained by sub-atmospheric pressure inside a concrete track having the shape of an inverted U, 1.3 m wide inside and 1.8 m wide, 0.8 m high outside (Figure 10.5). The cabin is 10 m long and comprises five modules of six seats each. When the off-peak condition of traffic allows it, two vehicles are coupled permanently.

The vehicle is propelled forward and backward by six linear motors of 200 daN thrust each. Taking into account the total load of 7000 kg the maximum acceleration is 1.5 m s^{-2}. This high value is compatible with an all seated passenger configuration and the acceleration is achieved on street cars without inconvenience to the passengers. The combination of widely separated suspension arms, controlled air pressure springs, shock absorbers and self damping suction cushions explains why URBA has eliminated all the well-known shortcomings of suspended vehicles (oscillations, wind sensitivity,

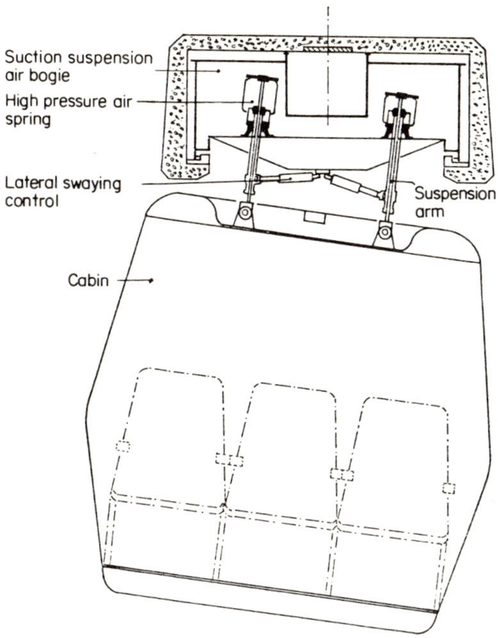

Figure 10.5 Transverse section of URBA 30.

curves control etc.). Maximum speed is 80 km h^{-1} and slopes of 10% can be climbed. Braking is achieved by d.c. supply to the linear motors providing a deceleration of 1.55 m s^{-2}. Power is supplied through catenaries. The system can be overhead, underground or on open cut. Stations, overhead or underground, are of standard design.

The interstation distance varies from 400 m to 1000 m. Along this type of circuit, discontinuous operation allows a commercial speed of 35 to 40 km h^{-1} to be reached, thanks to the acceleration and deceleration rates and the short stop-time, facilitated by the large number of doors (automatically controlled) in the vehicle (one door for six seats).

The interval between trains is of the same order of magnitude as on regular mini-train systems (60 to 80 seconds), compatible with termini operations and safety.

These intervals can be achieved with maximum safety as the braking distance is roughly 220 m. It should be noted that the vehicle has, in addition to the regular braking system through the linear motor, a mechanical brake for precise stop-positioning in the stations and a very efficient emergency braking system which operates by opening the air chamber to the atmosphere and landing on shoe brakes. The deceleration can thus reach 2.5 m s^{-2}.

The capacity of transport of this system is a function of the number of vehicles per train and the time interval (Table 10.13).

Table 10.13

URBA's passenger capacity

Time interval in seconds	Number of vehicles per train 1	2	4	6
240	450	900	1.800	2.700
120	900	1.800	3.600	5.400
80	1.350	2.700	5.400	8.100
60	1.800	3.600	7.200	10.800

10.5.3 *Innovations in URBA and their consequences*

The URBA system includes some important technological innovations. These innovations (air suction suspension, suspended position) are simple, limited, reasonable and precisely located on an area of maximum efficiency, the heart of the system, on a single new component of small dimension: the Dynavac air bogie (1.5 m x 0.7 m x 1.3 m, see Figure 10.1).

The sub-atmospheric system, Dynavac, was conceived in 1966 and for seven years has been the object of extensive studies and tests. As there are no random stresses such as shocks, vibrations and slipping, its development has been made easier. Computer studies of the behaviour of the air-bogie have greatly reduced the risks of development as is proved by the coherence between theoretical studies and experimental results (Figure 10.8).

The URBA linear motor has the same characteristics as are in the catalogues of industrial manufacturers.

The problems of this motor have nothing in common with units 20 to 30 times more powerful and several times faster. They are simplified by the use of the Dynavac principle, by the effectiveness of the initial choices and by the experience acquired in this area.

The suspended position, the effectiveness of which was proven in operation by the Monorail of Wuppertahl, brings real advantages over supported overhead systems. The advantages include a slimmer track, an easier supporting system and protection of the track against weather conditions. Suction suspension has enabled the design of a particularly simple and light switching arrangement to be effected (Figure 10.6).

The suspended position, the effectiveness of which was proven in operation by the URBA 30 brings to the user a quality of service quite unknown until today and which is absolutely necessary to deter the motorist from using his car: no noise, shock or vibration; high commercial speed; attractiveness of the suspended position (good visibility, maximum comfort); passengers all seated; one door for six passengers.

It fulfils exactly the specification for a system satisfying market needs as established in Sections 10.1 and 10.2.

It can be inserted easily into an urban environment because it allows 40 m radius curves and 10% slopes. It can be installed as a simple track or a twin-track. It can

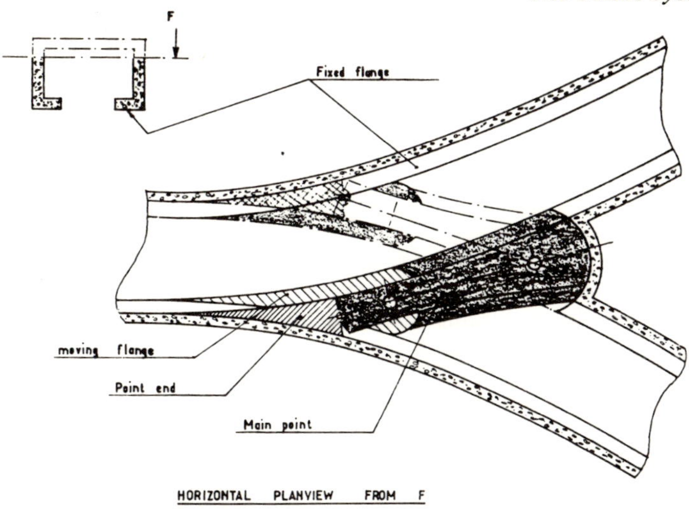

Figure 10.6 URBA's switching arrangement (first generation).

be elevated or it can be placed underground in small dimension tunnels (less than 6 m diameter, double track).

All these elements are summarised in Figure 10.7 which compares URBA to wheeled systems taking into account the relative weight of the various aspects to be considered and gives (shaded areas) the improvement margin brought by URBA.

10.5.4 *The development programme and methods*

Table 10.14 shows the whole methodical development of the URBA programme which has been followed for seven years. It explains why there has been no drawback during this development up to the final success.

Table 10.14

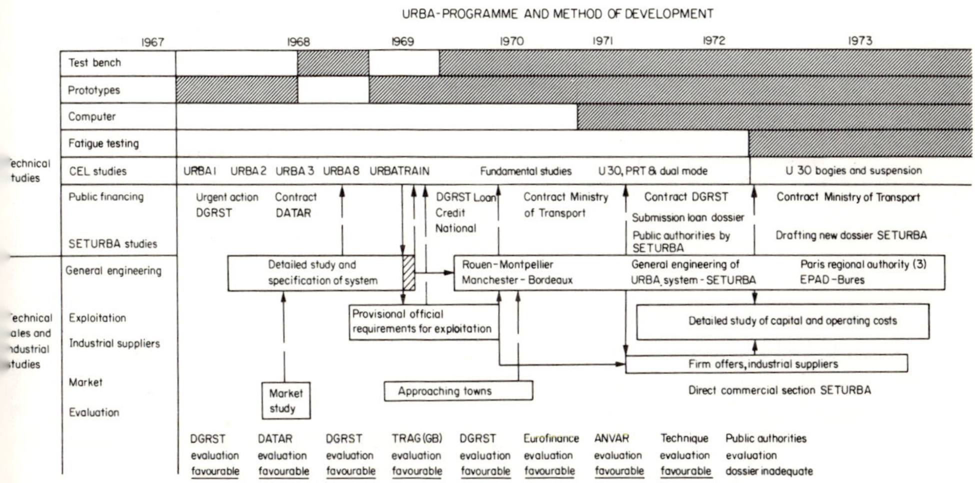

Successively, a series of five initial models were used to verify the air suction/suspension principle, to determine its laws, to define the main functions, to make a complete list of the problems related to an URBA vehicle, and finally to verify, for the first time in the world, the possibility of operating a train of air-suspended vehicles.

These five prototypes made it possible in less than three years, to choose the final technological scheme, defined in 1968, which has led to a family of homogeneous prototypes of which the basic design has not since changed.

This study was itself the consequence of a very elaborate specification. The value of this specification has been proved by the fact that, afterwards, it corresponded point by point with the specification proposed in 1971 by the 'Institut de Recherches des Transports' for new transportation means in medium-size towns.

Obviously, this development programme involved the most modern methods of development, the most sophisticated electronic means for measurement, testing and simulation.

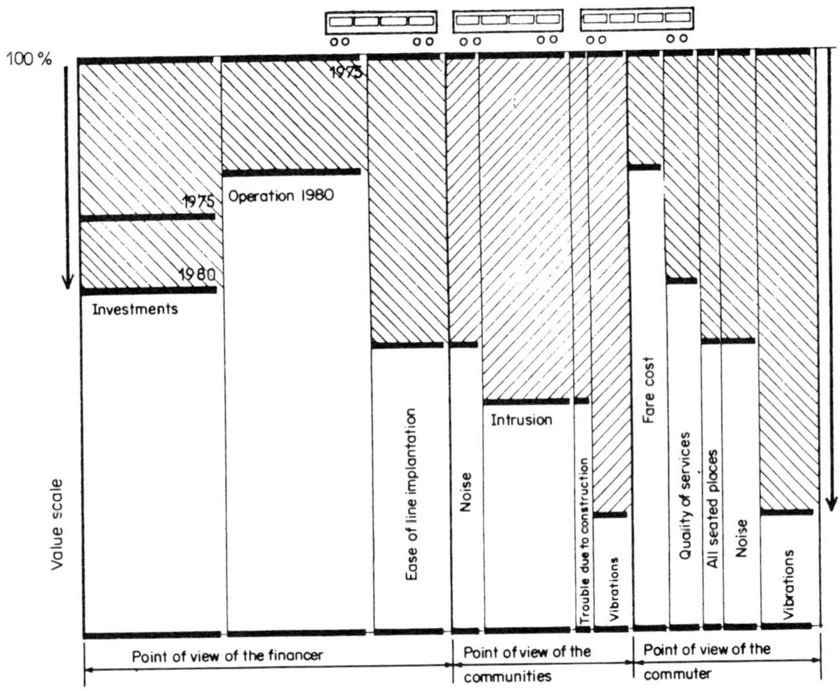

Figure 10.7 Diagram to show URBA's competitive position. The shaded region gives URBA's margin of improvement.

So, the URBA development has successively made use of:

Individual test benches to verify the basic elements, obtain the necessary experimental values of key factors such as the flow/pressure characteristics of the seals of the Dynavac cushions.

Sub-assembly test benches to obtain similar information as above at sub-assembly level and to study the interactions between the individual components.

. Test bench for a 'slice' of a vehicle and of the track.

. Finally, as it was possible to use electronic computers (no contact, slip or shocks, and generally speaking, no discontinuous phenomena) the scope of the development tools was completely changed.

10.5.5 *Computer simulation of experimental models*

The computer programme began with the operating conditions of the basic element of the machine: the air-bogie, the sub-atmospheric chambers of the Dynavac and the simplified interactions with the cabin. Having controlled the static and dynamic behaviour of the air bogie, the behaviour of the cabin was programmed, integrating a growing number of freedom parameters.

Thus, it became possible to re-inject into the air-bogie behaviour programme the results obtained from the cabin and to achieve a fairly exact 'image' of the total behaviour of the vehicle, suspension and air-bogie, in all possible conditions of operation, load line profiles or weather conditions. A final study of the stability conditions of the system ensures that even under the most difficult abnormal situation, no resonant problems or instability processes occur (Figure 10.8).

Naturally, during all this period, checking the experimental results with the mathematical equations played an important part. Results obtained through the computer were double checked on the test benches.

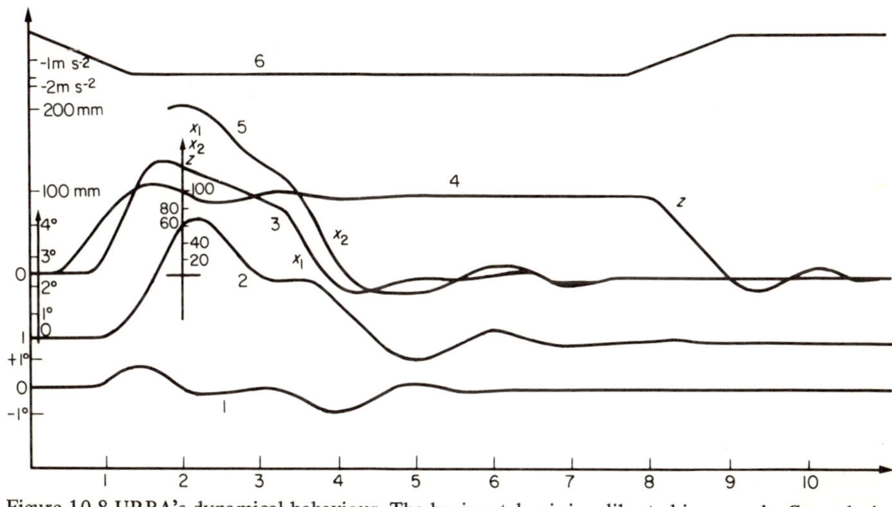

Figure 10.8 URBA's dynamical behaviour. The horizontal axis is calibrated in seconds. Curve 1 gives bogie's tilt angle; 2, Tilt angle of cabin; 3, Vertical displacement of cabin; 4, Forward displacement of cabin; 5, Lateral displacement of cabin and 6, deceleration.

Development on computer being realistic, the computer programme becomes a real development tool. After each test series and redesign, it becomes more precise and reveals possible imperfections in the behaviour of the machine. It is then fairly easy to introduce the corresponding change in the design and mathematical image and, according to the results obtained, to make a choice among several possible solutions for improvement. It is equivalent to designing and testing a great number of successively better refined prototypes at a much lower cost.

Mathematical images and prototypes design have a parallel evolution going further in the study of the behaviour of the machine.

Obviously, this method has practical limitations, as for example it cannot take account of some atmospheric phenomena, of corrosion or the impact of the future user on the system, but nevertheless this process allows a large number of potential problems to be eliminated.

Furthermore, the computer programme can be used to determine with great precision the stresses in the components under actual operation. It is then possible to reproduce the working conditions in simulated fatigue tests, allowing the measurement of the stress distribution and the critical stress, the elimination of stress concentrations and the fatigue tests of parts over a sufficient number of cycles.

A machine designed and tested like URBA on computer and separate test benches, reaches quickly and accurately the stage of a reliable machine, at the very first test in actual size.

As noted above (Section 10.4.2), this is one of the main reasons why it is easier and safer to develop an air cushion vehicle than a system with wheels and bogies, in spite of the enormous experience accumulated in these fields and the remarkable know-how of the specialists.

10.5.6 *Cities installation studies*

As seen earlier (Section 10.2) a key factor in successfully developing a transport system is to start with a precise specification taking into account the user's needs, the environment and the economic constraints. Whatever care is taken in doing so, such a specification is likely to remain on the vague side unless it is implemented with a number of specific exploitations. This is why URBA's coherent development programme has incorporated a number of detailed line installation studies carried out at the request and by the finding of a number of cities (Table 10.15).

Table 10.15

URBA installation studies

Name of City		Length of line	Slope > 5%	Length of tunnel (km)	Number of stations	Annual traffic (million) passenger	Peak flow one direction (per h)
ROUEN	(1971)	7.5			10	35	10,800
BORDEAUX 1.	(1972)	12.1	2	1.65	20	27	4,400
BORDEAUX 2.	(1972)	11.3			18	27	4,400
MONTPELLIER		10			17		6,000
MANCHESTER	(1970)	22			16		10,500
PARIS 1. GARE DE LYON ORLY	(1973)	14.6			5		1,500
PARIS 2. NATION ROSNY	(1973)	9.14		1.16	7		4,500
PARIS 3. STALINGRAD BOBIGNY	(1973)	7.35			5		9,000
PARIS 4. LA DEFENSE	(1973)	4			6	←	4,000

To make them more precise and faster a computer programme was developed which took into account exploitation factors such as line profile, station spacing, motor performance, curves, vehicle loading, wind conditions, and delivered exploitation, results such as: travelling time, time table of trains, energy consumption, maximum instantaneous power consumption, etc..

Furthermore, the range of requirements of such very different demands of traffic and installation led to a much more precise specification and development programme which led to a product much more likely to fulfil the needs of a wider array of cities.

Table 10.15 gives the list of cities studied and the main interesting results of the studies. It shows the very wide variations of the installation cost depending on the profile of the line and of the local site constraints.

10.5.7 *Industrial feasibility*

Most of the transportation projects so far have been proposed by R and D companies, who have focused their work on technical aspects.

The unhappy result was that the rare projects which went all the way up to a first demonstration line for commercial service were suddenly confronted with the problems of manufacturing the system. They discovered then only that some of the key components were not industrially available. It was the case for high speed electric motor/compressor units, for speed control systems of linear motors, for fully reliable automated traffic controls, etc.. The URBA programme avoided such shortcomings by running the research programme and cities installations studies (Section 10.5.6) in parallel with a thorough industrial procurement study: technical and commercial offers were obtained from industry for every component of an URBA system.

The process was painful at the time because it imposed on two occasions (traffic control and cushion compressors) a sacrifice of performance in order to use proven, reliable equipment.

% seat miles occupied	Commercial speed	Number of cars	Turn key construction cost (million F)	Overall exploitation cost (F)/year	Cost (centimes) per seat per km	Cost (centimes) per passenger
8.5%	30.6	124	75	8.8	3	26
6.5%	27.9	152	162	16.4	3.9	61
8.9%	24.2	125	119	12.3	3.9	44
		423	450	73.7		
	64.4	30	151			
	48.5	70	140			
	5.1	104	145			
			18.5 (track, piles, foundations, station)			

However, it has resulted in a system which could readily go into production without important departures from the present set of drawings. But this is now a problem of Government decision making.

10.6 Conclusion: new transportation, a problem of decision making

When looking at the last ten year's efforts to develop new transportation systems, one is compelled to note the number of attempts that have been unsuccessful. In some cases both industrial and government decisions have been wrong.

Why? Is it a question of finance, methodology, vested interests or environment? Finance is hardly a satisfactory answer. The development of a medium speed transportation system can cost anywhere from £2 000 000 to £4 000 000. This is of the order of 3% of the sum that a city like Paris invests *each year* in new line installation.

In fact 3% is precisely what an industrial firm, under competition pressure, would invest each year in innovation for survival. Paris RATP could therefore afford to develop at least five systems in five years time and pick up the best.

The only trouble is that public transportation being a monopoly does not have to face competition and therefore can afford *not to enter into new systems development.* The '*not invented here*' reaction is a very well known excuse for resistance to innovation.

There are surely good, solid reasons for it in transportation: the sheer size of vested interest, of industries and of infra-structures, the necessary compatibility with existing transport modes are all powerful restraints to any desire to change.

But one cannot help thinking that *railways would have never been born if the British Government of the time had nominated a Commission of horse carriage specialists to give the green light on an R and D programme presented by Stephenson!*

Methodology is a much more important factor than finance. The various innovation rules and methods which have been described in various parts of this contribution to this volume stem from daily innovative experiences of the electro-mechanical industry and from experience in teaching innovation methodology. For several reasons, often due to the pecularities of the transportation world, these methods and rules have been more or less ignored by many inventors, industries, public transport agencies and Governments involved in new transportation. The result could prove disastrous unless a dramatic reappraisal of the working methodology is made by the parties involved.

Leading electromechanical engineers and more especially successful innovators in that field should have more say in Government decisions, which are too often influenced by experts in concrete design who, by definition, ignore what makes machinery work or fail and who have no idea of what electromechanical development is.

In one extreme case where the Government's experts reported figures concerning key points that were generally wrong by a ratio of 2, sometimes 5 and on one given fact 10, all these points could have been checked by relatively elementary physics and engineering commonly used in the electromechanical industry.

Inventors on the other side need to become more market, development, industrially and exploitation minded.

A remarkable study of the decision making processes on new transportation systems in Europe and U.S.A. has been made by Professor Lucien SFEZ of Paris University and its conclusion was striking: every time he demonstrated the final choice to be the worst possible solution!

11

Three-Dimensional Engineering

E. R. LAITHWAITE

The greatest retarder to technological progress is probably the vast backlog of information already amassed and known to the researcher, who almost subconsciously accepts the work of his ancestors as basically 'correct'. Technology therefore tends to *evolve* rather than to advance, and evolution favours short-term progress and plays into the hands of the accountant, whose policies of 'balancing the books by the end of this year' have already been the death of several promising major advances, and once threatened such massive technological achievements as the Concorde aircraft.

When the 'redesign' of an induction motor is contemplated, it is assumed that the power will be supplied to the stationary part, that the conductors will be housed in slots, the ferromagnetic parts on both primary and secondary members will be laminated, the air gap will be small (rarely exceeding 2 mm) and within the active pole area, a cross-section perpendicular to the machine axle will be the same, wherever along the axial length it is taken.

It may come as a tremendous, revolting thought to the conventional induction motor designer that *none* of these features is an *essential* ingredient in the design of a highly-efficient, cost-competitive motor.

The arguments in favour of retaining all these features in a rotary machine of this type are powerful indeed. What is far less convincing is that when a radical change in basic layout is considered, i.e. the change from rotary to linear motion, the same fundamental structural features should be retained.

Habit forming

There are two driving forces, both economic, that dictate how a machine for the production of straight line motion should differ from that of a motor built to produce *torque*. These may be classified as follows:

1. Inherent physical effects due to the almost unavoidable increase in air gap incurred by 'going linear', and the edge effects at the front and back ends of the shorter member, force the designer into defensive measures to combat the detrimental results of the new topology.
2. The newly-found 'freedom' of three-dimensional space that allows the designer to venture into new shapes and physical layouts unimaginable to the designer chained to his topology by belief that rotary motion, largely spurned by living creatures, puts mankind in a class above the animals.

Unfortunately, human nature being what it is, conservatism has tended to prevail and the brave who chanced their arm were shot at from all sides by the greedy, the over-cautious, the jealous and the monetary advisers. Almost a quarter of a century of linear induction motor development was therefore largely pre-occupied with class (1) above. It might be given the name 'defensive' engineering, whilst by contrast, class (2) could be regarded as 'aggressive' in its approach.

What is quite fascinating in the development of the modern linear induction motor is the way in which features that were essentially being tackled with the defensive outlook led almost directly to the 'about-face' needed to adopt the class (2) outlook. Let it be said at once that this would never have been achieved but for the courage of those who were prepared to venture a few million pounds on a full-scale project from which there was to be no turning back, and in the face of total technological failure, the money had to be written off.

The first of the 'brave' were undoubtedly the Westinghouse engineers of the aircraft launcher 'Electropult' [1]. Zehden (1905 – US Patent No. 732312) and Emile Bachelet (1914) had had the good ideas, but never found the courageous men to finance them. The Electropult, on the other hand, had massive financial support, but the designers failed to appreciate that they had been 'let out of the topological cage', and limited their design to 'class (1)' thinking. They might be compared to fish that had been kept in a small rectangular pond for a number of years and then suddenly transferred to a large lake. For several hours fish subjected to such treatment will continue to swim around in a rectangular path, often accelerating at fantastic rates (as when frightened) but always pulling up to within a millimetre of the 'wall' of the imaginary pool whose size is identical to that of the one that had conditioned them. In biological terms, both the fish and the early designers of linear motors had been conditioned to a 'habit', and well-established habits are extremely difficult to eradicate. When the size of the project is that of a passenger transport system, the time-scale of the habit-demolition process tends to be measured in decades rather than, as in the case of the fish, in hours.

Whatever criticisms may have been, or may continue to be launched against the National Research Development Corporation in Britain, their name shall surely be listed among the brave who fell in the cause, for £4.5 million is a fair stake in a game of roulette – for high-speed transport in the 1960s could be likened to just such an adventure.

The progress in linear motor technology from the 1950s to the mid 1970s can indeed be likened to an adventure story in which a climber has the will and the drive to discover what is behind a seemingly impenetrable range of mountains. Many have set out to climb and many returned beaten – some not at all. As in John Bunyan's *Pilgrim's Progress*, that range of mountains had a name. It was called the 'Small Gap Range'.

Magnetic circuits

Among the physical properties of materials such as the tensile strength of steel, the electrical conductivity of copper and the gravitational pull of matter on matter, few are more powerful than the magnetic conductivity of the *absence* of material – the 'thing' we call 'space'. Space has a value of permeability, μ_0, lying between zero and infinity, whatever system of units we use. There are substances that are a thousand times more conducting, but there is not one that is a thousand times less. So there is no magnetic insulator. Couple with this the fact that 1000 μ_0 is not such a high permeability for a magnetic circuit as is the conductivity of copper for an electric circuit and the result is well known if not well appreciated by all electrical engineers. In 150 years of electrical machine development, the electric circuit has been the subject of hundreds of new shapes and thousands of patents. The electric circuit can be multi-turn, of small cross section and therefore easily shaped. The ingenuity of the engineer has produced complicated but highly effective ideas for the exploitation of the physical properties of copper.

Meanwhile, the magnetic circuit has remained a single turn, short and fat. Where a machine is required to suffer relative movement between primary and secondary electric circuits, the designer laments the fact that the air gap required by other considerations (e.g. harmonic losses, mechanical tolerance) often contributes over 50% of the whole reluctance of the magnetic circuit. The men who ventured to suggest linear motors for linear motion soon found that it was *more* difficult to keep mechanical clearances low in a linear machine, and those who sought to make the secondary conductor consist of a simple conducting sheet, so as to make a long track in transportation systems, threw away a factor of 10 or more by abandoning the technique of sinking conductors in slots, that had served the rotary machine designer so well. It was this factor of 10 that turned back many 'pilgrims' from the 'Small Gap Range' of mountains.

Magnetic pull

The first mountain range was crossed almost simultaneously in several places in the early 1960s. In the USSR the concept of the Magnetic Reynold's Number, and in the UK the creation of the Goodness Factor showed that what was lost in gap could be recovered in pole pitch, and a factor of 10 in air gap could be cancelled by a factor of only $(10)^{\frac{1}{2}}$ in pole pitch. For constant frequency, this meant an increase in speed of $(10)^{\frac{1}{2}}$ over those surface speeds encountered in rotary motor design.

But was not high speed a part of the 'promised land' behind the mountains? Alas, the men who stood at the top of range (1), only saw in the distance an equally formidable range whose name was Magnetic Pull. The rotary machine designer was already knowledgeable about this effect. Rotor and stator members of nearly all rotary motors are effectively iron-cored electromagnets that attract each other like the opposite poles of a permanent magnet. Such forces are like a tie bar in tension. If the rotary machine has *perfect* cylindrical symmetry, there will be no external effect. The radial pull will sum to zero for the whole pole surface. Yet so potent is this force that a very small eccentricity in manufacture – a preferred grain-orientation in the steel, a non-uniform distribution of copper in the end rings, or even a keyway – may be sufficient to make the unbalanced magnetic pull (UMP) dominate the design to the extent of fixing the shaft size to avoid the whirling speeds that might spell disaster.

In a linear motor, the whole (not merely the imbalance) of the magnetic forces are arithmetically additive. *Experience* in rotary machine design created the second mountain range, for it was well known that the magnetic pull (the tensile stress) was, in a 'good' design, of the order of 10 to 20 times the *useful* force, seen as a shearing stress in mechanics, so that a vehicle fitted with a linear motor that was seen as a rotary design simply split and unrolled, might easily double the effective weight of the vehicle on its suspension. More than that, the extra force would vary with supply voltage (which would need to be changed to control speed) and with the air gap clearance itself.

Double-sided sandwich motors

A narrow pass was seen in the second mountain range. At the expense of increase in effective air gap, the whole of the secondary conductor could be extracted from the slots and cast into a sheet to be placed between a *pair* of primary units. The primary units would then pull only against each other, but travelling as a pair they would constitute a portable magnetic circuit, leaving only the conductor to be distributed over hundreds of miles of track. The track cost is thus minimised simultaneously with the elimination of the magnetic pull. The transport Utopia surely lay through that pass in the second mountain range?

Maybe, but the pass itself was full of thorns. Aluminium (the obvious choice for track conductor on both conductivity/weight and conductivity/cost criteria) does not have the tensile strength of steel, and is therefore suspect for any welded rail proposals. What is worse, if the track plate fractures, whether at a joint or not, the lateral forces applied to the plate by the primary guidance system (at high speeds it would almost be imperative to guide the motor separately from the vehicle, or the air gap clearances required for mechanical safety would become prohibitive) would produce the lethal situation depicted in Figure 11.1, where the oncoming edge of the displaced primary would strike the track plate.

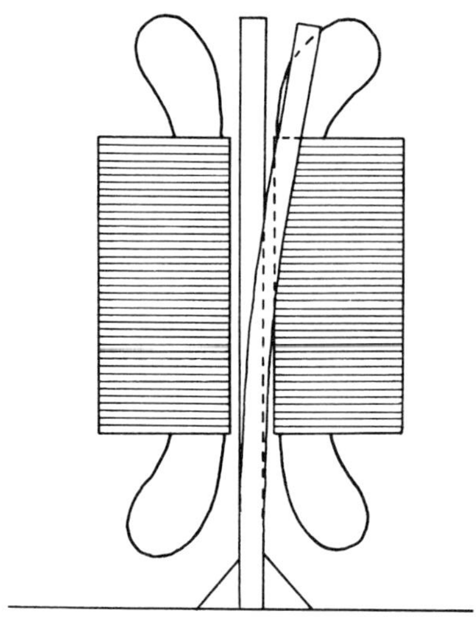

Figure 11.1 The effect of lateral thrust on a split track plate.

The problem of the big pole pitch

There were some who hacked their way through the thorns, some who merely looked through the gap in mountain range (2) and saw mountain range (3) that was, if anything, higher than either (1) or (2). Both electric and magnetic circuits are being stretched in the wrong direction as the demand for higher speed arises, provided that there is a requirement to use the normal industrial supply frequency (50 or 60 Hz). A speed of 400 km h^{-1} demands a pole pitch well over a metre. Figure 11.2 indicates that the *useful* part of the electric circuit corresponds to the track width, the wasted part to the pole pitch. The amount of overhang of end windings can be reduced by short chording, but the bigger problem is that of core depth for the magnetic circuit in both halves of the primary (in a double-sided motor) or in the primary *and* the track in a single-sided system, must be sufficient to contain the whole of the flux per pole.

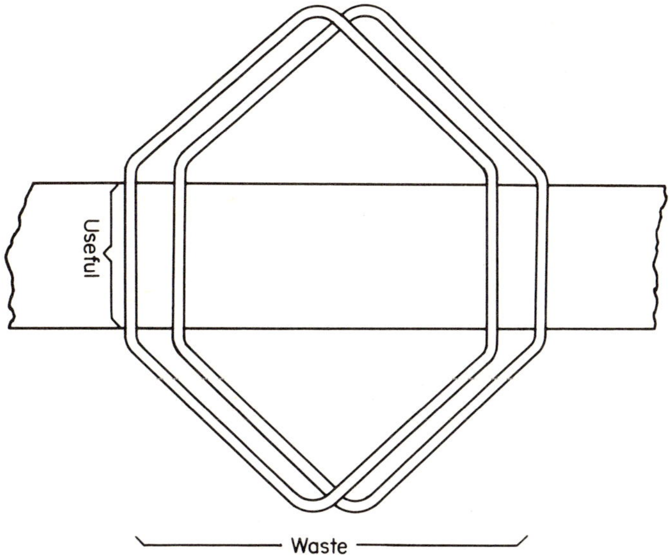

Figure 11.2 Useful and wasted parts of a conventional winding.

There were those who declared the third mountain range to consist only of clouds of water vapour. A change in frequency would solve the high speed problem. Some were bold enough to declare: 'Inverter system vital to rail traction by linear motor' [2]. The first estimates for the weight of inverter required for a 4 MVA machine varied between 13 and 17 tonnes, for a vehicle of nominal weight 30 tonnes. The third mountain range was more than clouds!

To increase frequency to increase speed is basically wrong, even though the slogan 'one cycle per second per mile per hour' coined in the early 1960s died hard. First, the Goodness Factor (G) is proportional to (pole pitch)^2 x (frequency). To double speed by doubling pole pitch is to multiply G by 4; to double f is to double G, and one needs that extra potential in G to cope with a larger air gap required for mechanical safety at double speed.

Secondly, the specific output per unit square of pole face is reduced. Whilst it is true that reduction in pole pitch reduces core depth, the slot depth at the higher frequency cannot be as great, for fear of increasing the primary leakage reactance to a level that would give the machine a power factor that was abysmally low. Inverters do not take kindly to a low power factor.

The transverse flux concept

A solution to the long pole pitch problem was to be found in the basic geometry of an electromagnetic system. Flux density B, acts on current loading J (ampere/metre) effectively only on components of B and J that are at right angles to each other. The resulting mechanical pressure, shearing pressure (a strange unit, but nonetheless correct – N m^{-2}) acts at right angles to both B and J.

Therefore, the 'conventional' linear motor, by which is meant the one derived from the rotary machine by the simple 'splitting and unrolling' idea, is most unsuitable for a high speed motor, as shown in Figure 11.3 (a), where both electric and magnetic circuits are unduly stretched. The lateral (shorter) dimension should surely be exploited for *one* of these circuits (Figure 11.3 (b)). The first to be tried was the

(a)

(b)

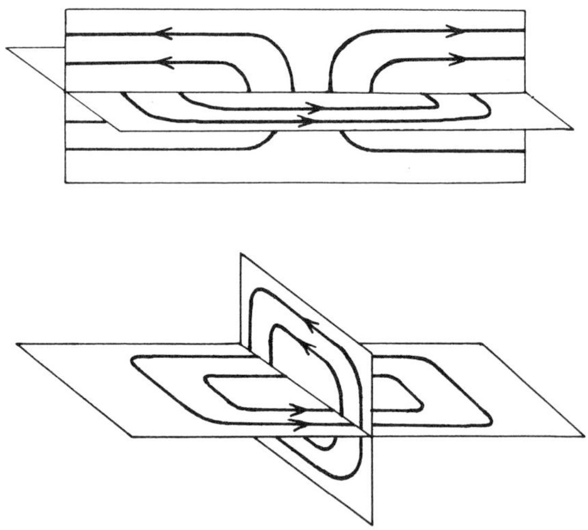

Figure 11.3 Electric and magnetic planes related to the direction of motion.
(a) The axial flux machine
(b) The transverse flux concept.

electric circuit, for this represented a return to a well-tried, if out-of-date technique known as Gramme-ring winding, of 1870 vintage. Figure 11.4 illustrates the disadvantages of such a system.

1. The size of the core in a large pole pitch machine makes the useless part of the electric circuit as long, if not longer than that of a surface winding's conventional end turns.
2. There is no improvement in the magnetic circuit and the total weight is no smaller.
3. The useless part of the electric circuit lies everywhere adjacent to the ferromagnetic material and the leakage flux is enormous, not merely increasing X_1 but demanding a further increase in core depth to accommodate it.

Full-scale models revealed that Gramme-ring motors for 1.5 metre pole pitches had power factors of the order of 0.05. Yet the solution was so near. The electric circuit is just able to cope with such a large pole pitch, if chorded to 1/3 (this

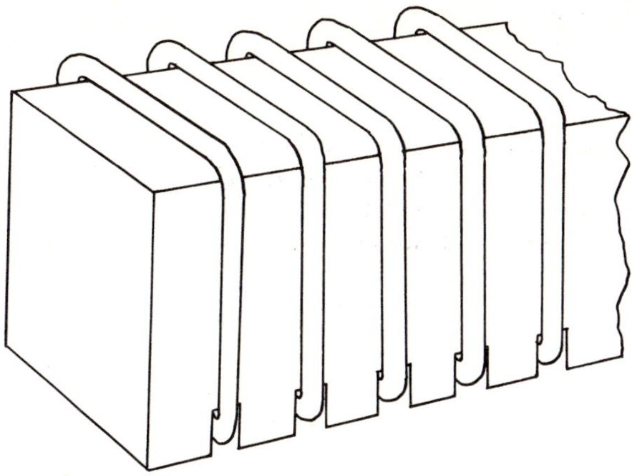

Figure 11.4 Gramme-ring windings on a motor with large pole pitch.

doubles the primary copper loss compared with a full-pitch winding). The magnetic, being by far the weaker of the two circuits, benefits enormously if placed transversely, as illustrated in Figure 11.5. The result of this exercise is nothing more than a pair of linear motors, side-by-side, with opposite poles next to each other throughout the length. Machines of this type have been named Transverse Flux Machines (TFM).

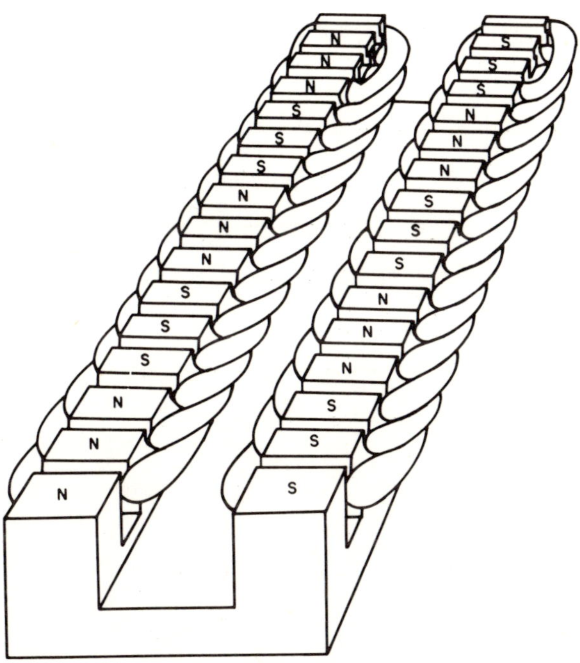

Figure 11.5 The TFM concept.

The construction shown in Figure 11.5 can be considered to be the 'primitive' TFM, for the leakage flux from conductors on the outside of the machine is very large. But as in transformer design, the 'E'-core followed closely the 'C'-core and the 'E'-core shown in Figure 11.6 brings out a whole new concept in linear motor design which is not limited to the induction machine alone. In the 'E'-core, the whole of the primary coils are embedded in steel and it becomes impossible to differentiate between 'useful' and 'wasted' conductor. More than that, if the primary coils are assumed to have triangular geometry in the longitudinal slots, as shown in Figure 11.7, and the angle

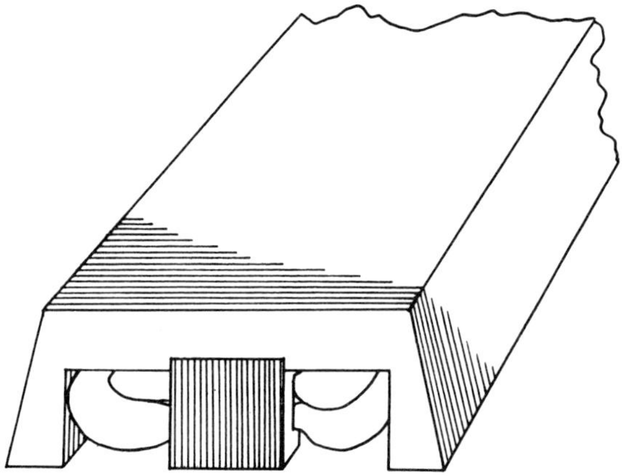

Figure 11.6 An 'E' core, 'hybrid' TFM (lamination in 2 planes).

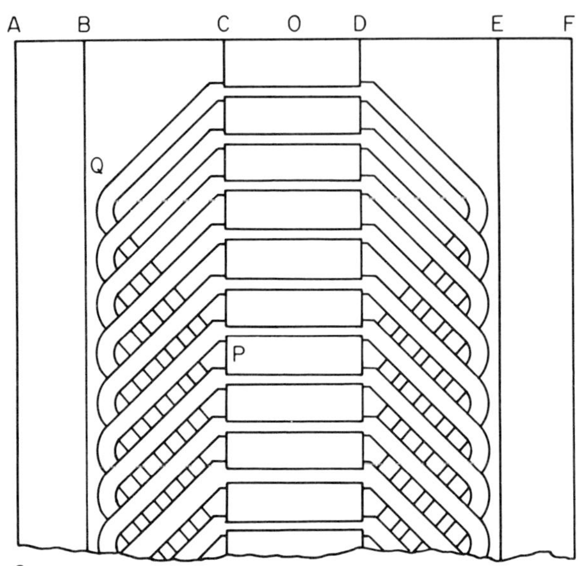

Figure 11.7 Plan view of an 'E' core TFM. The whole surface produces the same shearing pressure.

that each makes with the slot is 45°, for example, then the various parts of the motor face contribute to the thrust as described below.

The centre section CD is identical to that of a conventional motor, either linear or rotary. Let the shearing pressure in this area be P_s which is proportional both to air gap flux density B and to current loading J, the latter measured in the direction of motion AG. Since the left hand half of the central portion of the magnetic circuit CO returns its flux to AB, whilst OD returns its flux through EF, the portions AB and EF will produce thrust at the pressure P_s provided the secondary conductor extends beyond the edges A and F. In the longitudinal slot areas BC and DE, the current load-

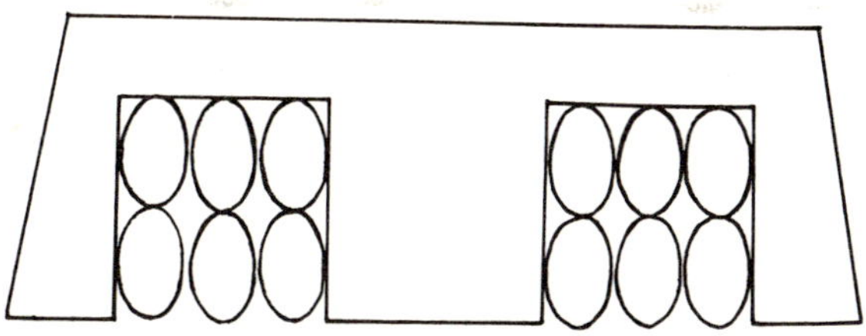

Figure 11.8 Cross section through the TFM shown in Figure 11.7.

ing conductors (PQ in Figure 11.7) is $\frac{2^{\frac{1}{2}}J}{2}$, and the whole layer is therefore theoretically capable of producing pressure $\frac{2^{\frac{1}{2}}P_s}{2}$ at right angles to PQ. But the top layer of conductors has current loading along PQ and the result is that the shearing pressure is $\frac{2^{\frac{1}{2}}P_s}{2}$ along PQ. The total result is that the shearing pressure is $2^{\frac{1}{2}}P_s\,(\sin 45°) = P_s$. Thus the whole width of the motor A to F produces the same pressure and raises the intriguing question, why have slots at all? (See Figure 11.8).

A system of coils shown in plan in Figure 11.9 (a) and in cross section in (b) lies on the face of a laminated primary slab facing a track of the usual steel-backed aluminium. The only disadvantage of such a motor is the increase in magnetising current incurred due to the removal of the primary steel teeth, but for very large pole pitches the slotless motor is at least 'in with a chance'.

One feature of the TFM has been deliberately deferred up to this point for fear of disturbing the logical argument relating to basic topology. But a glimpse at Figures 11.7 and 11.8 is sufficient to see that in some parts of the primary steel the laminations lie in planes containing the direction of motion, whilst other parts of the magnetic circuit are laminated as it were athwartships. This is done for ease of manufacture alone, but it is an example of the benefits of three-dimensional engineering. When an alternating flux passes through an iron block there are always two mutually perpendicular planes in which it can be laminated. Any plane that contains the common line of intersection of these two planes is also a possibility. The motor built according to Figures 11.7 and 11.8 allows the centre limb to be constructed exactly like a conventional linear motor, but with reduced core depth. The transverse punchings then fit on to this core. The arrangement is known as a 'hybrid TFM'.

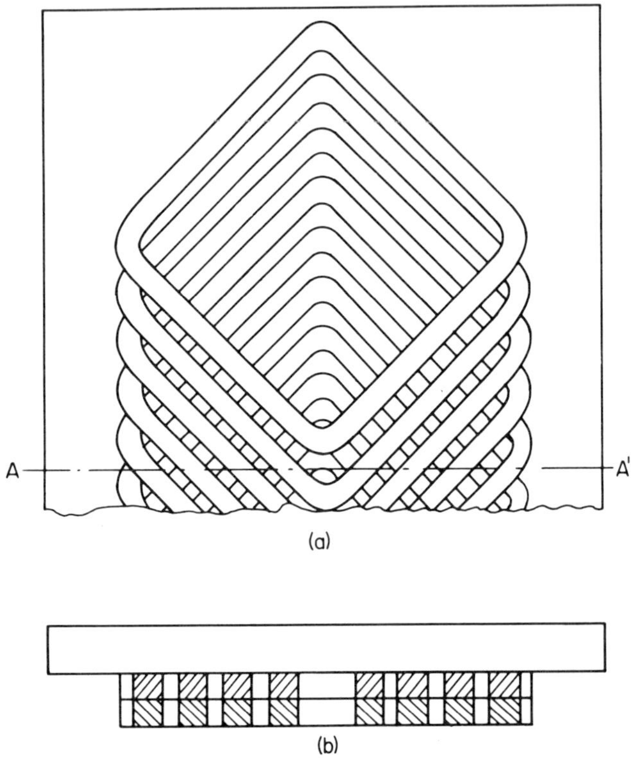

Figure 11.9 A surface wound TFM.
(a) Plan view
(b) Cross section through AA′ in (a).

If the track is to be laminated, this can either be done in planes parallel to the motion or athwartships, for flux paths in three dimensions, as shown in Figure 11.10, are possible. It will be shown in the next section however, that the secondary steel of a TFM is often so thin that lamination as a whole can be dispensed with.

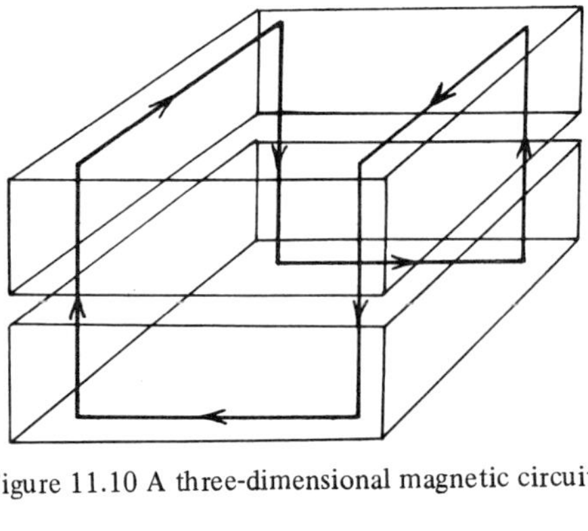

Figure 11.10 A three-dimensional magnetic circuit.

The tooth/slot ratio of linear machines

In 1969 it was established[3] that the real effect of the larger air gap imposed by 'going linear' and by using a sheet secondary was not so much that of lowering the performance compared with that of a rotary machine, but of necessitating a radical re-appraisal of the slot geometry, as had long been the case with fractional H.P. induction machines. It was established that the ratio of air gap flux density B_g to primary current loading J was dependent almost solely on the ratio of pole pitch to air gap length, thus

$$\frac{B_g}{J} = k\frac{p}{g} \qquad (1)$$

where k varies only in the ratio $5\frac{1}{2} : 1$ over a range of sizes from a fraction of a horse-power to thousands of horsepower. The ratio B_g/J is virtually the ratio of (tooth width)/(slot width) and in very small rotary motors where g cannot be reduced in proportion to p, for mechanical reasons, the effect can be seen in the use of teeth as narrow as can be punched without danger of the teeth breaking off.

Figure 11.11 shows a comparison between a stator punching (a) based on a

(a)

(b)

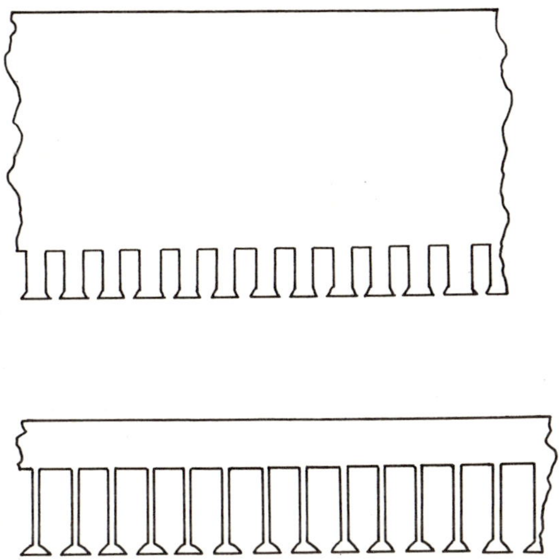

Figure 11.11 Comparison between the slot geometry of (a) a 'developed' rotary machine, and (b) a well-designed linear motor.

rotary machine design of the same pole pitch, power output, core length, etc. as the corresponding values in the linear machine. This was the technique adopted for a century following the invention of the linear motor by Wheatstone in 1845, and was the primary cause of the failure of industry to exploit the linear machine during that period. With such a design, the air gap of which might be 10–50 times that of the rotary 'equivalent', excessive magnetising current lowered the power factor and efficiency to the order of 0.2 and 30% respectively.

The *ab initio* design of the linear primary steel is shown in Figure 11.11 (b) where the effect of implementing Equation (1) has been to reduce the tooth width by a factor of 5 (from 50% of a tooth pitch to 10%), allowing the slot to be increased from 50% of a slot pitch to 90%, and therefore to allow a proportionate increase in slot depth, i.e. 9/5, making the increase in slot area, and therefore in J, $(9/6)^2 = 3.2$. The core depth is reduced in proportion to the reduction in tooth width, i.e. by a factor of 5, (flux doubling, that occurs due to end effects, affects the core depth in both designs to the same degree).

The output (proportional to $B_g J$) for design (b) is therefore reduced to 64% of that of design (a), per unit length of motor in the direction of motion. In a long pole pitch, conventional linear motor, since it is necessary to limit slot pitch to a maximum of 5 cm in order to avoid pole face losses, there may be as many as 12 slots/pole, necessitating a core depth of 12 tooth widths. The overall depth of the motor in Figure 11.11 (b) is therefore reduced by 3/7, and the machine can therefore be increased in length by 1/0.64 to make up for the reduction in output per unit area and the weight of core, teeth and slot conductors is still less than that of design (a). The extra weight of end-turns resulting from the increased slot section will probably make the total weight of (b) nearly equal to that of (a).

Whilst the above comparison appears to make the design of a conventional (longitudinal flux) motor feasible for a gap increase of 5 x 3.2 = 16 times that of a rotary machine (where the gap may be 2 mm), there are two factors that indicate the need for the TFM geometry for very high speeds.

1. 16 x 2 mm = 3.2 cm is not a sufficiently large air gap to accommodate aluminium of the order of 1 to 1.5 cm thick and leave a safe mechanical clearance.
2. The above example using 12 slots/pole, at 5 cm slot pitch makes p = 60 cm and corresponds to a synchronous speed at 50 Hz equal to 60 m s^{-1} (240 km h^{-1}). 400 km h^{-1} is the target figure for most high speed investigations currently in progress.

Any increase in core depth beyond 6 cm, indeed 6 cm itself, is a high price to pay for track core, remembering that the track length could easily exceed 1000 times the motor primary length. Using TFM construction, the track core can be reduced to 1 cm or less and still be run at sufficiently low flux density to allow the steel to be laid in solid sheets.

Single-sided motors

The discussion on TFMs and that on Equation (1) have not been dependent in any way on the choice between single-sided and double-sided sandwich designs. But the fact that the arguments resulting from Equation (1) when applied to the TFM result in designs with wide, longitudinal slots and therefore a wide motor overall, conflict strongly with the need to keep the height of the plate secondary in double-sided systems as small as possible for mechanical reasons. Researchers have been slow to accept that magnetic pull can be reduced enormously compared with that commonly measured in rotary machines and even more reluctant to admit that it might even be reversed!

The solution lies in further examination of Equation (1) and in adopting the *complete* expression for the normal force in an electromagnetic arrangement where, if G is reasonably high, equal and opposite *currents* face each other across the air gap. Opposite currents, unlike opposite magnetic poles, repel each other. The expression normally used by machine designers for the calculation of UMP is $B_g^2/2\mu_0$ per unit area. This expression is incomplete, the full equation for normal force F_n being:

$$F_n = (B^2/2\mu_0 - \mu_0 J^2/2) \text{ per unit area} \qquad (2)$$

When typical values for B and J from existing rotary motor designs are substituted into Equation (2), the conventional designer is seen to be justified in using only the first term of Equation (1) for it constitutes over 90% of the total. But in a linear machine design (for example the one used in the argument relating to Figure 11.11) B_g is lowered by a factor of 5 and J is raised by a factor of 3.2. If it is assumed that $B^2/2\mu_0$ constitutes 10 units of F_n and $\mu_0 J^2/2$ one unit (values appropriate to Figure 11.11 (a)), then the corresponding units for Figure 11.11 (b) are 10/25 and $(3.2)^2 = 10.24$, that is the $B^2/2\mu_0$ contributes one unit of F_n for 25.6 units of $\mu_0 J^2/2$, i.e. the first term is now less than 4% of the whole and a huge repulsive force exists.

When the argument is extended to TFM designs, even greater lifting forces are obtainable even though both primary and secondary members contain ferromagnetic cores. This makes possible the design of combined lift, guidance and propulsion systems, described later under the heading of 'Magnetic Rivers'. The very small component of F_n due to the magnetic pull between steel parts justifies the argument that suspension heights for electromagnetically suspended systems using TFM designs can be comparable with those proposed for cryogenic MAGLEV systems.

The 'great divide'

In 1973, an article in *Electronics and Power* [4] proposed a classification of all electrical machines (excepting transformers and electrostatic devices) hitherto undescribed. The distinction was made by focusing attention on the air gap, in this case to be seen as the interface between the moving and stationary parts, the region where the desirable shearing stresses, referred to earlier, are generated. Figure 11.12 shows the possible arrangements in which either electric current, permanent magnet, or merely steel can constitute one side of the interface.

The left hand side of the diagram is seen to include induction, synchronous, d.c. and all a.c. commutator machines. The right hand side includes reluctance and hysteresis motors, and all electromagnet/iron keeper arrangements. The need for such a classification arises in connection with the effect of the *scale* (in size) of a system.

In the magnetic machine world, force is produced in proportion to $\mathbf{B}^2$ per unit area. In all systems employing ferromagnetic material, the maximum value of $\mathbf{B}$ is always limited by saturation. The force produced by a magnetic machine therefore rises as the *square* of length. The weight of the device rises as the *cube* of length (assuming all dimensions are scaled by the same factor) and the system is analogous to that of the flight of birds, where lift is proportional to wing area, weight to volume and therefore length cubed. Scale up a sparrow to the size of an ostrich and it cannot fly. The force/weight ratio of magnetic machines increases as they are made *smaller.*

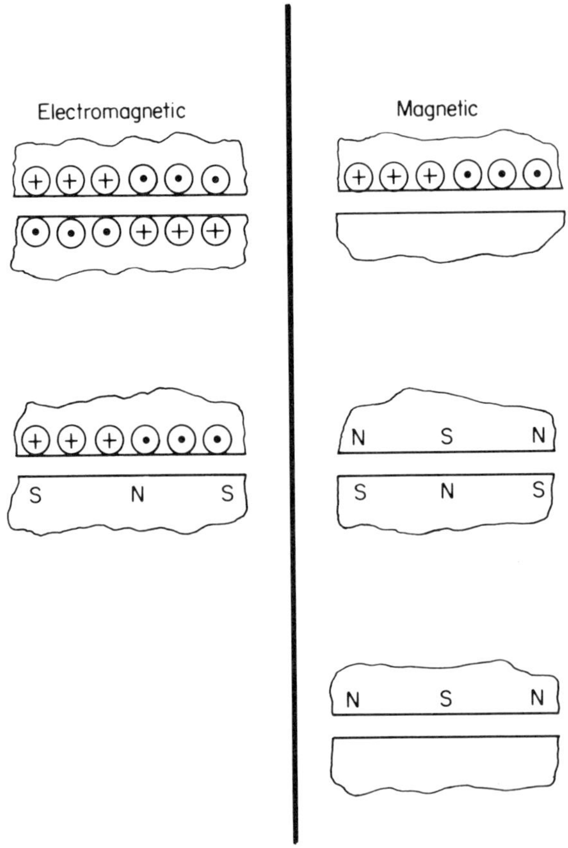

Figure 11.12 The division of machines into 'magnetic' and 'electromagnetic' groups.

The force of an electromagnetic machine is proportional to **BJ** per unit area. **B** has a fixed maximum as before, but **J** increases with increase in size, and by an amount greater than a linear increase so that the force increases by $(\text{length})^n$ where $n > 3$, and electromagnetic machines have a force/weight ratio that increases as they are made *bigger.* The left hand side of the Great Divide in Figure 11.12 is the Brobdingnag of *Gulliver's Travels*, the right hand side is Lilliput!

Levitation

Two methods of wheel-less suspension were developed in the 1960s both confusingly named MAGLEV. The confusion was due to the fact that one lies to the left, the other to the right, of the Great Divide. The one is easy to demonstrate at the model stage, and will not scale up effectively; the other is unconvincing at model stage but it needs great courage to invest £10 million in the belief that all will be well at full scale. The terms 'attractive MAGLEV' and 'repulsive MAGLEV' have been used to distinguish the two but this has only caused greater confusion in the minds of politicians and journalists for whom the words 'attractive' and 'repulsive' are easily interpretable as 'desirable' and 'undesirable'!

The terms 'tensile MAGLEV' and 'compressive MAGLEV' will be used here as being more descriptive, in an engineering sense, of the precise nature of the mechanism. In tensile MAGLEV the vehicle is 'hung' on a pair of ferromagnetic rails by electromagnets whose current is controlled by an amplifier that in turn receives its signals from a sensing device that effectively measures the distance between magnets and rail. A schematic diagram of one half of such a system is shown in Figure 11.13. A certain amount of induced current flows in the rail, producing drag forces on the vehicle but these are easily minimised by making the magnets and rail long and narrow, and steel, in any case, is not the best of electrical conductors.

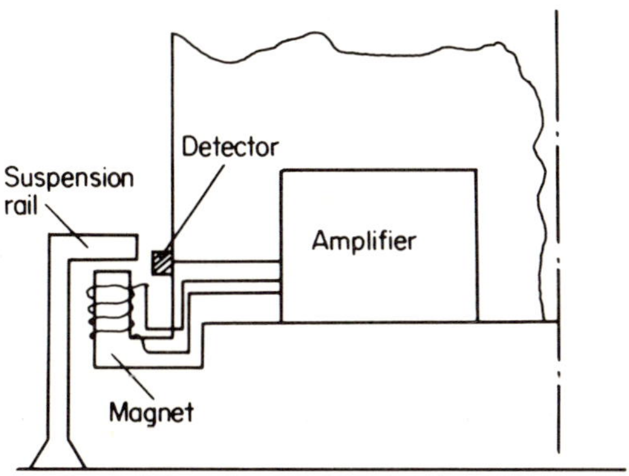

Figure 11.13 Basic layout of a tensile MAGLEY system.

The compressive MAGLEV consists of cryogenic (air-cored) magnets on the vehicle, kept at 4 K in liquid helium jackets, and capable of producing very high flux densities (of the order of 7 Tesla) downwards into the track. Unlike tensile MAGLEV, no lifting force is available when the vehicle is stationary and very little lift when moving slowly. The vehicle can 'take-off and land' as does an aircraft on wheels. The currents induced in an aluminium sheet laid flat on the track produce both lift and drag, the drag/speed curve being of the general form shown in Figure 11.14. It is, of course, a mirror image of an induction motor thrust/speed curve.

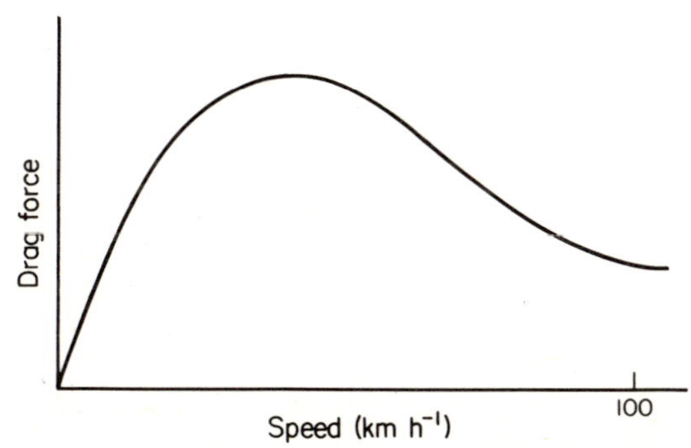

Figure 11.14 Drag/speed curve of a cryogenic magnet.

The 'Magnetic River'

The descriptive names of the products of high technology have a habit of being condensed into initials that spell a pronounceable name, or, if simply one or two words, have a habit of being shortened. With the best intentions of indicating that it was not a magnetic device but an *electromagnetic* one, the system about to be described was originally named the 'Electromagnetic River', but was very soon shortened to the title of this section and later to 'Mag. River'.

It owes its origin to developments in TFMs that were aimed first at power factor improvement and secondly at track cost reduction. But it was found that a primitive 'C'-core TFM would levitate and stabilise any conducting sheet of sufficient thickness and width at any height within an expanding trough as shown in Figure 11.15. The power fed to the primary determines the height lifted, and the stability at such a height depends on the levitated sheet being made the right width to 'fit' between the dotted lines shown in Figure 11.15.

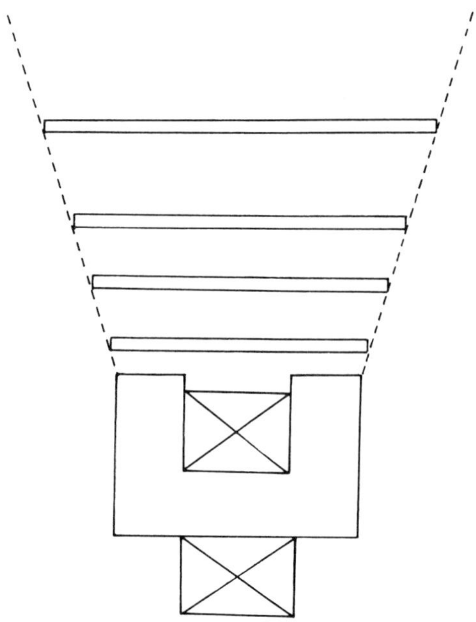

Figure 11.15 The expanding geometry for lateral stability from a 'C'-core Magnetic River.

At first it was thought that the lift was obtainable only at enormous cost in power input, most of which was lost as heat in the suspended plane. It was at this point however, that three-dimensional engineering began to force itself upon the researchers (rather than by any conscious efforts on their part!). In a conventional rotary machine the radial (magnetic) pull is of the order of 20 times the specific tangential (useful) force. If the user had to pay for such a radial force in the coin of a proportionate amount of power, no rotary motor could exceed 10% in efficiency. We know that this is not the case, and theoretically, zero (and practically very little) power is needed to set up radial force, for there is no radial motion and therefore no mechanical output is produced. The technology of the Mag. River was founded years before its invention on the reversal of domination between the terms of Equation (2) (i.e. $B^2/2\mu_0$ and $\mu_0 J^2/2$), so the levitation force is merely an 'unbalanced magnetic push' [5] exaggerated to its maximum by making it single sided and is therefore basically a *powerless* force. The same is true of the guidance forces. In three-dimensional engineering you only have to pay your 'entry fee' (i.e. primary I^2R loss and iron loss) *once.*

Therefore, the wisdom of the designer lies in the realisation that he can design two or more linear motors, obtain the normal power balance equation, Input = Output + I^2R loss + Iron loss, and have forces in directions perpendicular to the direction of motion at zero watts per newton! Of course there is a price to pay for this luxury. Whilst it is not in watts it is in reactive volt-amperes, and the higher the supported height the lower will be the power factor, due to what would be called in a conventional motor 'secondary leakage reactance'. The 'C'-core TFM which is admirably stable and adequately damped in pitch, roll, yaw, vertical bounce and lateral displacement when well designed (for such properties *are* within reach of the skilled designer) has excessively large primary leakage due to end windings along each side of the motor, but an 'E'-core TFM is basically unstable in roll.

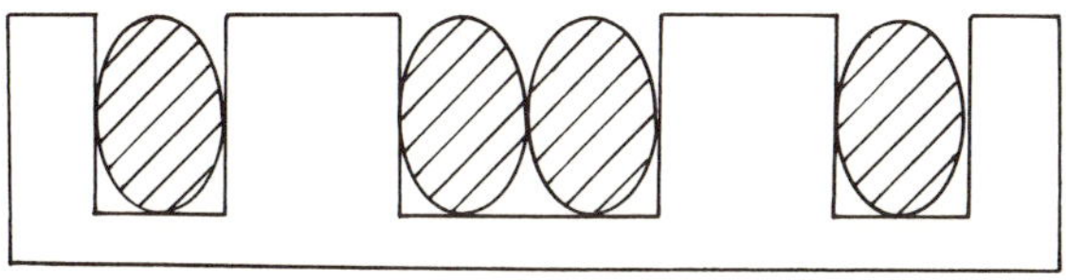

Figure 11.16 Cross section through a Xi-core Magnetic River.

Solution to the problem of combining the best qualities of both 'C'- and 'E'-core types was unbelievably obvious (as are *many* good ideas once they have been formulated!) The containing of leakage flux by 'side walls' in the 'E'-core can be copied in the 'C'-core without loss of even symmetry and the result was labelled a 'Xi-core', since the Greek letter Xi has four 'limbs'. A cross section through such a motor is shown in Figure 11.16.

Nor is there any undue complexity in what is essentially a two-dimensional winding on a flat surface. Linear motors should perhaps be designed 'per square metre' rather than 'per metre', not unlike the way that a tailor designs a suit of clothes and cuts it to fit, a concept first expressed in much earlier days in linear motor research [6]. The high speed linear motor of the 1980s could well resemble a closely woven mat only one or two centimetres thick, covering the entire under surface of the vehicle.

Hyperconductivity

A natural reaction to the proposals for Mag. Rivers so far expressed is that they must have very poor magnetic circuits since the secondary plate (in practice, of course, this is the *track* member rather than the levitated one) is not backed by steel. Steel *must* produce magnetic pull and/or instability. But examination of the earlier discussion on the subject of the Great Divide and of the new concepts in B_g/J ratio reveals that this need not be so. If the machine is *large* enough, induced current effects (compressive and stabilising) will increase, whilst magnetic pull effects (tensile and de-stabilising) will be reduced.

The effect may be illustrated with very simple equipment. Figure 11.17 (a) shows an iron-cored coil of outside dimensions 2 cm diameter, 1 cm long, fed with alternating current and placed over a sheet of aluminium 1 cm thick. At reasonable current densities the coil will be unable to lift itself clear of the surface, but it can be

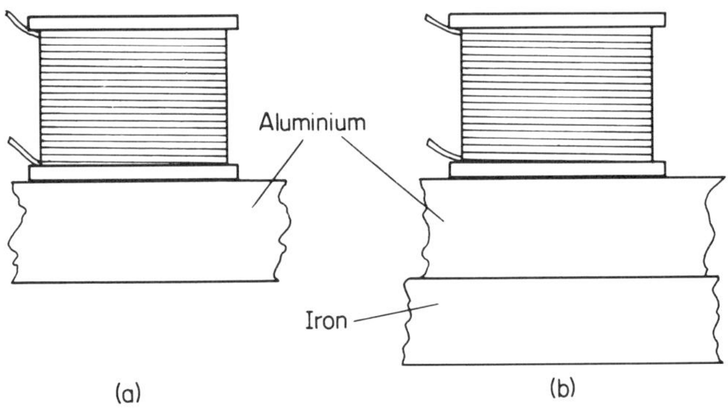

Figure 11.17 Normal forces on a single, iron-cored coil floating over a variety of metal sheets
(a) small coil over aluminium sheet will not lift off.
(b) small coil on aluminium sheet, steel-backed.

weighed, with and without excitation and the resulting lift force determined. It will be of the order of one tenth the weight of the magnet. If placed on a sheet of mild steel however, the attractive force will be found to be many times the weight of the magnet. Such an object truly belongs in Lilliput. Place it over the aluminium sheet when the latter is backed by the steel sheet as in Figure 11.17 (b) and a large attractive (tensile) pressure is obtained.

If now the experiment is repeated with a 20 cm diameter coil 10 cm long over the same sheets, the remarkable changes predictable from Equation (2) can be measured. The upthrust over the aluminium sheet alone is 25 N. The tensile pull on the steel sheet alone is 400 N. The vertical force over the composite secondary is a *lift* of 40 N. Such a lift could only be obtained using aluminium sheet alone, if its thickness was increased to 2.5 cm. The aluminium/steel combination suggests that if the steel is hidden beneath, the aluminium has somehow acquired a conductivity 2.5 times its normal value. The word 'superconductor' has had a very special meaning for a considerable time, so it seemed appropriate to refer to aluminium track, when steel-backed, as 'hyperconductor'.

It might be argued that in performing the earlier experiment with the small coil, the sheet thicknesses should have been scaled down also. This would only result in further deterioration of the device as a levitator with very little effect on its performance as a magnet in its literal sense.

Variation of vertical force with speed

Early experimental evidence supported by calculations showed that the normal force between primary and secondary of a flat linear motor could be tensile or compressive at standstill, depending on the values of B_g and J, but that as the motor gained speed, lift forces were reduced and tensile forces were increased. A typical curve is shown in Figure 11.18. At the time such information was beginning to come in, it was noticed that the front portion of a motor would exhibit a different normal force curve from the rear portion (shown dotted). This could give rise to pitch instability and there was an awareness of a motion that resembled that of a porpoise at play. It was known that increased values of G would aggravate such instability.

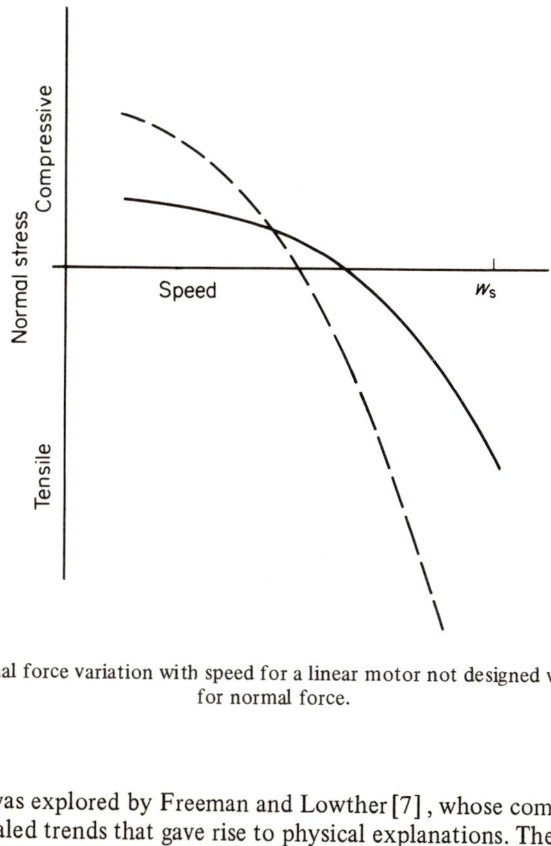

Figure 11.18 Normal force variation with speed for a linear motor not designed with special regard for normal force.

The subject was explored by Freeman and Lowther [7], whose complex theoretical programme revealed trends that gave rise to physical explanations. The family of curves shown in Figure 11.19 indicate that the larger the air gap, the flatter the curve of normal force/speed. To use a large gap is in accordance with the findings of researchers into compressive MAGLEV, the mechanism of which is closely related to one axis of the Mag. River. To use a large air gap and maintain a reasonably high value of G makes it imperative that the pole pitch be large and that the advantages of the short magnetic circuit of the TFM be fully exploited. The Mag. River is an 'all or nothing' exercise. It is high technology for a fast-moving society. It belongs only in Brobdingnag.

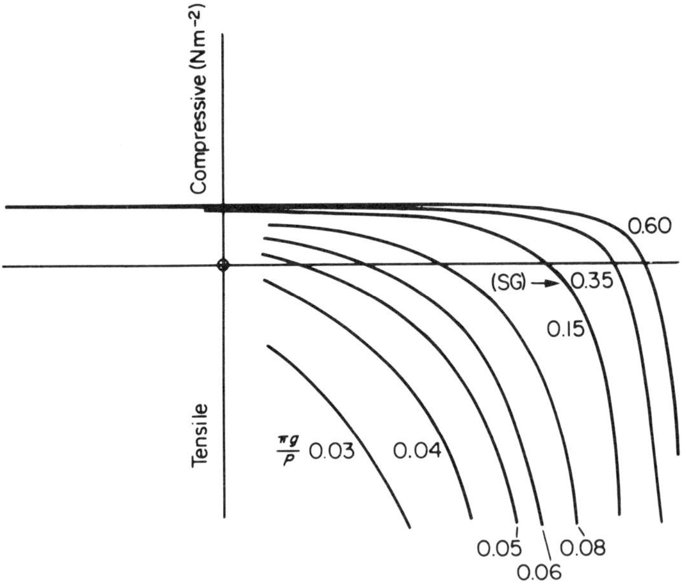

Figure 11.19 Variation of normal force/speed characteristic with air gap length (due to Freeman and Lowther).

Electromagnetic Joints

Another example of the possibilities of three-dimensional engineering is to be found in solving the problem of making track joints in the conducting plate, whether it be used in a compressive MAGLEV system, in a Mag. River, or simply in a linear induction motor for propulsion only. The nature of the problem is illustrated in Figure 11.20. In Figure 11.20 (a) the current paths in a conducting secondary correspond closely to those in the primary from which they originated. Any constriction to the flow of such currents, such as slots cut in the positions shown in Figure 11.20 (b) can only result in the deviated current finding no primary current to 'mirror' over the region AB, whilst the primary winding in the same region is free to surround itself with purely magnetising flux, with the result that the impedance of the primary coil so disposed will be highly reactive.

When the slots widen into a complete break, as in Figure 11.20 (c), the secondary currents are reduced to rather pathetic little 'pools'. (Pathetic because the end portions of the primary winding face magnetising reluctances in the primary steel and the whole of the primary that straddles such a joint is virtually on open circuit, always assuming that the value of G in the absence of a joint is high, i.e. the designer has done his job well.)

Even this effect would not be disastrous were it not for the fact that a coil that straddles a joint is connected in series with all the other coils of the same phase, so that the effect of a joint is equivalent to placing in series with the primary a reactance equal to the magnetising reactance of the machine per phase, divided by n, if there are n coils/phase in series. Cutting the plate at an oblique angle only prolongs the duration of the low performance. Aluminium does not have the tensile strength of

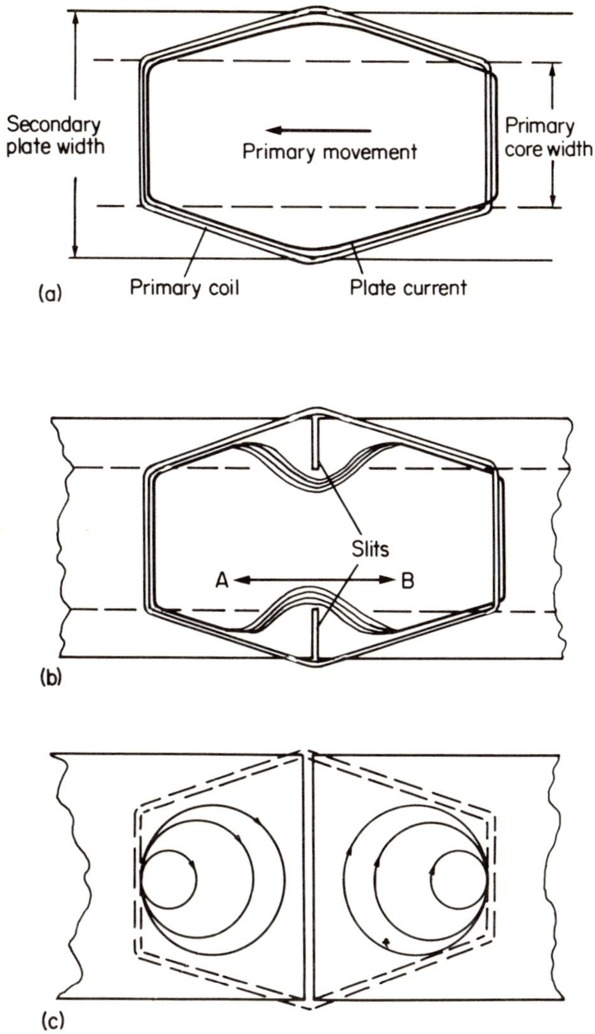

Figure 11.20 The problem of expansion joints in aluminium sheet tracks.
(a) Current paths in an unbroken sheet.
(b) Current path deformation due to slots.
(c) Current paths in the presence of a break.

steel, nor the same coefficient of thermal expansion, so the conducting plate needs to move over the steel with changes in temperature and to have expansion gaps between relatively short sections like those used on railway tracks before the days of the welded rail technique.

But there remains a dimension (the vertical) still unexploited in a track joint. Figure 11.21 shows an expansion joint (greatly exaggerated in size in order to display the current paths clearly). Both aluminium sheet and steel backing are bent downwards

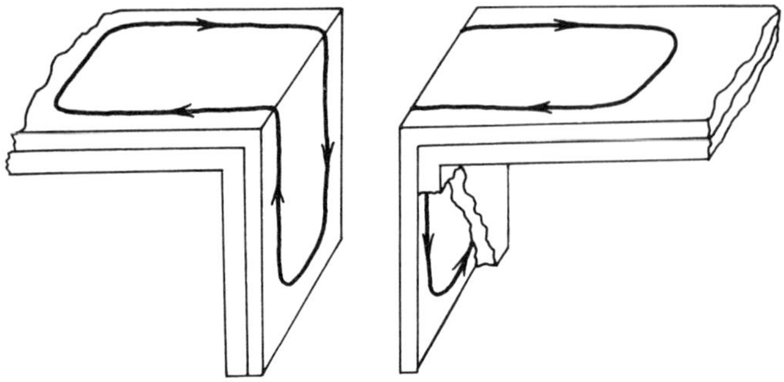

Figure 11.21 An electromagnetic track joint.

to face each other. The highly desirable current paths on the track surface continue in vertical planes, as shown, but the 'magic' lies in the fact that oppositely-directed currents face each other in vertical planes at all points. The vertical part of the joint is no less than a 1:1 current transformer with an enormous section of conductor in each 'winding' (single turn). It has a good magnetic circuit and can handle 100 000 A with ease. The price of this joint is small. A half pole pitch of vertical drop suffices almost to eliminate the primary transients that would occur with a surface plate only. The three-dimensional joint achieves the apparently impossible surface-current configuration shown in Figure 11.22. Three-dimensional engineering of a linear motor might well have spin-off in the world of rotary motors since it enables such important breakthroughs as the three-dimensional joint to be realised.

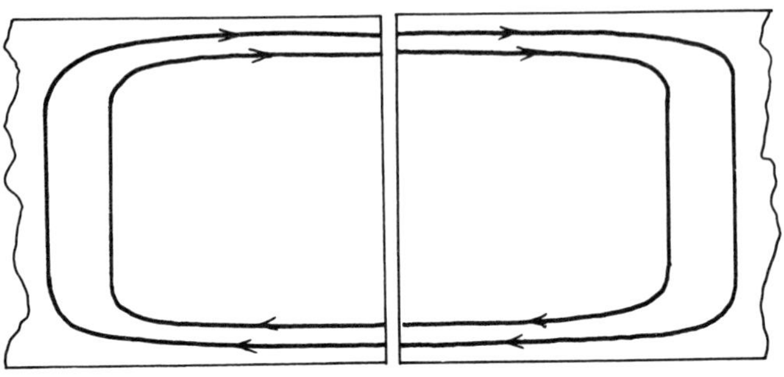

Figure 11.22 Plan view of plate currents on each side of a 3-dimensional joint.

By courtesy of British Rail I am permitted to show test results over track joints obtained with a linear motor of synchronous speed 80 km h^{-1} mounted on a railed vehicle and travelling at full load speed (Figure 11.23). The track consisted of a 5.0 mm thick aluminium sheet, 32 cm wide, for a motor pole face 19.5 cm wide. The track core iron was the same width as the pole face and 7.0 cm thick. The folded joint, A, had a total clearance of 7.0 mm and the vertical portion was 10 cm deep. For a pole pitch of 0.275 m the vertical portion was not as great as $\frac{1}{2}$ a pole – thought to be the best dimension. Nevertheless, the reduction in all the undesirable effects of an ordinary butt joint B is seen to be considerable.

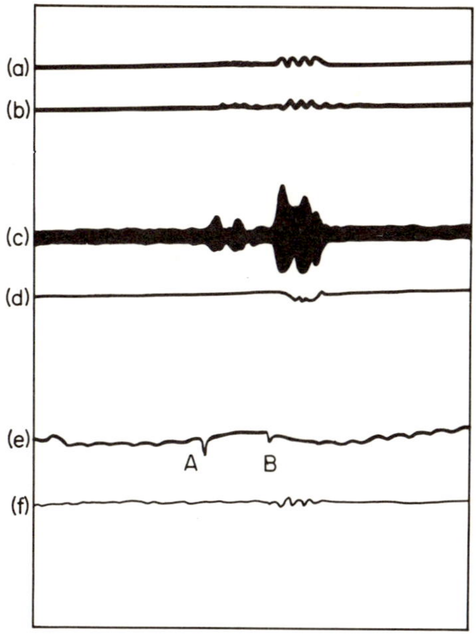

Figure 11.23 (a) Y-phase power
(b) B-phase power
(c) Pitch force
(d) Total vertical force
(e) Gap identity
(f) Phase current
(*Courtesy of British Rail*)

Emergency braking – failing 'safe'

The progress of science has always been a process of groping forward in the dark, looking for a glimpse of a pattern in experimental results, hoping for an inspired guess, testing and rejecting this or that theory, always in a 'half light'. When the paper is published and the material has been put in logical order, it looks as if that was the brilliant manner in which it was conceived, and papers are filled with sentences that begin, 'Clearly it can be seen that . . .', 'Obviously the optimum . . . ', 'It follows at once . . . ', and so on. Rarely was it ever 'clear' or 'obvious'. Research is done mostly 'from the back'. So at least it was with the emergency braking system devised for the ill-fated Tracked Hovercraft RTV (Research Test Vehicle) 31 that achieved 170 km h^{-1} against a 30 km h^{-1} headwind, starting and stopping under the power of one conventional (i.e. not TFM nor Mag. River) linear induction motor within a distance of approximately 1 km (Figure 1.17).

Having thus admitted the existence of the 'young ambition's ladder' (that took three years to climb), the reader will perhaps tolerate the obviously logical approach by which the ascent can be made in a couple of pages. If a 30 tonne vehicle is travelling at 400 km h^{-1} and suffers total disconnection from its supply there are clearly several things that need doing fairly quickly:

1. Its suspension and guidance must be continued at least until it can 'touch down' on its landing wheels.

2. Its speed must be reduced to landing speed as rapidly as is compatible with passenger comfort.
3. Something approaching 18 million joules must be got rid of, without the heat sink catching fire.

The one commodity that is in plentiful supply is energy – there is an embarrassing surplus of it. The only heat sink big enough to cope with it lies, not in the vehicle, but on the track. An induction motor can provide large enough braking forces if its excitation can be continued and if it is run either (1) in a 'plugging' condition where its direction of motion is opposite to that of its travelling field (this *requires* a large power supply) or, (2) in an induction generator condition, where the speed of its field is less than that of the motion proper (this *generates* a large power output). Combining (1) and (2) must be simple. On power failure the linear motor will be immediately divided into two halves. The one half will run as a generator and supply all its output to the other half run as a plugging motor. The whole motor will be fully excited. Lift and guidance will continue. The whole motor will produce maximum braking force.

The only remaining question hinges on the fact that an induction motor is incapable of supplying either its own magnetising current, or any other reactive volt-amps that leakage flux produces. But a synchronous machine is excellent for this purpose. The only thing a synchronous machine cannot do is to go on running without a power source. The situation is beautifully summarised in a well-known nursery rhyme:

Jack Spratt could eat no fat,
His wife could eat no lean,
And so you see, between them both,
They licked the platter clean.

A rotary, synchronous machine of very modest proportions may be made to resemble, electrically, a huge capacitance of several Farads. Such a machine, installed in a vehicle, might be permanently connected to one half of the linear motor and used to drive the fans of the air-conditioning system, the pumps of the refreshment refrigerator and like auxiliaries. In emergency it becomes the 'pacemaker' (a most apt analogy with the medical device of that name), the very heart of the stricken linear motor. The only action that is necessary as the result of total disconnection from supply is the reversal of two phase leads to the second half of the motor, operation in fact of the change-over switch S in Figure 11.24.

The operation of the 'emergency brake' proceeds as follows. On loss of supply, neither the vehicle nor the synchronous machine will stop rapidly. So long as the latter revolves at full speed, full excitation will be maintained. Switch S is operated. The small synchronous machine is initially doing all the 'giving' and no 'receiving'. Its speed must fall rapidly, taking with it the value of excitation frequency. A few per cent reduction will suffice to turn section A into a super-synchronous generator. The rotary machine's speed is now stabilised and entirely dependent on vehicle speed, for section A will always sense ground speed and adjust power flow so that its 'heart' receives the power it needs to keep supplying the lifeblood (reactive VA). Voltage is largely self-regulating. A 50% drop in speed means a 50% drop in both voltage and frequency. The magnetising reactances of both A and B sections receive

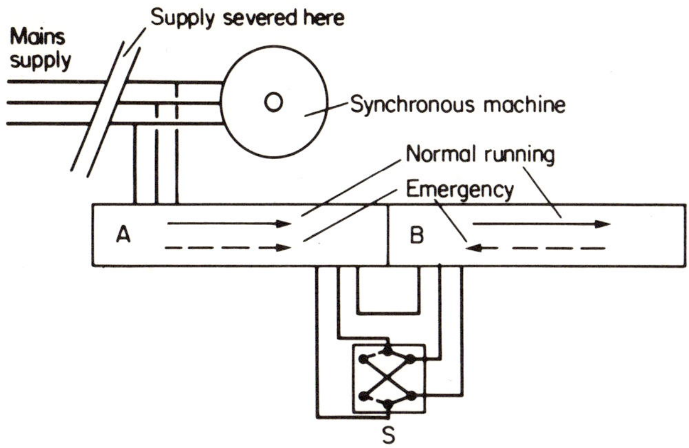

Figure 11.24 The fail-safe emergency braking system schematic.

full current. There is therefore full flux (approximately). Such a system will continue to be excited until the speed has fallen to a few per cent of full speed (provided the value of G for maximum frequency was high). The plugging motor gives the same lift as it did when motoring (see Figure 11.19). There is virtually no control gear at all to complete the emergency landing programme. It is one of the many marvels of the phenomenon we call 'electromagnetic induction'.

References

[1] 'A wound rotor motor 1400 ft. long', *Westinghouse Engineer*, 1946, **6**, 160.

[2] Farrer, W. (1974) 'Inverter system vital to rail traction by linear motor', *Electrical Times*, 20 June, 10–11.

[3] Laithwaite, E. R. and Barwell, F. T. (1969) 'Applications of linear induction motors to high-speed transport systems', *Proc. IEE*, **116**, No. 5, 713–724.

[4] Laithwaite, E. R. (1973) 'Magnetic or electromagnetic? The great divide' *Electronics and Power*, **19**, No. 14, 310–312.

[5] Freeman, E. M. and Laithwaite, E. R. (1968) 'Unbalanced magnetic push', *Proc. IEE*, **115**, No. 4, 538.

[6] Laithwaite, E. R. (1957) 'Linear induction motors', *Proc. IEE*, **104A**, No. 18, 461–470.

[7] Freeman, E. M. and Lowther, D. A. (1973) 'Normal force in single-sided linear induction motors', *Proc. IEE*, **120**, No. 12, 1499–1506.

12

Advances in Transport Without Wheels—The Future

A. J. ELLISON

12.1 Introduction

In this chapter the various forms of surface guided transport which have had, or are having, studies on them carried out, or which have been proposed, are described. Other chapters have given details of the studies, tests and future proposals (excepting those of the Soviets) of the various countries interested in the subject of future transport, namely, the United Kingdom, the United States, Canada, France, Germany and Japan.

Since the increase in world recession, some of the major governmental support for research and development in the area of future transport has been cut, especially since certain somewhat misguided developments have failed to come up to their sponsors' unreasonable expectations. The result has been to put off for some years the introduction of certain other more promising systems. Nevertheless, development is bound to continue and ultimately, in one way or another, to be successful. The countries which adopt the most imaginative approaches and which provide appropriate facilities for engineers both academic and industrial to innovate and conduct trials, are likely, not only to have transport systems ahead of those in the rest of the world, but also to have saleable techniques and hardware which the others will be obliged to buy if they wish to catch up.

Advantages of wheel-less transport are primarily smoothness of ride, quietness, low stresses due to the load being distributed over a large area, light, cheap, supporting structures due to those low stresses and less possibility of impulsive loads. Quietness is certain to be of increasing importance as greater emphasis is placed on quality of life. A smooth ride is also bound to be given an increasing value for the same reasons. The tendency will be towards a general reduction in the factors which increase wear and tear on the nerves of the population and there is likely to be a rapidly growing demand for quiet, smooth, rapid, reliable, automatic means of transport, free of pollution, tasteful and inconspicuous. The various forms of vehicle supported only by a magnetic field are bound to have advantages, with respect to all of these factors.

There are basically two types of magnetically supported vehicle, where the support is due to either attraction or repulsion. Attraction systems necessarily use dynamically controlled d.c. electromagnets or a.c. electromagnets, perhaps combined with the propulsion means. Systems using repulsion may use one of three possible methods, the force of repulsion being produced by permanent magnets, by a.c. excited electromagnets over a conducting secondary member, or by rapidly moving d.c. excited electromagnets over a conducting stationary secondary member. A.c. excited electromagnets may be arranged to be supplied by polyphase currents and produce the propulsive and guidance forces of the vehicle also. D.c. excited electromagnets will probably use cryogenics so that extremely high currents may be used with no continuous expenditure of power in the windings. Such d.c. windings might also be used as the field windings of linear synchronous machines for driving the vehicle, the polyphase windings being arranged along the middle of the track. There are advantages and disadvantages associated with all these systems and these will be discussed as the systems are described.

It is important to note the confusion which sometimes arises in discussions of this subject as a result of the introduction in the USA of the word MAGLEV to indicate magnetic levitation. Users of this word have frequently a very restricted definition of it in their minds (if, indeed, they define it to themselves at all) and consider it rather vaguely to cover only cryogenic levitation (Section 12.4) and the simplest magnetic attraction system (Section 12.2.1). But there are several other methods of suspending vehicles by a magnetic field, especially, and perhaps most importantly, by the Magnetic River system (Section 12.5) in which the levitation and propulsion means are combined. It is probably best not to use the word MAGLEV at all, but to state exactly what system or systems are meant.

12.2 Electromagnetic attraction systems

12.2.1 *Dynamically controlled d.c. electromagnets*

Vehicles using servo-controlled electromagnets have been developed at the University of Sussex and by MBB (Messerschmidt-Bolkow-Blohm) and KM (Krauss-Maffei) in Germany. The electromagnets operate on magnetic structures – rails – incorporated in the track, that is, the vehicle is prevented from moving downwards by magnetic attraction between electromagnets on the vehicle and steel rails fixed to the track structure. The magnets may be excited by either alternating current or by pulses of direct current, the latter being preferred as explained below.

This system of suspension of a vehicle is clearly unstable. If the strength of the m.m.f. producing the field remains constant (or is of constant r.m.s. value if produced by alternating current) then a tendency for the vehicle to descend will increase the air gap length between magnets and rails and weaken the field. The force of attraction will thus be reduced and the vehicle will fall. If the vehicle tends to rise, the air gap length will be reduced and the vehicle will move upwards as a result of the increased field strength, ending its travel by making mechanical contact with the rails. The current flowing in the coils of the suspension magnets must therefore be tightly controlled, any slight movement of the vehicle downwards being immediately detected and the current increased. Conversely, any movement of the vehicle upwards must result in a rapid reduction of the current.

The control is not easy and provides a difficult problem in non-linear dynamics. Alterations in the length of the air gap must first be detected: there are several ways of doing this ranging from observation of the relative mechanical movement of the adjacent parts of the magnets to observation of flux changes i.e., of changes in the inductance of the suspension magnet or other magnet windings. Different methods have been tried by different researchers. As mentioned, the problem is non-linear in several ways: alterations in one parameter cause changes in the others. Also, flux is not proportional to magnetising current because of magnetic non-linearity (saturation) of the magnetic structures. In the simplest suspension, where one magnet is placed at each of the four corners of the vehicle, alterations in the forces produced by the various magnets as a result of differing air gaps at the four corners will lead to deflections of the vehicle chassis, i.e., to twisting of the vehicle, which is necessarily made as light as possible. In addition, for reasons of cost the supporting magnetic rails and their supports to the ground must also be made as light as possible, the rails being supported only at intervals and not continuously. Clearly there will be deflections as the vehicle moves along and the system will have associated with it a family of natural vibratory frequencies. There will thus be electromechanical resonances excited at different vehicle speeds and loading conditions.

The closed loop system to provide the controlled current (normally pulses of direct current) to the suspension magnets must have associated position and velocity feedback loops (acceleration feedback is also useful) and sufficient current capacity to provide very large magnetic forces quickly under the worst dynamic conditions, that is, the suspension must be stiff enough to maintain the air gap within very close tolerances as a result of gusts of wind from various directions, track irregularities and the other variables mentioned. The power to such a vehicle, both for driving it and for the suspension, will normally be conveniently of 3-phase alternating current. For a d.c. suspension it must therefore be transformed to an appropriate voltage, rectified and controlled. The equipment to do this is necessarily heavy and expensive. Such a suspension system requires quite small power to hold the vehicle steady while it is stationary: of the order of 0.1 W per N. (This is less than the power for an equivalent air cushion system.) However, the size and weight of the power electronic equipment is decided by the worst dynamic conditions and so the figure just quoted is scarcely relevant. The heavier the onboard power equipment, the less passenger carrying capacity is available.

Regarding vehicles which have been built and run, these are the subjects of other chapters. The MBB/KM developments in Germany are the most developed and Transrapid (the joint company formed by MBB and KM at the instigation of the German government) have a vehicle capable of carrying people which has run at 200 km h^{-1} (and has an air gap of about 10 mm).

One disadvantage of the d.c. servo-controlled electromagnetic suspension system is the drag force due to eddy currents induced in the stationary magnetic structure and which necessitates extra drive power to overcome it. Appropriate design can reduce the drag force to an acceptable value and it could, in any case, be made very small by laminating the magnetic rails (with a large increase in cost).

12.2.2 *A.c. electromagnets*

The a.c. suspension system is not apparently a serious contender. It has a superficial attraction in that it may be 'tuned' to give a stable suspension. This means that the supply circuit parameters can be adjusted so that an increase in air gap length will lead to an increase in magnet current, and vice versa. The 'tuning' is between the reactance of the magnet coil and external capacitance. Unfortunately, the normally decreasing suspension force with increasing air gap length can be made to increase by this method for only very small changes in air gap length and a real system requires control over greater excursions of air gap length than appear to be practicable by this means. So far as the writer knows, it is not therefore being seriously studied further, earlier work having clarified its impracticability.

Vehicles suspended in this way must be driven also by a non-contact means and the linear induction motor is normally the most appropriate. It may be of the single- or double-sided type. The latter was originally adopted for the same reasons that it was chosen for propulsion of the other systems, that is, it provides a propulsive force without unbalance. A single-sided motor has an unbalanced force between the stationary secondary member and the primary excited member. That force is approximately perpendicular to both and varies with their mutual velocity, being for most conventionally designed machines a repulsive force at lower speeds and an attractive force at higher speeds including the running speed. Changeover occurs somewhere around half way between standstill and synchronous speeds. It was soon appreciated by British investigators (Tracked Hovercraft Ltd.) that a double-sided motor was too dangerous for high speed travel. The primary members on the vehicle needed to be placed on each side of the vertical secondary plate along the centre of the track, that is, the vehicle straddled the secondary plate. If the plate fractured, perhaps due to ambient temperature changes, then the vehicle could run into the sharp end of the oncoming plate which might enter the vehicle, with unfortunate consequences to the passengers. The British workers therefore adopted in 1967 a single-sided configuration with the secondary member bolted safely and smoothly to the track. Other countries, as will have been seen in this book, now have a similar policy. The single-sided motors used to propel vehicles controlled in the way described above may conveniently use the stationary magnetic member as the secondary conductor with or without an additional surface plate of higher conductivity (normally of aluminium). Any unbalanced forces must then be accommodated by the suspension system.

12.2.3 *Interaction between suspension and propulsion systems*

The interaction between the forces due to the suspension and propulsion systems is exceedingly important and must be quantitatively considered at the design stage. An example of where this was not appreciated is provided by the intermediate transport proposals for the Toronto authorities made by KM and won by that company against wide competition. Their proposed vehicles were suspended by a magnetic attraction system (Section 12.2.1) and were to be propelled by a single-sided linear induction motor. The suspension and attraction systems were apparently designed quite separately. When the vehicle was floating very satisfactorily supported by the magnetic suspension system, the single-sided linear induction motor was switched on to drive it forward. The large levitation force produced by the motor could not be accommodated by the suspension system, which was driven hard over to make contact with the track. The vehicle was driven forwards in that condition, accelerating until (as is well known to the UK workers) the levitation force changed to an attraction force. The suspension was then driven to the limit in the other direction. That major technical difficulty, together with certain other unsatisfactory features, including anticipated difficulties with track junctions, led to the KM company's withdrawal from the contract.

12.2.4 *Combined electromagnetic attraction and drive*

Various suggestions have been studied for producing both levitation and translation, with lateral stability, from a single-sided linear motor arrangement and developments might be expected to show a great simplification and a large cost reduction. One arrangement that shows these features was demonstrated by Rohr in the US in 1972[1] and used the attraction forces of the linear motor which provided the propulsion, but development appears now to have been discontinued.

12.3 Permanent magnets in repulsion

Considerable advances in the development of improved materials for permanent magnets have now made their use quite practicable for suspension of vehicles[2]. Particularly important has been the development of ceramic ferrites and cobalt/rare-earth magnets. These magnets are manufactured in brick form and have pole faces on the largest sides of the bricks. The vitally important feature of such magnets, which is relatively new, is their enormously high coercivity. The permanent magnets of 30 years ago could not be made in flat brick form, or used in repulsion, as they would have become rapidly demagnetised and useless. This is not the case with the modern permanent magnets, which can be produced in long 'rails' with the north pole all along one flat side and the south pole along the other. Such a rail can be used in repulsion with a similar rail placed just above it and several pairs of such rails running in parallel along a track can be used to support a vehicle, the required lift forces being obtained by arranging sufficient rows at an appropriate spacing.

It can be shown (Earnshaw's theorem) that any system of permanent magnets so arranged in repulsion cannot in principle be designed to float stably, that is, to produce the suspension for a levitated vehicle without other aid. The system requires lateral stabilisation, which can be achieved mechanically, for example, by the use of wear-resistant rollers running on the track sides. Lateral stabilisation can also be achieved electromagnetically by controlling the currents in electromagnets reacting on a vertical magnetic structure incorporated in the track design, forming a horizontal arrangement of the vertical suspension system described in Section 12.2.1.

A vehicle suspension using permanent magnets has many advantages:

(a) No energy is required to achieve the lift forces.
(b) Lighter vehicles result from the elimination of equipment required for other means of support.
(c) There is a freedom in track design to suit various forms of transport, e.g. vehicles may be supported above, or suspended below, the track.
(d) No maintenance is required.
(e) There is no noise or vibration.
(f) As the magnets are of insulating material, there are no drag forces resulting from eddy currents.
(g) Economical and inherently safe operation follows from (a).

Problems that could be caused by the attraction of stray metallic objects can easily be overcome by suitable track design and in any case the danger is remote for elevated tracks.

It seems reasonable to assume that as the materials for the modern permanent magnets are amongst the most plentiful in the earth's crust, such magnets could be produced relatively cheaply on a large scale and the relatively high present cost brought down to a low enough level for their use to become practicable.

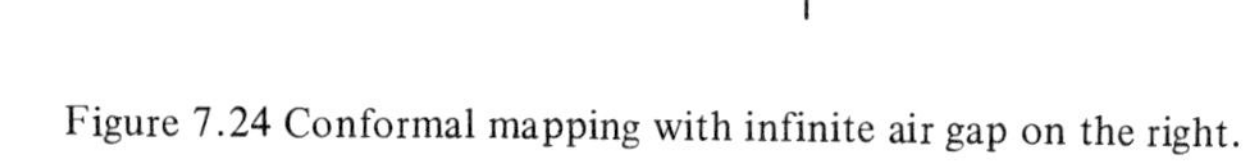Work on permanent magnets for transport has been carried out in the UK, in the USA and in Germany. Small trial vehicles have been built. Constant improvement of the magnetic material and the development of high-performance magnets make this form of suspension important to bear in mind.

12.4 Cryogenic magnets

If a permanent magnet is moved at high speed over a conducting sheet, currents will be induced in that sheet. By Lenz's law the currents will be in such a direction as to tend to destroy their cause, that is, a flux will be produced in the opposite direction from that inducing the currents and a force of repulsion between the magnet and the conducting sheet will be the result. There will be a drag force on the magnet so that work is done on the magnet, in proportion to the ohmic losses in the conducting sheet, in addition to that required to push the vehicle through the air. The ohmic losses will depend on the conductivity of the sheet and on the depth of penetration and the magnitude of the currents. The higher the speed, that is, the shorter the time a particular part of the sheet experiences the flux from the magnet, the less time there is for the flux and resulting current to penetrate but the higher the voltage induced in the surface layers.

Permanent magnets are too weak to produce levitation of a vehicle in this way but if they are replaced by cryogenic electromagnets, giving an enormously increased magnetic flux, then a perfectly practicable system of vehicle levitation is possible. It is proposed, in this system, that current-carrying superconducting loops arranged in the vehicle (like very strong permanent magnets but requiring to be permanently supercooled) would induce currents in discrete loops or conducting sheets on the track as a result of the motion of the vehicle. At a certain velocity (depending on the design) the repulsive forces would be sufficient to support the vehicle. Lateral stability would be achieved in one arrangement by means of vertical loops at each side of the track, producing restoring forces when the vehicle moved out of alignment. Large stable suspension heights of 15 to 30 cm have been predicted. It is not anticipated that onboard refrigeration would be needed for journeys of the expected lengths.

Cryogenic suspension is in the experimental stage and no full scale performance figures are yet available. As will have been seen in earlier chapters, considerable effort is being devoted to its development. It is fairly clear where the difficulties and advantages might be expected. With such large suspension heights the suspension will be 'light' and the vehicle exceedingly 'bouncy'. Evidently, considerable additional damping will be needed and this will probably have to be 'active', that is, powered, to give a smooth stable ride. The levitation will not occur until the vehicle reaches a certain 'take-off' speed (which will occur at quite a small proportion of full speed) and so wheels will be necessary to enable it to reach this speed. One may expect difficulty in slowing down smoothly to rest as a result of rapidly increasing drag force (as the main propulsive force to overcome air resistance falls off) which reaches its peak and then drops off very quickly to zero at zero speed. Specially designed control will be needed to cope with this characteristic.

Turning now to the propulsion of a cryogenically levitated vehicle, one obvious way would be to use an onboard single-sided linear induction motor, the existing track conducting sheet for levitation doubling as the secondary conductor. This would not appear to be very practicable as the magnitude of air gap required for such a drive would have to be much smaller than the suspension height, in order to avoid excessive need for reactive power to magnetise the gap (resulting in a low power factor), together with dynamic control of the length of gap. One early arrangement suggested used a double-sided linear induction motor with a central secondary conductor, but this is now considered too dangerous (as explained in Section 12.2.2). The prime contender for the propulsion of cryogenically levitated vehicles is the linear synchronous motor. The linear induction motor just mentioned requires all its power to be picked up and, for a vehicle of the highest speeds envisaged (about 400 km h^{-1}), this would be of the order of 7 or 8 MW at an inconveniently low power factor. Methods of accomplishing this satisfactorily at the speeds for which these vehicles are intended do not yet exist and remain a major problem. The linear synchronous motor proposed would use an 'active' track, that is, a 3-phase winding would be arranged down the centre of the guideway and fed from track-side sub-stations at intervals. (A large number appears to be necessary.) The power would be switched to a section of the track as the vehicle approached and switched off after it had passed. The field winding of the linear synchronous machine would be carried on the vehicle and also supercooled so that the very large field current necessitated by the large air gap would be arranged with no ohmic losses. The power factor could be near unity. An ingenious arrangement would be to use the same cryogenic coils for excitation and levitation, but appropriate design would not be easy and most present designs use separate coils. An active track would obviate completely the at present intractable problem of satisfactory power pick-up at 400 km h^{-1}. An important feature of a linear synchronous machine using cryogenic field coils would appear to be that no iron would be needed to form the magnetic circuit, as in a conventional machine, the result of the very large m.m.f. available for the correspondingly large air gap. The absence of ferromagnetic material would cheapen the track (which would be expensive in any case), but efficient magnetic shielding would be needed for the passengers.

Even assuming that a linear synchronous machine of this kind could be designed and made to work satisfactorily at a cost not entirely unacceptable, there remains the problem of accelerating the vehicle from standstill. (A synchronous machine produces force only at the fixed synchronous speed, decided by the pole-pitch of the 3-phase track winding and the frequency of the supply.) One way would be to accelerate it as a linear induction motor, an appropriate secondary conducting sheet or winding being arranged on the vehicle. Synchronisation would then have to be carried out as the vehicle approached synchronous speed. Alternatively, acceleration might be carried out if the vehicle could be raised to a low speed by some other means, by raising the supply frequency from a low value, synchronisation taking place at low speed. This would necessitate a variable frequency supply at the track side. Variations of speed during normal synchronous running would also necessitate variations in the supply frequency. The cost of such arrangements would appear to be very high and the complications great.

12.5 Combined levitation, guidance and propulsion (the 'Magnetic River')

The 'Magnetic River' is so called because it is like a river of water on which pieces of wood may float and be carried along, with a small 'slip'. The magnetic river is a flowing magnetic flux pattern on which a conducting sheet may float and be carried along also with a small slip and with guidance forces coming into operation when it departed from the central position. The vertical forces of a linear machine, like the radial forces of a rotary machine, absorb no power and, in principle, might support a vehicle for the expenditure of no power. However, as the vertical forces of a 'normal' linear induction motor change from 'up' to 'down' at about half full speed and, in addition, do this at different speeds at the leading and trailing ends, it must be specially designed to produce lift at all running speeds. This special design might reduce the efficiency and therefore require extra power.

The magnetic river became possible with the invention in the UK of the transverse flux linear induction motor. In this arrangement U-cores were arranged with the plane of the U perpendicular to the direction of travel of the secondary conducting sheet. This not only necessitated a low depth of iron core independent of the pole pitch of the machine and therefore economical in cost, but also enabled an appropriate width of conducting sheet equal approximately to the outer dimension of the U-core across the top of the open U to be used and stably floated. A strip of iron behind the conducting sheet increased the flux and the levitating force.

Since the U-core arrangement of magnetic circuit was invented – or rather re-discovered, because most electricity meters had used that arrangement for many years and some of the earliest electric machines also used it – various other arrangements of magnetic circuit have also been examined which carry the flux in planes roughly perpendicular to the direction of motion of the vehicle. In all, consecutive magnets, or groups of magnets, carry windings connected in each of the three phases of a 3-phase system so that a pattern of north and south poles is produced which travels in the direction of motion of the vehicle, that is, the flux is transverse but the direction of movement of the pattern of poles is axial.

The magnetic river has been described earlier in this book (see Chapter 11) and it is not proprosed to give more detail here but only to make a few general remarks. The first is that demonstration model vehicles have used a wound track, that is, a series of U-cores have been given a 3-phase winding to produce the magnetic river, and a plate of aluminium has been floated on it. The model vehicle has been attached to the aluminium plate. Such demonstrations are very impressive, especially if two assemblies of U-cores are used and placed at a slight angle end to end; the model vehicle successfully negotiates the irregularity at speed. By having the two halves connected to produce magnetic rivers flowing in opposite directions such a model vehicle can be arranged to fly continuously from end to end of the assembly. In such demonstrations it is usually explained that in practice a full size vehicle would have the primary windings, producing the magnetic river, mounted underneath it, and the secondary conducting sheet would be placed along the middle of the 'track'. A strip of mild steel, which is relatively inexpensive, placed beneath it, would improve the magnetic circuit and increase the lift (and because the machine is of the transverse flux type, this plate could be relatively thin and therefore low in cost). This also can be demonstrated readily.

In considering such a system for practical use for high speed future transport several factors may be of importance. The first is that the device is electromagnetic, and therefore gets better as it gets bigger. This is very encouraging. Possible difficulty with this particular magnetic river system at the highest speeds is in the collection of the necessary power (of the order of 6 or 8 MW at a relatively low power factor) at the top speed. As the vehicle is not in mechanical contact with rails like an electric train, three 'rubbing' contacts would be needed for current collection. Even if three pantographs were mechanically practicable it would be exceedingly difficult, if not impossible, to maintain contact with over-head conductors at top speed (say 400–500 km h^{-1}). Exceedingly straight conductors would be needed, probably with dynamically controlled pantographs, and the problem is not an easy one to solve. It must also be remembered that the vehicle itself will be moved by gusts of wind and track irregularities. Instead of pantographs, track-side conducting rails might be used and the rubbing contacts arranged on both sides of each rail so that movements away from it on one side would increase the contact pressure on the other. Even if the vehicle were imagined to take a completely straight course, the forces on the rubbing contacts due to irregularities to be expected in the rails would at top speed be very large. Difficulties might be expected with this system also. A third possibility is to collect the current via a plasma, that is, by continuously running electric arcs between rods projecting from the vehicle and flat plates. The rods could be fed out as they burned away. Clearly, irregularities in the distances between rods and rails would not be so critical with such a system but the difficulties of maintaining stable arcs in the presence of random gusts of wind and other variations at top speed would be formidable. (Some research is being carried out into such a system.)

With the primary member of the magnetic river system on the vehicle there is also the problem of control. Series/parallel and other arrangements of windings could be used to provide some control in steps. Variation of the applied voltage by means of reactors (or, more inefficiently, by resistors) would be possible at the cost of an on-board weight penalty. There is also the problem of safety when the power supply is accidentally cut off for any reason. A system using an onboard synchronous generator to excite part of the linear motor, feeding the other parts as an induction generator has been patented and would appear practicable for maintaining levitation and braking power until the vehicle stopped and settled down on skids on the track.

There is no reason why the current collection difficulty could not be overcome for the magnetic river as for the cryogenic system by using a live track. If a 3-phase 'meandering' winding were laid on a steel plate and covered with an appropriate protective layer the secondary member could be on the vehicle just as in a small demonstration model. Appropriate orientation of the windings and shaping and orientation of the secondary member would be needed to produce magnetic river operation. In order to control speed, either the primary winding frequency could be varied at the track-side substations, as for the linear synchronous machine drive (described above), or the secondary member on the vehicle could be made into a polyphase winding (like the secondary member of a slip ring induction motor) and resistors inserted in the phases. There is, however, a new and possibly useful way (Prov. Patent Application No. 3312/76) of controlling the speed of a linear induction machine without auxiliary resistors (or reactors) and without any need to alter the frequency of the supply; and there are other incidental advantages. The method in its essential features (and when applied to linear machines) was invented at The City University, London, and involves splitting the wound secondary member of the linear induction machine laterally into two equal parts and connecting the windings in series by flexible leads. The two parts may then be moved with respect to each other along the centre line of the vehicle. If a single secondary member is imagined to be split in this way and one of the halves moved through a distance equal to a pole pitch, then the voltages induced in the phase windings of the member that has been moved will have been altered in phase by 180° and the voltages will be in anti-phase with the voltages in the member that has not been moved. Clearly, the two sets of voltages will cancel, and no current will flow in the secondary windings on the vehicle. There will be no propulsive force and the vehicle will be at rest. As the movable half is adjusted in position relative to the other half the net voltage and current of the secondary phase windings will increase and the propulsive force be applied to the vehicle. The two halves of the secondary member do not have to be in juxtaposition for full force: they may be an even number of pole pitches apart. The two halves will always contribute equal components of the propulsive (and vertical) forces. There is, in addition, the possibility of adjusting the halves independently in vertical position and in angle relative to the track and this feature might be useful for stabilisation.

It is important to remember that adjustment of the speed of a 'magnetic river' arrangement by adjustment of the current in the windings will still depend on secondary loss and will also change the levitation force. Separate single-phase levitators may be needed to compensate and such levitators would make a contribution to the propulsive force of a vehicle (as single-shape linear induction motors).

12.6 Testing: drums, discs and tracks

The uninitiated will be impressed (as mentioned below) by an efficiently running model if they are not aware that (sometimes) a full scale version of the model could be quite impracticable. Tests on models are very important provided it is understood how the results of tests on models may be scaled up to give information useful for the design of larger devices: the models must be designed with this scaling up particularly in mind. This topic is of great and growing importance, especially in a time of financial stringency. It is vital for all investigators to understand the laws of nature relating to how some systems become better as they get bigger and others become less effective.

The ultimate test is the full scale test and there comes a time in every development when this is essential for further progress – and the provision of the facility can take several years. In the case of a very high speed magnetically levitated system, test rigs are useful, especially of the disc and drum types and several very large drum rigs are in existence. However, there is a lack of realism about a disc or drum rig. For tests on a disc rig, a linear machine must be built in the form of an annulus; a drum rig requires an arc-shaped primary member. In the UK at present, it is being suggested that a large drum rig should be built, the idea behind it being apparently that this is a cheap way of carrying out full scale tests, especially heat runs. This is not in fact true and all the leading workers in the field in the UK are opposed to the expenditure to build such a drum rig of a sum larger than would have been needed to establish Earith as a national full scale test and co-ordination centre.

The disadvantages of a large drum rig are as follows. Such a drum would be bringing round to the stationary vehicle member of the machine, under test, secondary conductors becoming continuously hotter and more unrealistic as the test proceeded. A real machine on a track would have continuously presented to it cool, fresh track conductor. A drum rig would therefore be useless for heat runs. An arc shaped stator would have a parallel air gap at only one length of gap. (A full scale track could be used for any length of gap.) The forces on the member in all six degrees of freedom would therefore be very unrealistic and normal movement during operation would give rise to quite unrealistic corrective (or other) forces and torques.

A real test track would present realistic wind and weather conditions: gusty winds would be of great importance. A real test track would have irregularities and resilience with a family of natural frequencies together with damping factors which would be of enormous importance in connection with the motion of a vehicle. For tests of systems which required it, a test track would need a realistic current collection system: a system for a drum rig would be quite unrepresentative of reality. The UK researchers were unanimous in their views that the realistic Earith test centre was vital to progress and they had already, or had planned for early completion, their own laboratory scale drum and disc rigs for small scale model tests. It is interesting to note that French investigators have a large and apparently redundant drum rig. German investigators are building full scale track (but now of a reduced length). Tracks exist also in the USA and Japan.

12.7 Politics and the future

The breathing space which has resulted from the costly errors, and the world recession, has provided the opportunity for increased emphasis on thought, research and testing, and reassessment before the next long steps forward. There is some danger of mistakes being made as a result of a lack of appreciation of certain physical facts by politicians and perhaps especially by some of the scientific and engineering advisers of politicians. In the UK, the introduction of the so called Rothschild doctrine, in which Government money for applied research is not made available unless a 'customer' says he wishes for the development, had led, *inter alia*, to the shutting down and subsequent destruction of the Earith test facility (originally of Tracked Hovercraft Ltd.) which would have been of inestimable value to many researchers in the UK and in Europe, to the aircraft industry and to other Government Departments. (The retention of Earith as a joint Universities/industry test and research co-ordination centre was unanimously recommended in 1973 by the House of Commons Select Committee for Science and Technology, which represents equally both sides of the House and which interviewed or received reports from everyone, Ministers, engineers, academics and transport authorities, who had any relevant information and views to contribute.) The destruction of Earith went a long way towards destroying the clear lead which the UK appeared to have over the rest of the world in the magnetically levitated transport area.

Regarding the form which future very high speed guided land transport systems are likely to take, Government and transport authorities are deflected from investment in the development if there are wide differences of opinion between researchers. They are, especially, put off by mistakes. The efficient interchange of information is vital to researchers in different countries and books such as this, and international conferences, are of growing importance. A clear understanding of certain relevant facts of nature is also important. For example, as mentioned in Section 12.6, some systems become more effective as they become bigger, others less so. The uninitiated will be over-impressed by an efficiently running model if they are not aware, and it is not pointed out to them, that a full-size version could be quite impracticable.

Researchers in the UK will never forget the derogatory remarks made by a Minister about the excellent THL full-size experimental vehicle: he appeared to consider that it should not be taken seriously because it contained no seats. (It was, of course, carrying a load of automatic monitoring and recording equipment.) Engineers often do not appreciate the need for proper and careful regard to psychological considerations when Government support is needed for an important project.

A difficulty experienced by researchers in several countries but especially in the UK concerns the time scale of need for possible new systems. It is perfectly clear to any deeply thinking person in the UK (and in other countries similarly) that technically excellent though the advanced types of wheeled passenger train are, and however appropriate is their use to the utmost of the existing infra structure of railway tracks, there is bound to come a day when they are obsolete. This happens with every technology, a good recent small example being the replacement of slide rules by electronic calculators. Research and development of the various alternatives to wheeled trains and other guided vehicles must be carried on now and at a sufficient level to keep an advanced country like the UK in the forefront of transport progress, not only with respect to that country's own transport systems but also with due regard to future exports. Otherwise the lead will pass to other countries (as happened in the UK with the textile industry because the industry did not keep up to date; the Swiss watch industry is at present in grave danger). In addition, some developing countries which are not at present short of foreign currency may well wish to proceed immediately to certain types of 'advanced' transport if they can be quite sure that what is offered has been realistically tested and is as reliable as it is possible to make it. Exports of highly developed and complex engineering devices requiring a high input of brain power and 'know how' are vital to the survival of a small developed modern country.

Another important aspect of this problem of time scale relating to the development and introduction of new transport systems and indeed to any such new long term developments relates to the life time of Governments. No government – and this applies especially to the UK at the present time – looks much further forward than the next general election. Responsibility for the more distant future, in particular now in the UK, what will succeed the Advanced Passenger Train, is vested in no one but private individuals, most of them in universities, and occasionally in a more enlightened public corporation, which might set aside a very little of its resources for such research and development. However, when almost all public corporations lose money most of the time, their aims are bound to be concentrated within a fairly short time scale. And it is difficult or inexpedient for them to advocate publicly Governmental support for more advanced work in view of the likelihood that their own support from the same source would be cut as a result of the direction of funds elsewhere.

12.8 Conclusions

It is hoped that some of the features of wheel-less guided land transport systems still requiring research and development effort have been brought out in the earlier chapters and re-emphasised in this final chapter. The importance of adequate research before operational installation has been highlighted by the Toronto difficulties which have resulted in some disillusionment in the minds of politicians in several countries. Combined with a shortage of development money due to world recession, this has in turn led to severe cut-backs in the scale of work on innovative transport systems in a number of countries, especially in the UK, Germany and the US (but not, apparently, in Japan, where increased public concern over the noise caused by wheeled trains has led to an even greater emphasis on the need for magnetically suspended vehicles).

It has also been recently appreciated by many that the transport systems of a country, as of a town, must be considered as a whole and the emphasis is turning to 'balanced transport', in which the full development and use of existing infra structure is also emphasised, besides the need for adequate interchange between different systems and the gradual and planned introduction of new systems.

Consideration of the whole transport of a country has to include future improvements in communications, which will render some travel of diminishing importance, and must include the transport by pipe networks of solids as well as liquids and gases, which will probably render unnecessary some of the present rail and road transport of raw materials, especially of hydrocarbon fuels.

Transport systems cannot be considered in isolation from general town and country planning, as the building of a new metro or other system encourages rapid property development along the line of the new system.

The advanced passenger trains in more than one country are completely in tune with the present times and are designed to make the utmost use of the existing infra-structure of railway lines. They can run at a higher speed than can present trains round the relatively sharp radii to be found on the tracks designed for the lower speeds of the past and which, without the body tilting feature, would give radial accelerations too high for passenger comfort. The light body construction will also reduce wear and tear on the rails, as will modern developments in suspension and wheel profiles. Developments of conventional systems have also been introduced into 'rapid transit', where complete computer control is of increasing importance, reflecting the need for less labour-intensive transport systems and the increased safety resulting from the elimination of most of the human element. Developments of this kind are bound to continue.

The emphasis on quality of life today is also indicating strongly the need to reduce the flow of motor vehicles, with their noise, fumes and over-crowding in the centres of large (and smaller) towns. It is now being widely appreciated that the building of more and 'better' roads, which soon become equally choked with cars, is no solution, and is destroying the beauty of towns besides the quality of life in them. It is likely that light guided wheeled or levitated systems, probably completely automatically controlled, will relatively soon be introduced into the larger towns. Along the lines of what was in mind for Toronto, these will be midway between a bus and a train, and may have some of the characteristics of taxis.

It may be stated finally, without much fear of contradiction, that the transport systems of our towns and countries will be revolutionised within the lifetimes of most of us working in the field. The lessons of the recent past have now been learned and the principles for the next moves forward understood by most, if not all, of those interested. When the world financial recession draws to a close and prosperity is with us again, the thoughts of politicians, as of the ordinary citizens, will turn to the implementation of some of the innovative systems which are available. We must be ready for this day.

If we can regard the whole body of researchers into high-speed, wheel-less transport as one single team with a common problem and a common purpose, then we must be ever mindful that if one of us fails, then at least, in part, we *all* fail, in the minds of politicians, journalists and the public. Only success succeeds.

References

[1] Ross, J. A. (1973) 'ROMAG transportation system', *Proc. IEEE*, **61**, (5), 621–630.

[2] Bahmanyar, H. and Ellison, A. J. (1974) 'The application of permanent magnets to the suspension of surface-guided vehicles', Proc. of Conf. on Linear Electric Machines, October 1974, IEE Conference Publication No. 120.

Note: Additional information concerning many of the devices and systems considered in this final chapter is to be found in the publication referred to in reference [2] above and in Ellison, A. J. and Bahmanyar, H. (1974) 'Surface guided transport systems of future', *Proc. IEE*, 121, No. 11R, November 1974, IEE Reviews.

Index